PATTERNS OF HUMAN HEREDITY

"ME 'N JACKSON ARE EXACTLY THE SAME AGE. ONLY HE'S DIFFERENT. HE'S *LEFT-HANDED!*"

PATTERNS OF HUMAN HEREDITY

AN INTRODUCTION TO HUMAN GENETICS

James R. Brennan
Bridgewater State College

Prentice-Hall, Inc., Englewood Cliffs, New Jersey 07632

Library of Congress Cataloging in Publication Data

BRENNAN, JAMES R., date
Patterns of human heredity.

Includes bibliographies and index.
1. Human genetics. I. Title.
QH431.B664 1985 573.2'1 84-6798
ISBN 0-13-654245-X

Editorial/production supervision: Virginia Huebner
Interior design: A Good Thing: Howard Petlack
Cover design: Christine Gehring-Wolf
Manufacturing buyer: John Hall

Cover Photographs (James R. Brennan):

Top: Two different stages of contraction of human chromosomes in mitosis—white blood cells.
Middle left: Normal red blood cells and two white blood cells. The larger white cell is a polymorphonuclear leukocyte with a "drumstick" (condensed X-chromosome).
Middle right: Blood smear from Rh positive blood showing hemolysis due to Rh antibodies.
Bottom left: Cancerous tissue from human bronchus.
Center bottom: Blood smear of sickle-cell anemic showing sickled red blood cells.
Bottom right: Human cheek epithelium cell from the lining of the mouth.

Printed in the United States of America

10 9 8 7 6 5 4 3 2 1

ISBN: 0-13-654245-X 01

Prentice-Hall International, Inc., *London*
Prentice-Hall Australia Pty. Limited, *Sydney*
Editora Prentice-Hall do Brasil, Ltda., *Rio de Janeiro*
Prentice-Hall Canada Inc., *Toronto*
Prentice-Hall of India Private Limited, *New Delhi*
Prentice-Hall of Japan, Inc., *Tokyo*
Prentice-Hall of Southeast Asia Pte. Ltd., *Singapore*
Whitehall Books Limited, *Wellington, New Zealand*

This book is dedicated to everyone who takes the time to read it. It is particularly dedicated to any readers from a very special F_1 who might like to know more about blue eyes and lefthandedness.

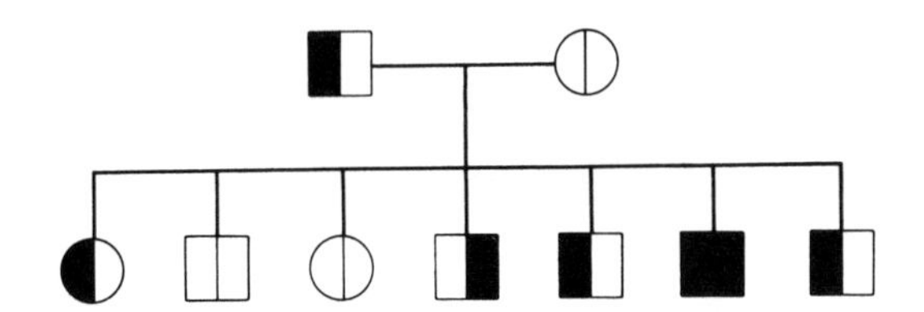

CONTENTS

8

GENETIC DISEASES / 200

9

FAMILY PEDIGREES AND ANALYSIS OF HEREDITARY PATTERNS / 219

10

POLYGENES AND QUANTITATIVE INHERITANCE / 235

PREFACE

Patterns of Human Heredity has been written for those who want to learn about the genetic basis of human life. It is hoped that it will provide a useful understanding of the hereditary mechanism at work in humans. Inherent in such an understanding should be a view of commonly shared genes within families and populations. Unique individuality arising from a background of vast diversity is a concept that deserves scientific support on this tense and crowded planet. All who profess to be educated today must know about the fundamental basis of individual differences.

The science of biology has expanded so greatly that it has become a contentious matter to design a "general" approach for college students. Opinions vary among professionals about what can and should be included now that so much must be left out. A commonplace approach in recent times has been to teach a one-semester course in "principles" and then allow nonmajor students to opt for an introductory course in a specialized area. I hope that students with diverse interests will see the merit to an exploration of human genetics in this context. It is hard to imagine a more interesting and relevant subject for a plethora of endeavors. Where "value oriented" courses are called for, this topic is significant.

The text is deliberately didactic and organized along lines that provide the most logical learning sequence for a novice human geneticist. Because we often learn best by repeating the historical development of a subject, this aspect has been emphasized.

The organizational sequence of topics is not likely to satisfy all others teaching human genetics. While the historical sequence is important and could be the best way for a beginner to approach the subject, the mathematics of Mendel seems much more meaningful after the gene has been defined chemically and physically. A good understanding of the reproductive process also provides a much clearer perception of the basics of genetics. Thus, these topics are placed ahead of a detailed look at Mendel's principles and examples of human genes. Chapter 7 could be dealt with immediately after Chapter 1. In this way Mendel's story of basic genetics could be studied first, followed by a chemical explanation of the gene (Chapter 4) and then the story of chromosomes (Chapters 5 and 6). If some prefer to do chromosomes prior to the chemical basis, the text is flexible enough for that option.

I prefer to discuss that important tool of human genetics, the pedigree chart, after all the basic genetics has been done (including sex linkage), at least for nonmajors. Again, an instructor could deal with Chapter 9 along with Mendelian genetics, or even do the first part of Chapter 9 before Chapter 7.

Instructors should feel free to change sequences. I believe that the text will be adaptable to such changes when it is necessary to fit a course structure.

The book will serve a most important function if it leads the reader to explore further the vast amount of knowledge that has been accumulated by the legions of international workers. Humans are quickly becoming the most well-known organism genetically, due to widespread interest and elegant new techniques. Every day exciting and astounding facts are added to the existing catalog of information.

Science in general, and biology in particular, is often viewed by students as studies centered on words. Many times memorization and regurgitation of terms are substituted for an understanding of concepts and ideas. Unfortunately, if one is to develop concepts and discuss them with others, a reasonable vocabulary of precise scientific terms is essential. To remind students of this, boldface type is used to draw attention to words that will be helpful in dealing with the concepts presented.

Likewise, a fairly short glossary (by biological standards) appears at the end of the text. Some of the more frequently used and significant terms that are printed in boldface type have been included. Terms that are likely to be encountered only once or twice in the text, or that do not carry broad significance, have not been included. Specific disease names or detailed structures and chemicals are also not in the Glossary. The definitions provided are brief and simple, limited in their utility to the context of this subject. More thorough and technical definitions can be found in dictionaries or more detailed texts.

By no means is an exhaustive list of genetic traits included here. It would not serve the intended purpose of this fundamental approach, nor would it be possible. Once the basic ideas have been mastered and a vocabulary developed, students can find information about the genetics of many other human traits simply by involving themselves in some detailed research.

At the end of each chapter a short list of additional readings is provided. Some are general, some very specific; some are easy to read, some fairly rigorous. The list only samples some topics; as always, they will lead on to even more sources of information.

The Review Questions at the end of each chapter are there for the purpose of getting students involved directly. They are not easy and sometimes a direct, undebatable answer cannot be found. The questions may require a student to delve into additional sources of information and to spend some time thinking about responses. This thinking will often lead participants to a topic covered later in the text. Whether assigned or not, the questions are integral parts of the text that should expand each student's understanding considerably.

Each chapter ends with a summary. When all work on a chapter is finished, including any additional reading and problem solutions, students can profit from using the summary to review the material. If an attempt is made to expand and embellish he concise summary with as many details as can be remembered, a student is likely to view the material as a unit and not as a series of disjunct facts and ideas.

I am indebted to Bridgewater State College and The Open University in Milton Keynes, England, for providing good colleagues and a quiet, intellectual atmosphere where thinking is stimulated. Mostly my thanks are due to people who shared their thoughts and time. Cathy Lauwers, Diane Peabody, and Walter Morin have read manuscript and made valuable comments. Claire Devincentis has turned my unreadable scratchings into neatly typed pages. The sketches of historical personalities by Christine Polivogianis added greatly to the descriptions. Thanks to all who have helped, directly or indirectly!

JAMES R. BRENNAN
Bridgewater, Massachusetts

CHAPTER

1

THE ORIGIN OF A SCIENCE

Who wer as lyke as one pease is to another.
John Lyly, *Euphues* (1578).

As long as humans have been able to make thoughtful observations of themselves and their characteristics, there have surely been questions about inheritance of various traits. It is embarrassing sometimes to a child to stand up to comparisons made by relatives and friends in regard to real or imagined similarities and differences between parents and offspring or among siblings. This comparative process must have taken place in early periods of human history, with no adequate explanations for the mechanism of heredity and variation in offspring.

However, some basic, sophisticated knowledge of hereditary patterns was possessed by certain cultural groups of humans early in history. A familial association of red-green colorblindness was known to the ancient Greeks, while the Hebrews in the second century A.D. showed a rudimentary understanding of the association of hemophilia to sex determination. However, such isolated bits of knowledge did not constitute a body of concepts that could be distinguished as a science.

The development of a valid explanation for the patterns of **heredity** and **variation** was impeded because early observers tried to fathom simultaneously the mechanism determining inheritance for many characteristics in a single individual. An adequate explanation could only be obtained easily by considering one trait at a time. Science is sometimes a deceptive field of endeavor: The simplest explanation, disdained by great thinkers, is often the only valid route to true understanding. Genetics thus eluded the powers of human understanding until one man carefully considered the patterns he found in one of his favorite living things. Using the powers of mathematical logic in his train of thought, he was able to analyze in a fashion that brought forth an explanation.

THE BEGINNING

For many years prior to 1865, biologists attempted without success to understand the principles of inheritance. Many prominent scientists observed and studied various characteristics of living organisms in an effort to interpret the mechanism involved in the passage of traits from one generation to the next. However, no satisfactory explanations were proposed until 1865 when an Austrian monk named Gregor Mendel (Fig. 1.1) offered a well-documented theory at a meeting of the Brunn Society for the Study of Natural Science. His work, compiled after eight years of careful research, was clearly organized and logically designed to explain the process of inheritance, but Mendel was years ahead of the scientific world. Little was understood concerning reproduction and cellular structure, and the significance of his work was not clear to his fellow scientists. Mendel's research made little impression on those present at the meeting, and only slight recognition was given to the carefully assembled data. The paper that Mendel presented in two successive readings in February and March of 1865 was published in the Proceedings of the Brunn Society and appeared in print

Figure 1.1. Gregor Mendel (1822–1884), whose careful research with pea plants revealed the basic mechanism of heredity.

in 1866, but it was overlooked by Mendel's contemporaries. Other scientists saw the publication in later years, but they also missed its significance.

The biological world at the time was preoccupied with a new theory by Charles Darwin and Alfred Wallace. Their work concerned **evolution** through **natural selection,** and it stimulated much scientific thought, debate, and controversy. So startling were the ideas of Darwin and Wallace concerning the common ancestry of living things, there is little wonder that the biological world overlooked the unimpressive experiments of an obscure monk. Darwin and Wallace were right in their basic theory that living things change and adapt with time and circumstances. Unfortunately, they had no idea what the mechanisms of inheritance were which permitted such a phenomenon. Ironically, Mendel's work contained an explanation for the way in which evolution is directed, but the relationship would not be recognized by biologists for many years.

In 1900 three biologists from different parts of the world independently rediscovered the principles Mendel had outlined and recognized the value of his work. Carl Correns in Germany, Hugo de Vries in Holland, and Erich von Tschermak-Seysenegg in Austria had conducted original studies concerning the mechanism of inheritance and noted the original work in their reports (Fig. 1.2). Technically, the science of **genetics** was born with Mendel's first report in 1865. In reality, its birth

Figure 1.2. Clockwise from upper left: Erich von Tschermak-Seysenegg (1871–1962), Hugo de Vries (1848–1935), and Carl Correns (1864–1933) who all independently rediscovered Mendel's principles of heredity in 1900.

occurred in the year 1900. In the decades that followed research into the study of heredity proceeded rapidly, leading to one important discovery after another. Seldom does such a clear-cut and well-documented origin exist for a field of study.

Science rarely proceeds smoothly, in effect, with a simple sequence of discoveries that logically lead to more complex ideas. Even though the seeds for an explanation of evolution through natural selection were present in Mendel's work, Darwin's theory was thoroughly debated and discussed without consideration for the mechanism of inheritance, often with rather weak arguments on both sides. So it was also with Mendel's ideas. Biologists of his time primarily observed and described what they saw and they were unable to apply mathematics to biological phenomena. Mendel himself knew practically nothing about cell structure and chromosomes. In his work even the process of reproduction was vague and basic descriptions of the process were almost nonexistent. No doubt Mendel would have been able to analyze more accurately and understand more thoroughly if he had known more about **cells, chromosomes, fertilization,** and formation of **reproductive cells.** Likewise, if Darwin and Wallace had known and understood the principles outlined by Mendel before proposing their evolutionary ideas, their theory of evolution through natural selection might have been more solidly based.

Genetics has had many contributors to the understanding of the simple, underlying processes, and it is somewhat unfair in a short discussion to select a small number of the most important workers in the field. However, the parade of biologists who followed Mendel is long and varied, and it is not difficult to name some who contributed significantly to the origin of the science.

The new science of biological inheritance came to be known as **genetics**, a name applied by William Bateson of England in 1906. He developed this name by using the word **gene**, a term that had been coined by Wilhelm Johannsen, a Danish botanist, in 1902. Mendel had referred to the units of inheritance simply as **factors**. He implied that some sort of particle was transmitted from generation to generation, but could not define it. He did not understand that cells were involved. Neither Johannsen nor Bateson knew what type of particle was involved, but this did not prevent genetic experiments or the naming of the unseen units.

William Sutton, an American graduate student, and Theodor Boveri, a German researcher (Fig. 1.3), knew something about the internal structure of cells and the process of reproduction. They were involved in studies in another new and rapidly developing science, **cytology**, the study of cells. They were familiar with the developments in genetics and recognized a similarity between genes and objects found in living cells called "chromosomes." Both suggested in 1902, from their own separate studies, that the genes are located on the chromosomes. (The evidence suggesting this theory is discussed in depth in Chapter 2).

The idea that genes are present on the chromosomes was proven through the efforts of T. H. Morgan and his coworkers starting in 1911. Most of their work was done using the common fruit fly *Drosophila* as experimental material. *Drosophila* proved to be an excellent organism for genetic studies. It has come to be the most completely studied and well-understood multicellular organism in regard to inheritance. These early studies in **cytogenetics** (cytology and genetics combined) con-

Figure 1.3 William Sutton (1877–1916) (left) and Theodor Boveri (1862–1915) (right) both independently recognized that chromosomes were the physical bases of Mendelian genetics.

firmed not only that genes are on chromosomes, but that there are many more than one gene per chromosome.

No description of the founding of the science of genetics would be complete without mentioning the work of William Bateson and R. C. Punnett. These early geneticists repeated and supplemented Mendel's work using sweet peas. They then studied poultry intensively in light of this new information concerning heredity. In all, they applied the principles discovered by Mendel to nine different genera of plants and four different genera of animals between the years of 1903 to 1913. Their work truly established genetics as an important and useful area of study.

INITIATION OF HUMAN HEREDITY STUDIES

It is difficult to determine when the early geneticists realized that their embryonic science had strong implications for the understanding of human patterns of inheritance. Possibly Mendel saw a parallel between his peas and humans. Certainly some of his followers did after the rediscovery of his work in the early 1900s. No matter what organism serves as experimental material for a biologist, no matter what phenomenon is being studied, sooner or later the comparisons to humans must develop.

Medical science is the area most concerned with an understanding of human biology. Therefore it is within the medical establishment that most of the advances in understanding of human heredity have taken place. Because of the small number of offspring produced by individual human parents, as well as the cultural and moral restrictions imposed on the human mating process, humans are not suitable for the type of experimentation and analysis done on pea plants by Mendel. The most commonly analyzed human genetic traits are those associated with serious defects encountered in a medical context. Development of an understanding of **human genetics** in the past has therefore been slow and dependent on information obtained from other living organisms.

The painstaking process involved in understanding human genetics was illustrated in 1900 when Karl Landsteiner, a physician in Vienna, discovered the blood typing system and the **ABO** transfusion incompatibilities between **sera** and **red blood cells.** Although this discovery came to the biological world simultaneously with the rediscovery of Mendel's work, it was not until 1925 that Felix Bernstein, a German, was able to describe the correct hereditary mechanism controlling this basic human genetic trait.

The "father" of human genetics is usually considered to be Archibald E. Garrod, a British physician (Fig. 1.4). In 1901 he described 11 cases of **alkaptonuria,**

Figure 1.4. Archibald Garrod (1858–1936), recognized as the father of human genetics for work with his so-called "inborn errors of metabolism" such as alkaptonuria. He was the first to demonstrate genetic control of biochemical reactions in the body.

a **metabolic disorder** in which the compound alkapton accumulates and is excreted in large amounts in the urine. Garrod recognized that this condition resulted from a defective metabolic process and he classified the disease as an "inborn error of metabolism." Aware of Mendel's recently discovered work and noting similarities, he demonstrated that alkaptonuria was due to a single **recessive** gene which could then be associated with one enzyme.

By 1909 when Garrod's book, *Inborn Errors of Metabolism*, was published, he recognized that the features of alkaptonuria were caused by the accumulation of alkapton which resulted from a failure of the body to break it down. He was the first to identify a human gene according to the new Mendelian system and the first to recognize that genes can control biochemical processes which are mediated by **enzymes.**

Like Mendel, he was well ahead of his time, almost stating what has come to be known as the "one-gene, one-enzyme hypothesis." Not only did his work give rise to the field of human genetics, it also gave birth to **biochemical genetics.**

Because Garrod was a medical doctor, he was able to correlate a distinct, discontinuous medical trait with a familial inheritance pattern. The study of medically disabling genetic patterns was of great interest and significance in view of the predictability shown by Mendel for genes. Even today the most intensely studied human inherited traits are those that are disabling. Diseases are first recognized and most completely understood within a medical context. Thus, some of the most active efforts made in human genetics involved medical researchers and practitioners who were able to combine their work with genetic analyses.

The origin of human genetics studies occurred very close to the rediscovery of Mendel's work. In general, rapid and important advances took place in genetics, while human studies lagged somewhat. Landsteiner and Garrod had laid the groundwork for an understanding of human heredity at the cellular and molecular levels, but human traits were not obviously Mendelian in their inheritance patterns and humans were not easily studied in regard to genetics. A realistic understanding of human genetics evolved as the control of proteins by genes developed in the 1950s. An understanding of blood proteins and changes resulting from mutations opened up new vistas for study.

Blood proteins, enzymes, and finally the immune reaction based on proteins led to human applications of the rapidly expanding science of **molecular biology**. In the 1960s, new techniques of chromosome analysis provided the tools needed to associate specific genetic effects with chromosomal changes. An understanding of the molecular structure of the hereditary material in the 1950s and a detailed understanding of the means of controlling physical characteristics through the control of protein synthesis in the 1960s not only provided understanding but began to provide medical solutions to heretofore insoluble problems. Whereas bacterial cells provided a basic understanding of genetic mechanisms in the 1950s and 1960s, cultured human cells began to give answers specific to human chromosomes and genes in the 1970s. Infinitely more complex than bacteria and of considerably different structure, human chromosomes continue to be studied with exciting new results and astonishing prospects for manipulation and control.

Genetic mechanisms determining the characteristics of humans have been revealed only after slow, painstaking studies of families showing distinctive patterns of inheritance. In the 1940s studies of ***erythroblastosis foetalis***, a serious blood disease in infants, went on simultaneously and independently from experimental studies of various *Rhesus* monkey blood types. Monkey blood types and *erythroblastosis* were found later to be due to the same phenomenon, resulting from similar blood chemicals. Thus, a genetic basis did not suddenly become apparent, but the divergent work of many investigators led to its discovery as an important feature of human genetics.

Research interest in human heredity is currently centered on many diseases and debilitating conditions which have been revealed to be hereditary during the last three or four decades. In well-known situations like **phenylketonuria**, a genetic metabolic disease that results in serious brain damage and retardation, and **erythroblastosis foetalis**, treatments have been devised which circumvent the genetic determination of characteristics. Thus, it has proved advantageous to identify such problems very early in an individual's life to allow medical remedies to be utilized.

There are inherited diseases for which no distinct treatment is available, and an important medico-genetic area of endeavor has centered around means of identifying individuals who may transmit genetic defects. As a result, counseling about options that may be available in such situations has become an important aspect of human genetics.

Because of the practical implications of such important areas of research, medical science has devoted much attention in recent years to the investigation of hereditary defects. A large amount of knowledge of human genetics has been assembled and a pattern for the understanding of other genetically controlled features has developed. As more genetic features are revealed, the search for cures for those that are detrimental goes on. In addition more efforts are being devoted to disseminating such knowledge to the public.

GREGOR MENDEL, THE FIRST SUCCESSFUL GENETICIST

One can only wonder whether Mendel really understood the importance of his unnoticed research report in 1865. The fact that his work looks simple and unspectacular in comparison to our current knowledge in the field does not lower its significance. Mendel, through his genius for painstaking study and careful assemblage of data, was able to determine the orderly operation of a process that seemed chaotic to others.

Gregor Mendel was uniquely well prepared for his discoveries. Because his father was a farmer, he had a basic understanding of agricultural practices. He knew about contrasting characteristics among pea plants and was familiar with methods of culturing them, as well as methods of making **viable crosses** between two chosen parents.

Mendel was also a mathematics teacher. He was able to analyze mathematically the numerous pea plants he grew in the monastery garden and visualize them as segments or proportions of a large family. Even though he had failed to graduate from the University of Vienna after attending for two years, he was able to plan and carry out a study of immense biological significance.

Mendel spent eight years experimenting with common garden peas before announcing his theory. His two years of study in natural science and mathematics at the University of Vienna had been good preparation for his work. He kept his experiments simple and proceeded systematically to increase their complexity only after he understood the simplest. He examined one inherited trait at a time, a method not employed by his predecessors.

Mendel may still be praised for his choice of experimental material, as the selection of the garden pea had several advantages. Seeds were easily obtained from seed dealers and several varieties were known. Some of these varieties may still be purchased from seed catalogues. The pea plants grew rapidly and Mendel could make observations on relatively large numbers of plants in a small space. Also, the flowers of the pea plant are so constructed that they stay closed during the reproductive process. Thus, no **cross-fertilizations** occurred unless Mendel wanted them, and an undisturbed plant produced both types of reproductive cells involved in seed production for the next generation.

Mendel probably did not know much more about the process of reproduction than that some material for the next generation was contributed by each of the parents. However, Mendel indirectly contributed much to an understanding of the reproductive process; a better understanding of reproduction prior to his report would have made his work more understandable.

MENDEL'S PRINCIPLES

Based on Mendel's early work, it was apparent that certain principles govern the action of inheritance. These are often called Mendel's laws, but it is wise in biology to assume that laws or principles have exceptions and can be applied only under specific conditions. Indeed, over the years, many exceptions to these principles have been found. In general, the principles apply to most inherited characteristics and they provide a starting point for the study of those characteristics that show complications. The discovery of these basic principles provided a real understanding of the mechanism of inheritance. Because Mendel did not call them laws or principles, one can debate about how many basic ideas in his work should be so classified. The biological world might have understood and accepted his work if the humble monk had been more flamboyant and summarized his important findings in an orderly "one, two, three" fashion. In fact, to understand best the significant aspects of the discoveries, Mendel's work should be analyzed in this manner.

Mendel, first postulated that the appearance of a living thing was determined by some sort of **factor** for each visible characteristic. These factors were passed on to the organism by its parents, retained in its body throughout life, and passed on unchanged to its offspring. Because each organism has two parents, it seemed logical that two factors for every characteristic, one from each parent, would be found in the body. As noted earlier, the name ultimately given to these factors was **genes**, but Mendel simply called them factors. This first principle is sometimes called the law of **paired unit factors**. This law, not defined as such by Mendel or by many biologists, is simply used here to emphasize that Mendel recognized the basic idea that organisms possessed determiners of traits in pairs, one contributed by each parent.

Second, it was recognized that, even though both parents contributed a factor for a given characteristic, one of these was often masked or hidden in the offspring. Mendel knew it was present, because when he used such an offspring as a parent, the hidden factor could be passed on and appeared in the next generation. For example, when a tall pea plant was crossed with a dwarf pea plant, the resulting seeds all grew to be tall plants. However, if two of these tall plants (now each carrying a hidden gene for dwarfness) were crossed with each other, some of the seeds grew to be dwarf plants. It was obvious that a dwarf factor from each new parent had been passed on to these second generation offspring.

To explain this masking effect, Mendel assumed that there were two forms of each gene. One of these overshadowed the other form when they both occurred in the same organism. The overshadowing factor was called **dominant**, and the hidden one was called **recessive** (Fig. 1.5). The two different forms of the same gene are now called **alleles**, from a term used by William Bateson, *allelomorph*. This overshadowing action of one member of a pair of genes has been said to follow the law of **dominance**. The implication which is important from this so-called law is that an organism can carry a gene from one parent and pass it on to its offspring, without showing any visible evidence that it possesses the trait.

It was then apparent to Mendel as a third feature of inheritance that only one of the paired factors in an organism could be passed on to the next generation. If this did not occur, the offspring would receive two factors from each parent and have four factors for each characteristic. Thus, the law of paired unit factors would not be true. It was obvious then that the paired genes would have to be separated from each other in the reproductive process prior to the production of new offspring. This principle has been called the law of **segregation**. Mendel was correct in assuming that the paired factors carried by an organism were separated (segregated) and passed on, one at a time, to offspring.

Finally, when Mendel understood the inheritance of a single factor, and the laws governing such action, he was able to consider two or three factors simultaneously. In his experiments he found that different pairs of genes acted independently of each other. Each pair went on segregating during reproduction, in a fashion that showed no preference for one gene from a pair in combination with one from an independent pair. This last principle is known as the law of **independent assortment**. As stated, one of the great problems of Mendel's predecessors was their attempt to consider several characteristics at once. Because Mendel understood the independence of separate pairs, he was able to explain their simultaneous action rather simply.

The well-known law of simultaneous independent events states that the combined **probability** is equal to the product of the two separate probabilities. Because Mendel was able to apply this mathematical principle to his studies of living plants,

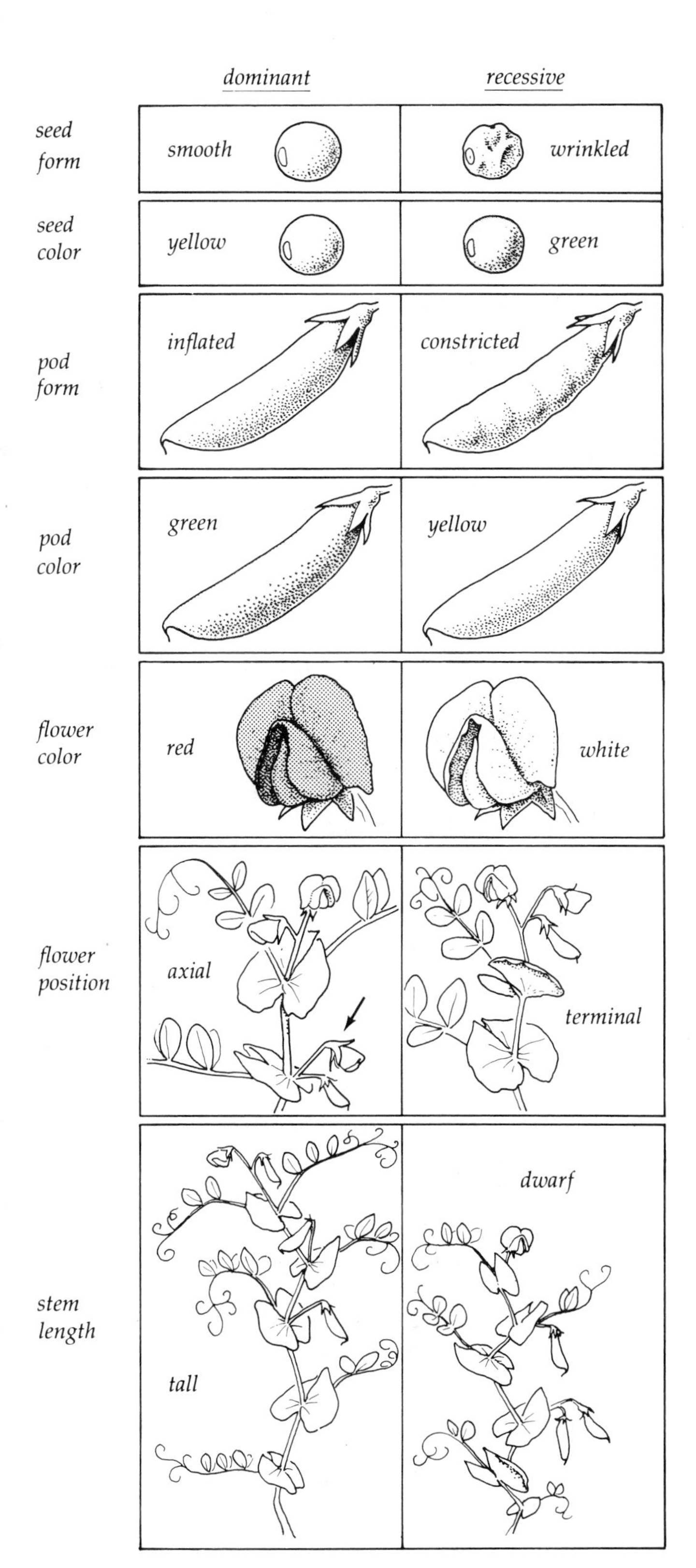

Figure 1.5. The seven pairs of dominant-recessive characters in the garden pea used by Gregor Mendel in 1865 to 1866 to demonstrate the mechanism of heredity. (From BOTANY: AN ECOLOGICAL APPROACH by William A. Jensen and Frank B. Salisbury. © 1972 by Wadsworth Publishing Company, Inc., Belmont, California 94002. Reprinted by permission of the publisher.)

he correctly predicted that the hereditary factors were independent of each other in the reproductive process.

Possibly his greatest achievement was the realization that the whole process of reproduction was based on **chance**, as is coin tossing or dice rolling. With this factor in mind, he was able to analyze or predict in terms of numbers of offspring receiving hereditary traits due to chance. Once he knew the separate, independent chance for certain traits to be exhibited, he was able to combine these and analyze a series of traits simultaneously in the same organisms. Mendel's predecessors were neither able to look at single traits carefuly nor to combine the knowledge of several single traits into a broader picture.

As Sutton postulated, and Morgan and his coworkers showed, more than one gene is located on a single chromosome. If two genes located on the same chromosome are selected for study, because they are connected together, they cannot assort independently of each other in the production of reproductive cells. Such a situation is called **linkage** of genes. This represents an important exception to independent assortment, but Mendel did not record any linked genes in his research. He selected seven characteristics of the pea plant for study, without knowing that the pea plant has only seven pairs of chromosomes.

It has often been said that Mendel was lucky not to have encountered linkage in his studies. In fact he did report on two-gene pair crosses with genes located in the same chromosomes in three cases. In all these cases the genes are far enough apart in the chromosomes that random breaks and refusions between the genes result in data that look exactly like independent assortment. This **recombination** process in chromosomes has been labeled **crossing-over.** Linkage and crossing-over were demonstrated only a few years after the rediscovery of Mendel's work. Both processes will be studied further in Chapters 2 and 5.

Actually, the gene controlling pod shape and the one controlling plant height are close enough together in the same chromosome that they should have caused unpredicted results in a two-factor cross. It is interesting that Mendel reported no such cross in his work. Could he have noticed a complication in such experiments and simply shelved the question for future research? If this is true, it may attest to the fact that Mendel was an astute simplifier of complex scientific phenomena. Realizing that a complexity he did not understand was involved in linked genes, he may simply have gone on to other factors that revealed the mechanism of inheritance clearly. Mendel may have possessed the valuable quality of refusing to be burdened by the complex until the basic principles were uncovered.

Mendel's great mind may well serve as an example for science (Fig. 1.6). All advances in science do not require the guidance of a superintellect. Sound, logical thinking is the basic requirement for scientific activities and is absolutely critical, in the field of genetics. An attempt should be made when studying heredity to look for underlying mechanisms that are based on a small number of basic principles. Even the exceptions are best explained by a clear understanding of the relatively simple principles. As Mendel, one must attempt to explain things in a simple, logical fashion, proceeding from basic underlying ideas to complex, broad patterns. Genetics often appears difficult and confusing, but there is always a thread of logic underneath that can be grasped and followed through to a meaningful answer.

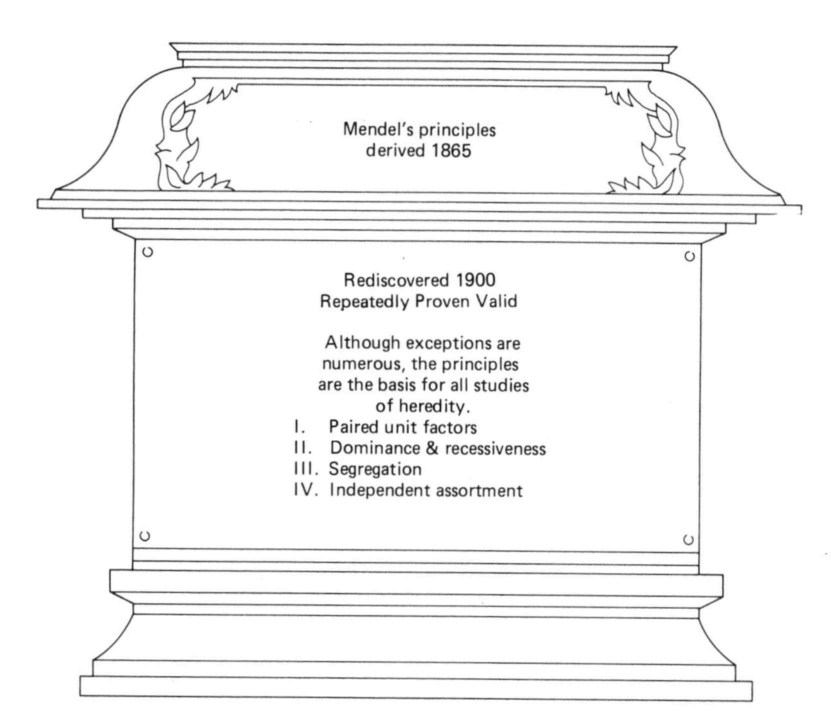

Figure 1.6. The principles of heredity discovered by Mendel in the garden pea serve as the foundation for all other genetic studies in sexually reproducing organisms.

ADDITIONAL READING

BOYES, B.C., "The Impact of Mendel," *Bioscience* (1966) 16, 85–92.

DODSON, E. O., "Mendel and the Rediscovery of His Work," *Scientific Monthly* (1955) 81, 187–95.

DUNN, L. C., *A Short History of Genetics*. New York: McGraw-Hill, 1965.

GARROD, A. E., "The Incidence of Alkaptonuria: A Study in Chemical Individuality," *Lancet* (1902) ii, 1616–20. (Reprinted in Boyer, S. H., IV, *Papers on Human Genetics*. Englewood Cliffs, N. J.: Prentice-Hall, 1963).

______*Inborn Errors of Metabolism*. London: Henry Frowde; Hodder and Stoughton, 1909. (Reprinted by Oxford Univ. Press, London, with a supplement by H. Harris, 1963).

HARDIN, G., *Nature and Man's Fate*. New York: Mentor, NAL 1959.

ILTIS, H., *Life of Mendel*. Translated by E. Paul and C. Paul. New York: Hafner, 1966.

KING, R. C., *A Dictionary of Genetics* (2nd ed.). New York: Oxford University Press, 1972.

MENDEL, G., 'Experiments in Plant-Hybridization," in *Classic Papers in Genetics*, ed. J. A. Peters, Englewood Cliffs, N. J.: Prentice-Hall, 1959, pp. 1–20.

STUBBE, H., *History of Genetics* (2nd ed., 1965). English translation by T. R. W. Water. Cambridge, Mass.: The M. I. T. Press, 1972.

STURTEVANT, A. H., *A History of Genetics*. New York: Harper & Row, Pub., 1965.

SUTTON, W. S., "The Chromosomes in Heredity," in *Classic Papers in Genetics*, ed. J. A. Peters. Englewood Cliffs, N. J.: Prentice-Hall, 1959, pp. 27–41.

REVIEW QUESTIONS

1. Cite as many reasons as possible that support the statement, "Mendel was ahead of his time, therefore his work in genetics was not understood by his fellow biologists."
2. Some biologists believe that Garrod was ahead of his time and therefore the significance of his work was missed in the early 1900s. In what ways was Garrod ahead of his time?
3. How do Mendel's principles relate to sexual reproduction or help to explain how sexual reproduction is advantageous to a species?
4. Using coin tossing as an example of chance (the chance is one-half it will land a head, one-half it will land a tail), explain how the coin might serve as a model for Mendel's principles of paired unit factors and segregation. How is coin tossing analogous to gene transmission?
5. Assuming that two coins can be tossed simultaneously, explain how this might serve as a model for Mendel's principle of independent assortment. How can the mathematical predictions for different combinations (two heads, head on the first coin, with a tail on the second, and so on) be determined using the mathematical law of simultaneous independent events? How many different kinds of results can be obtained when tossing two different coins?
6. State reasons for the idea that chance is the basic underlying factor on which Mendel's principles are based. In what way is chance a factor in the genetic mechanism associated with sexual reproduction?
7. Restate Mendel's principles so that they will apply specifically to humans.
8. Why were pea plants much better research materials to work out the principles of heredity than humans?
9. Suppose that a human trait is known to be genetically determined. How would you explain its mode of inheritance in light of the fact that most individuals with the trait have two parents who do not show it?
10. Based on Mendel's law of segregation, what proportion of a child's genes are shared in common with his or her mother? With the father? Using the same logic, what proportion of the child's genes are derived from a single grandparent? What proportion of a person's genes are common to those of a sibling (brother or sister)?

CHAPTER SUMMARY

1. Prior to 1865 there was no organized body of knowledge of the mechanism involved in biological inheritance and variation. In that year Gregor Mendel, an obscure Austrian monk, presented a well-documented theory explaining the results of his research with garden peas.
2. Mendel's work was overlooked until 1900 when it was seperately rediscovered by three investigators, Hugo de Vries, Carl Correns, and Erich von Tschermak-Seysenegg. Each had duplicated and verified portions of Mendel's work before discovering his 1866 publication.
3. William Sutton and Theodor Boveri independently recognized in 1902 that Mendel's hereditary factors, the genes, were carried on chromosomes. This theory was proven conclusively and extended to other important considerations by T. H. Morgan and his colleagues using an important organism for genetic studies, the fruit fly.
4. Careful studies of human genetics began to develop in the early 1900s using Mendelian principles. Archibald E. Garrod, the so-called father of human genetics, centered his early studies around inborn errors of metabolism such as alkaptonuria.
5. There are genetic bases for many human diseases and debilitating conditions. As a result, a great portion of the activities in human genetics is devoted to analyzing, identifying, and treating genetic diseases.
6. Gregor Mendel, because of his understanding of arithmetic logic, was able to analyze inheritance patterns and make mathematical predictions about future generations. He carefully verified his predictions with garden peas. Because of his analysis, he discovered four principles that are the basis for understanding all biological inheritance mechanisms: (a) Hereditary traits are controlled by paired unit factors (genes); (b) Two forms exist for each unit factor, one dominant and one recessive; (c) The paired unit factors segregate from each other during reproduction and are passed along singly to offspring; (d) Different factors affecting various traits are independent of one another.

CHAPTER

2

CELLS: THE FUNCTIONAL UNITS OF LIFE

What am I, Life? A thing of watery salt
Held in cohesion by unresting cells,
Which work they know not why, which never halt . . .
*John Masefield, Sonnets (1916)**

It is important to remember that in a living organism all activities are based on the functioning of individual cells. A human body should not be considered a mass of individually operating cells, for there is organization and direction within each entire organism. However, despite such organization, the overall function of a tissue or group of cells reflects the activity of each individual cell. The individual cell is the unit of life and is responsible for the activities of the whole organism. Thus, to understand the underlying mechanism in inheritance, it is necessary to consider the details of cell structure.

HISTORICAL ASPECTS OF CELL STUDIES

With the appearance of glass lenses, scientists found they could investigate two unknown areas: The distant stars and planets became clearer when viewed with a telescope, and the extremely small details of structure in nearby objects became visible with a microscope. Cell study is based on the development of the microscope for, without a means of magnifying living organisms, their cellular structure is invisible.

A father and son team of lens makers, Hans and Zacharias Janssens, made the first microscope in Holland about 1590. Galileo Galilei, noted for his work in physics and astronomy, was a lens maker working in Italy about 1600. He heard of the Janssens's achievement while working with telescopes and he built his first microscope in about 1610. Galileo's interest was in viewing distant heavenly bodies, however, and he did not spend much time with his primitive microscope.

Regardless of the early appearance of such an important tool, few biologists spent time examining living material with microscopes. Much was yet to be learned by examination with the naked eye and this new technique was difficult.

Among the prominent early microscopists was Anton van Leeuwenhoek who lived in Delft, Holland, between 1632 and 1723 (Fig. 2.1). Van Leeuwenhoek's hobbies were microscope making and the study of tiny living things, and although he made more than 250 microscopes, he never sold one. Some were given to friends even though his microscopes were primarily for his own use. He was proud that no one except himself had ever looked through some of his best instruments. His instruments were hardly better than a modern magnifying glass. However, the details he saw were truly amazing: His drawings indicate that he observed bacteria and their motions. Even today such observations are difficult for a student with a good **light microscope**. Van Leeuwenhoek was probably the first to observe and record the structure and movement of human sperm cells in 1677. Little was learned about the internal details of cells, but the pioneering efforts of van Leeuwenhoek and others proved the importance of microscopy to biology.

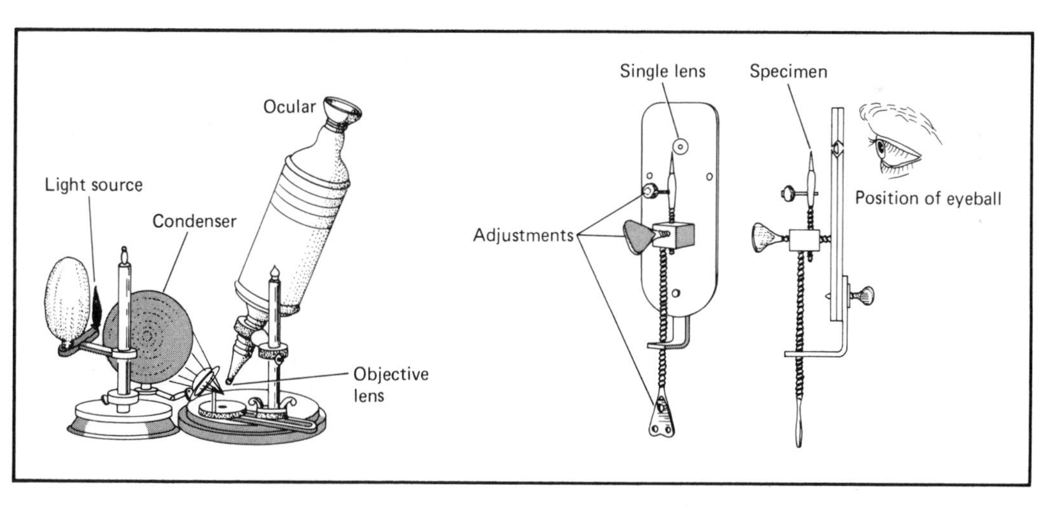

Figure 2.1. Diagrams of the microscopes used by Robert Hooke (left) and Anton van Leeuwenhoek (right). Although Hooke's microscope was used 50 or 60 years before van Leeuwenhoek's, it was a compound microscope with two lenses and was mounted on a stand. The simple, hand-held microscopes made by van Leeuwenhoek required good manual dexterity for manipulation of specimens and apparently extremely good eyesight for viewing. (Figure 2-1 (p. 9) from THE SCIENCE OF LIFE by Robert Day Allen, Copyright © 1977 by Harper & Row, Publishers, Inc. Reprinted by permission of the publisher.)

In 1665 William Hooke of England, sliced a piece of cork with a sharp blade and looked at the thin piece of material with a microscope (Fig. 2.2). His microscope looked more like today's compound microscopes than the simple lens of van Leeuwenhoek. Hooke saw the chambers left by cork cells after they had died and their living contents had disintegrated. To Hooke the remaining cell walls resembled the structure of a honeycomb; he used the Latin word **cella** to describe these small "rooms" in the honeycomb. **Cell** has since been used to describe the individual units of life in living bodies.

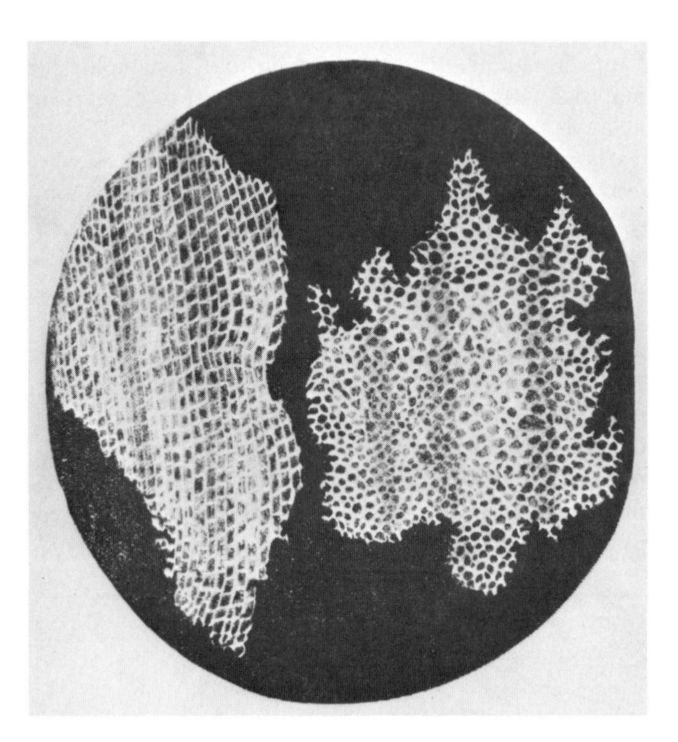

Figure 2.2. Diagram made by Robert Hooke and published in his 1665 *Micrographia*. Shown are side and top views of the sliced cork, the honeycomb structure which led Hooke to name the small compartments cells. Actually, cork is composed of the cell wall remnants of dead cells. (Courtesy THE BETTMANN ARCHIVE)

In 1838 to 1839 two German workers, Matthias Schleiden working with plants and Theodor Schwann working with animals, recognized that the structural and functional unit of life was the cell. Their observation has been recognized as one of the important steps in understanding the true nature of life and they are credited jointly with the founding of the **cell theory**. It was emphasized in these early years that organisms were constructed of individually operating units of life, the cells. It is now known that the total organism shows balance and integration in the activities of its many cells and, for this reason, the **organismal theory** has replaced the earlier view. Even though the whole organism is considered to be more significant than the sum of its parts, an individual cell is the functional unit of a living plant or animal. A thorough knowledge of its structure is essential to explain the functional aspects of the hereditary process.

In the early 1950s a pioneering group of biologists utilized the **electron microscope** with its higher **resolution** and very high **magnification** to study cells. Although experienced biologists, they quickly learned that the techniques of material preparation were much more involved and laborious than the standard techniques of light microscopy. With much effort and time, ways were developed to learn much about the tiny features of cells that were invisible in a beam of light.

The research preceding the pioneering discoveries with the electron microscope was equally significant. It became possible in the 1930s and 1940s to fractionate cells by breaking them open in various ways and then running the fragments for long periods of time in ultra-high-speed **centrifuges**. Based on the size and density of different cell components, the fragments sedimented at various levels in the centrifuge tube. Thus, certain cell parts could be highly concentrated and studied intensively. Some components were characterized functionally and according to size before the techniques of electron microscopy had revealed their true structural nature.

Many were responsible for these significant achievements in centrifugation techniques and electron microscopy. In 1974 the Nobel Prize for physiology or medicine was awarded to Albert Claude, Christian deDuve, and George Palade for their discoveries concerning the structural and functional organization of the cell. There is little question that the work they and others did to open new vistas of biology was as important and difficult as the original work with light microscopy.

Another technique of research that has proved to be of immense value to biological study, including investigations of gene action, is known as **tissue culture**. In the process, cells are removed from a living organism and maintained artificially in laboratory containers of various sorts (in vitro) by supplying nutrients and environmental conditions that are a close approximation of those in the original body. The technique was first used in 1909 by Ross Harrison of Johns Hopkins University and then developed to a fine degree in 1912 by Alexis Carrel of Rockefeller Institute with chick heart cells.

Human cells have been maintained as important reference cultures by some laboratories for many years. The cells have been utilized as basic research material for all sorts of studies and have been extremely important, due to the constancy of their responses. It is even possible to remove and grow human cells from an individual to study chromosomes and the proteins which they produce.

Many interesting experiments have been conducted with human cells in vitro, and an overwhelming amount of basic information is likely to be gained in the years ahead. Nightmarish scenarios that are only a short step from the present can be constructed if one considers that such cultures are really **clones** or separate and distinct extensions of the person from whom they were originally isolated. Although it can be considered only for the distant future when the techniques of tissue culture are highly perfected, it might be possible to induce the development of a complete human from the isolated culture cells.

The accumulation of basic information about living organisms and the genetic mechanism controlling them has been dramatic since the time of Mendel. While there was a microscope in the monastery at Brünn, Mendel knew little of the fundamentals of cell structure that are common knowledge to high school students today. By the beginning of the twentieth century the light microscope had revealed a wealth of information to the understanding of genetics and provided a basis for future study.

Very rapidly after about 1950, sophisticated electron microscope studies, centrifugation experiments, tissue culture techniques, bacterial and viral work, and elegant biochemical research all combined to shed extensive light on the hereditary mechanism. A most detailed understanding of the nature of the gene and the way it controls cells and whole organisms was developed. Even more elegant and revealing studies continue into the present; the progress that has been made is nothing short of awesome.

NATURE OF THE LIVING SUBSTANCE

The living material of which a cell is composed has been the object of many studies. This substance presented an even greater mystery to early workers. The name selected for the living material was **protoplasm**, a term often still applied to all the living material in a cell. However, the term is not precise, as a living cell includes different structures and aggregations of material within its living area that were not seen using primitive microscopes. There is still a need to talk about protoplasm; it should not be considered a static mass of material, as it shows a large number of dynamic activities.

To define cellular substances chemically, a chemical analysis of the materials could be made using the analytical methods of chemistry and physics. A long list of chemicals, organic and inorganic, has been compiled for protoplasm and it varies somewhat from one organism to another. Although life does have a chemical basis, such an analysis holds little meaning, and new protoplasm cannot be manufactured by mixing such a recipe of chemicals in the laboratory. It is important to realize that the single chemical found in the largest amount in protoplasm is water. It has been called the "universal biological solvent," and many of the materials that enter into living activities are dissolved in it.

While water is the most abundant chemical found in living cells, it is not the most important in terms of association with distinct life characteristics. Carbohydrates, fats, and oils serve as energy sources and as storage compounds essential to life activities, but these roles are also not peculiar to life systems. **Proteins** are the cellular chemicals that truly distinguish animate from inanimate entities. Living cells have the capability to build in precise fashion a vast array of different proteins. Proteins not only distinguish living from nonliving, they may also distinguish individuals from one another. It is probably true that no two humans (except indentical twins) produce exactly the same array of proteins.

Many structural chemicals making up the various distinct components within a cell are protein and proteins are often found to constitute an important part of the supporting, strength-producing elements in cells. An even more important role is that many proteins serve as **enzymes**. All activities and chemical reactions within a living system are due to these chemicals. Enzymes are produced by living cells and without them there would be no life as known. Breakdown of food materials and release of energy is dependent on their presence. Thus, any cellular activity, such as movement or growth, requires enzymes. While these enzymes are usually complex, they all are basically proteins. They act as mediators expediting all the chemical reactions characteristic of life.

It is no wonder that proteins are found abundantly in protoplasm. As structural components and as enzymes, they can be considered the most characteristic chemical type in a living system. Sometimes a single protein may function simultaneously as a subcellular structural material and as the specific portion of an enzyme. For an understanding of heredity, one should know something about the process of protein formation in cells. The basis of specific differences among diverse living things, no matter how slight, is found in the ability of cells to produce different types of proteins, which in turn control the characteristics of the organisms.

CELL STRUCTURE

To learn in great detail about all the structural and functional aspects of cells is clearly beyond the scope of this text. However, some areas of **cytology** will be investigated to implant the important concepts of cells into the broader picture of living organisms. All higher forms of life are structured along an architectural plan with cells as the individual structural and functional units. Furthermore, it is important to remember that each individual cell is constructed along a general scheme that follows very similar patterns from one cell to the next. Within each cell are found discrete structures called **organelles** (“tiny organs”).

By understanding the nature of individual organelles within cells, one can understand better the overall functioning of a single cell and, eventually, the living features of whole organisms. Because this text is concerned with the hereditary nature of humans, only a few of the structural and functional aspects of cells will be covered.

The protoplasm of a cell is infinitely more complex than it appeared to the early microscopists and much more complex than some simple definitions may indicate. Within the living portion of a cell there is a complex set of **membranes** and a whole host of organelles.

One of the most significant structural arrays of molecules found in cells is that which constitutes membranes. There are many different membranes in a single cell separating various components from each other and these membranes are not all exactly alike. They do, however, share a common structural basis—a double layer of **phospholipid** molecules (fat molecules with phosphate structures attached).

The phospholipid molecules of a membrane are structured in some as yet undiscovered fashion by the cell into a neatly ordered and closely packed arrangement consisting of two layers of molecules. This tightly organized double layer of water-insoluble molecules should prohibit the passage of anything dissolved in water, but the membrane is built so that passage of various substances is not only permitted, but controlled in a selective way. The passage and control mechanism appears to reside in proteins that are embedded in the double phospholipid layer in an integral way. Some proteins are also attached to the membrane surface, providing strength and elasticity. Most knowledge of membrane structure is still based on indirect chemical or physical features, not direct microscopic observation, as the membrane and its molecules are below the resolution of even the best electron microscopes. Locations of membranes are easily determined within cells, however, and it is possible to see how they are involved in the structure of organelles.

The organellar structure of cells was virtually unknown to early workers using light microscopes, because such tiny features could not be resolved by their instruments. Without the electron microscope, modern knowledge of subcellular fine structure would be very little advanced beyond the features recognized in the early twentieth century. Extremely high magnification techniques have revealed many interesting structural features. Organelles have been shown to be membranous, particulate, fibrous, or even tubular in conformation. Various chemicals are involved in the make-up of these structures, but most often (almost invariably) proteins are important constituents.

Early workers could clearly distinguish two distinct regions within the proto-

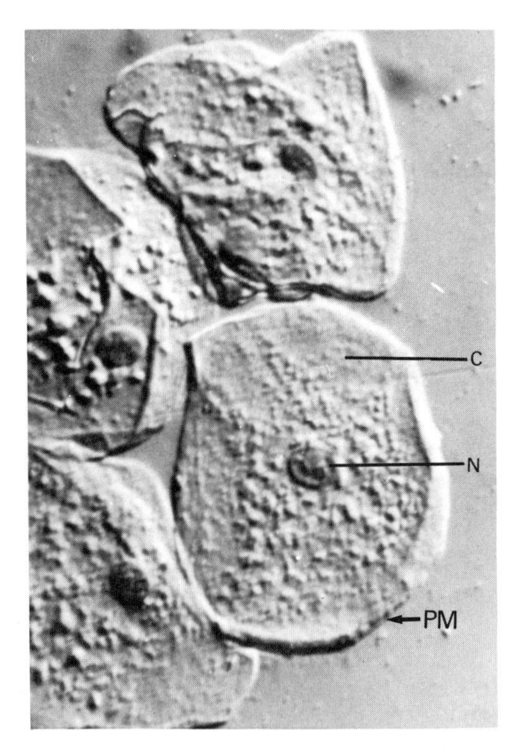

Figure 2.3. A photomicrograph of human cells (Mag. = 450×) removed from the lining of the mouth. A light microscope was used and not many details are obvious. N—nucleus, C—cytoplasm, PM—plasma membrane.

plasm (Fig. 2.3). There was found in each cell a relatively large spherical shape that generally remained constant in size and structure. This body was called the **nucleus**. The remainder of the protoplasm was referred to as **cytoplasm**.

By using various stains and careful microscopic study, the nucleus was found to contain a constant number of threadlike structures that are all replicated faithfully during the process of nuclear duplication. Due to their attraction for certain stains, they were called **chromosomes** (*chromo-*:colored, *-some:*body). Their number was found to differ among different organisms but remained constant for each separate species of organism studied. As described earlier, Sutton and Boveri recognized that these structures are the carriers of **genes** and are responsible for the inheritance of traits. Their chemical and physical structures will be covered in Chapters 4 and 5 but their importance as the carriers and ultimate determiners of genetic characteristics should be emphasized from the outset. At the moment it is worth noting that these long threadlike structures are composed of **DNA** (deoxyribonucleic acid) and **protein**. One physical feature that becomes obvious at certain times is a constriction at one point along the chromosome's length. This constriction is called the **centromere**, although it is not necessarily at the center point.

The chromosomes are embedded in a liquid within the nucleus which has been rather loosely called the **nuclear sap**. One or more small spherical structures called **nucleoli** (singular: **nucleolus**) are usually included in the nucleus, and the entire nucleus and its contents are separated from the cytoplasm by a **nuclear envelope**. Some photographs of the nucleus made with electron microscopes show that the nuclear envelope is a double-membrane structure containing numerous openings extending through both inner and outer membranes. In addition, some electron photomicrographs reveal that the outer membrane is continuous with a highly folded and compressed system of membranes extending throughout the cytoplasm (the **endoplasmic reticulum**). The exact significance of these openings or the connection to a system of membranes in a living cell is not completely clear. It is evident, however, that the possibilities exist for efficient communication between nucleus and cytoplasm. Although the pores in the nuclear envelope appear to be freely open to both sides, studies have shown that some sort of control mechanism exists which effectively permits or retards passage of certain substances.

Within the cytoplasm an array of structures is found, showing a number of different modes of organization. The size range of these organelles is fairly large, but only a few can be seen well with an ordinary light microscope (Fig. 2.4). (The maximum usable magnification of a light microscope is about 1000 times normal size.) To understand their structure, an electron microscope must be used.

The use of an electron microscope began in 1932, but, as mentioned previously, biological applications were not successfully explored until the 1950s (Fig. 2.5). When it was found that a beam of electrons might be used instead of a beam of light,

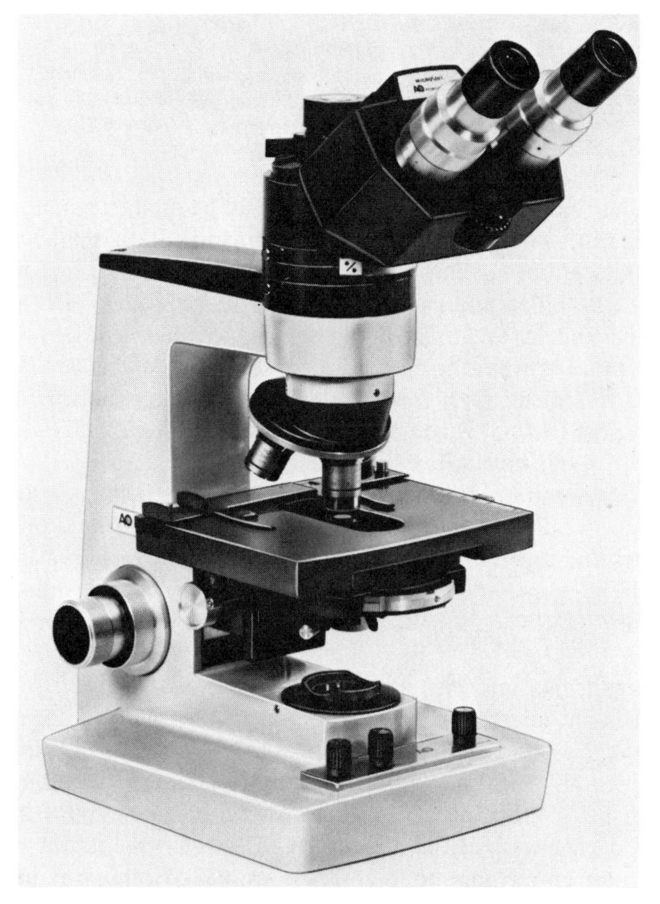

Figure 2.4. A modern, advanced student microscope, providing ease of operation, clarity, and good resolution. (Courtesy American Optical, Instrument Division)

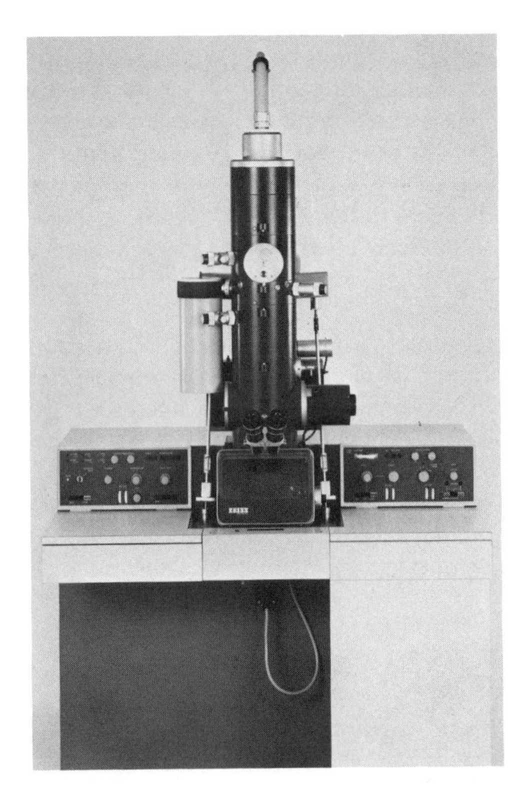

Figure 2.5. A modern electron microscope, providing simple control features and high resolution ability for biological materials. (Courtesy Carl Zeiss, Inc., New York)

a revolution in cell study occurred. The beam of electrons could be used to magnify objects up to about 1 million times normal size. This scale of magnification is equivalent to using a telescope to view unseen stellar bodies billions of miles away. Unfortunately, living cells have to be killed and sliced very thinly to be studied. In the preparation process various chemicals are utilized, and one cannot always be certain that the organelles are the same as when living. In any case the organelles are no longer functional after such treatment. Over the years, however, a large number of consistent observations have led to what is considered a valid understanding of the ultramicroscopic structure of living cells (Figs. 2.6 and 2.7).

Among the organelles found in the cytoplasm which are of genetic interest are tiny granular-appearing objects called **ribosomes**. These objects are composed of a number of protein molecules and several molecules of a chemical called ribonucleic acid **(RNA)**. Ribosomes will be discussed in Chapter 4, because they are an important part of the hereditary mechanism and are directly involved in protein manufacture.

Suspended in the cytoplasm of all cells are many small, oval-shaped objects that are barely visible with a light microscope. They are about 100 times the size of the ribosomes and electron microscopic studies reveal them to be double-membrane structures (one membrane is totally enclosed within the other). The inner membrane has many folds which form shelflike or fingerlike structures that project into the inside of the organelle. Such organelles must be mentioned in even a brief survey of cell structure because of their importance to life. They are called **mitochondria**

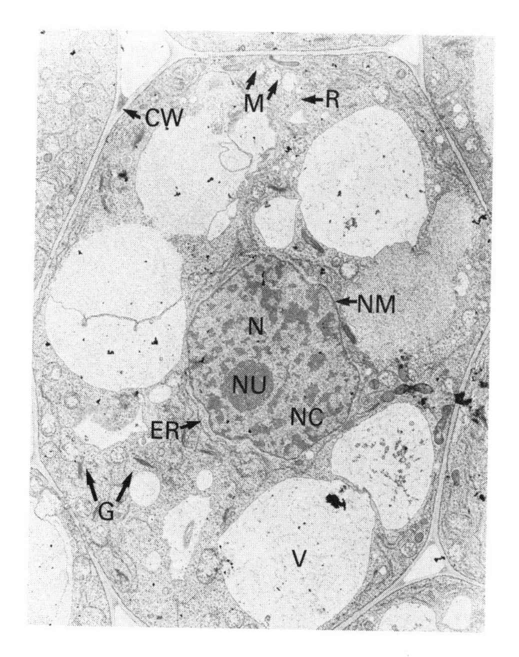

Figure 2.6. A photomicrograph of a plant cell (Mag. = 3000×) from an onion root. An electron microscope was used and considerable detail is apparent. C—cytoplasm, N—nucleus, NU—nucleolus, NM—nuclear envelope, NC—nuclear sap containing elongated and tangled chromosomes, M—mitochondrion, R—ribosome, ER—endoplasmic reticulum, G—Golgi body, CW—cell wall. Shown but not mentioned in the text are V—vacuoles (large droplets of water surrounded by membranes). The plasma membrane is not obvious because it is pressed tightly against the cell wall.

(singular: **mitochondrion**) and their structure includes a large amount of protein. They are an excellent example of the dual role of proteins. Many of their proteins serve a structural role in the membranes of the mitochondria and also act as important enzymes while attached to the membranes in a neatly organized sequential array. **Respiration,** a process which consists of food breakdown and the release of energy for all cell activities, occurs within the mitochondria. While the mitochondria are not autonomous, they have their own separate hereditary mechanism and are responsible for producing some of their own characteristic substances.

In the cells of animals and some plants, distinct central bodies or **centrosomes** may be seen with a light microscope. Such bodies appear to be minute spheres with a light halo surrounding them. Studies with electron microscopes have revealed that the centrosome is a small congregation of short **microtubules**. Careful observation reveals that there are two **centrioles** in each centrosome. Each centriole is made up of nine short triplet microtubules. Only one centrosome is found in each cell until cell division is initiated, when the structure splits, allowing each centriole to relocate at opposite ends of the dividing cell. The function of such an organelle is not known, but, when present, its division always precedes the division of the entire cell and the two centrioles serve as focal points in the chromosome-moving apparatus of the dividing cell.

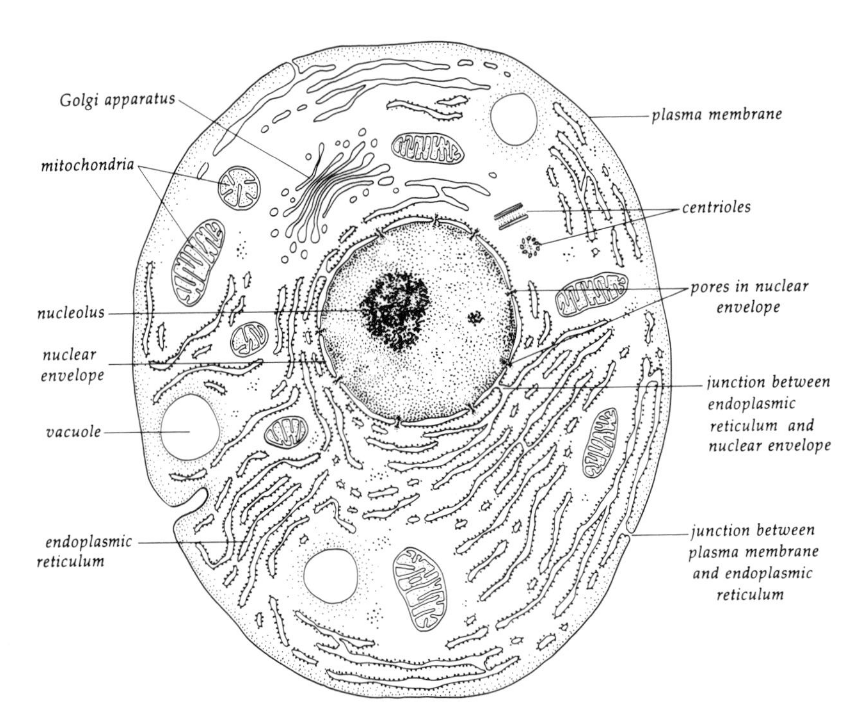

Figure 2.7. Drawing of a generalized animal cell, including many of the features that can be seen with an electron microscope. (From BIOLOGY OF THE CELL by Stephen L. Wolfe. © 1972 by Wadsworth Publishing Company, Inc., Belmont, California 94002. Reprinted by permission of the publisher.)

All these organelles are considered part of the living protoplasm. Although the cytoplasm in which they are embedded may appear to be a smooth homogeneous material when viewed with a light microscope, it contains a maze of folded and compressed membranes. Actually, most organelles in the cell are surrounded by a distinct membrane. In addition the entire outer surface of the cytoplasm is covered by a membrane called the plasma membrane. This is an important layer, for substances entering or leaving a cell must pass through it. As discussed, membranes have the ability to control selectively what passes through in either direction. Membranes always form enclosed compartments and are extremely important to the maintenance of specific mixtures of substances within these areas.

When viewed with an electron microscope, the cytoplasm is found to be permeated by the endoplasmic reticulum, the system of membranes connected to the outer membrane of the nuclear envelope. Often the ribosomes of a cell are found attached to its surface. It is well known that protein manufacture is dependent on ribosomes and that their association with the endoplasmic reticulum enhances the process in some cases. Some protein molecules are actually inserted through the membrane of the endoplasmic reticulum as they are formed by the ribosomes. As a result they are located within the compartment formed by the endoplasmic reticulum, separated from the cytoplasm by the membrane.

A smaller accumulation of folded and compressed membranes, somewhat similar to the endoplasmic reticulum in appearance but distinct from it and lacking ribosomes on its surface, is generally found in cells. This appears to be a small, irregularly shaped object when stained with silver salts and viewed with a light microscope. It is called the **Golgi apparatus**, a peculiar name which commemorates the discoverer of the structure. Functionally it is involved in the secretion of materials from cells, since it always shows the highest degree of development in situations where cells are actively secreting various substances, such as digestive enzymes from cells lining the intestinal walls. There is good evidence that proteins which were originally sequestered within the endoplasmic reticulum are moved to the Golgi bodies. Here they are sometimes modified chemically, often having carbohydrate portions attached to the proteins. The Golgi bodies "package" these molecules within small membrane-bound sacs destined to join the cell membrane and thereby secrete the contents to the outside of the cell. Such membrane-bound droplets are called **secretory vesicles**.

Some cells secrete a layer of nonliving material around their outer surfaces through the action of synthetic processes within the cytoplasm. A major difference between plant and animal cells is the presence of a nonliving layer or layers of cellulose impregnated with other chemicals on the outside of plant cells. This rigid enveloping layer is called the **cell wall** and it accounts for the strong, rigid form of plant bodies. In humans a very important cell type that also secretes a strong layer of insoluble salts around its outer periphery in a manner somewhat analogous to plant cell walls forms the bones.

The basic unity of cellular structure among all plants and animals, originally recognized by Schleiden and Schwann early in the nineteenth century, is an important foundation for the study of genetics. It explains the essence of life and provides a necessary concept for consideration of the role and mechanism of genes in human characteristics. Without an understanding of the cellular nature of all organisms, a study of their heredity and variation can have little meaning.

Table 2.1 summarizes by name and function the various components of cells that are important in the consideration of human heredity.

CELL DUPLICATION

To understand the process of heredity, the nature of cell division must be understood. The major function of cell division in an organism is growth through an increase in the number of cells. Traditionally, the process is called **cell division,** but it involves much more than a simple splitting of one cell to form two. Cell division is a multiplication process in which certain cell parts are precisely duplicated. Through this precise duplication mechanism, each cell in a plant or animal body receives exactly the same genes as the original single cell received from its two parents.

It would be impossible to divide a cell and produce two normal cells through some microsurgical technique. By cutting a cell into two halves, some key parts would be divided unequally and the cells produced probably could not survive. Prior to nor-

TABLE 2.1. Summary listing of various components of typical cells.

Cellular Components			Descriptions and Functions
Protoplasm	Nucleus	Nuclear envelope	Double membrane controlling the entrance or exit of materials.
		Nuclear sap	Suspending medium (liquid) of other nuclear components.
		Nucleolus	Aggregation of RNA molecules that will be utilized in constructing ribosomes.
		Chromosomes	Long threads of DNA and protein that are physical carriers of genes.
	Cytoplasm	Ribosomes	Small particles composed of RNA and protein that are essential to protein manufacture.
		Endoplasmic reticulum	Extensive internal membrane system to which ribosomes are attached and in which some newly constructed proteins are accumulated.
		Golgi bodies	Flattened membrane arrays that serve to package manufactured chemicals for secretion.
		Vesicles	Small membrane-bound droplets or accumulations of material.
		Centrosomes (centrioles)	Tiny bundles of microtubules that serve as polar centers of chromosome attraction in dividing cells.
		Mitochondria	Double-membraned organelle that is site of food breakdown and energy release.
		Plasma membrane	Membrane that surrounds the cell and controls entering and exiting materials.

mal cell division, these key parts usually divide to permit a splitting of the cell. Thus, the **centrosome** of a cell divides and the two new centrosomes move to opposite ends of the cell prior to the actual division of the cell. It may be that many cell organelles, such as Golgi bodies, divide before cell division occurs. Such processes involve duplication and then separation of the two new organelles, although this is hard to follow microscopically. Most important to the procedure of creating two viable cells from one, of course, is the precise duplication of the nucleus and the **chromosomes** contained therein. As the carriers of the two sets of genes, chromosome duplication and separation are of utmost importance to the survival and functioning of the two new cells. Thus, a biologist studying cell division concentrates on the nuclear phenomena which can be observed.

Mitosis

As emphasized earlier, the important directing parts of a cell, the chromosomes, are contained within the nucleus. These chromosomes exist as long, thin threads of protein and deoxyribonucleic acid. Each chromosome contains the directions for manufacture of many important proteins, all essential to the normal life of the cell. The nucleus must be carefully divided prior to the division of the cell to insure that all chromosomes and their genes will be present in each new cell. The division of the nucleus has been given a special name—**mitosis.** Because of its importance, a student traditionally spends more time studying **nuclear division** than all other aspects of cell division.

Mitosis involves the duplication of all the chromosomes and their equal separation and distribution to each of the two new cells. Such a process involves much movement and rearrangement within the cell. Because all cells exhibit the same general course of movements, mitosis has been divided into a number of easily identified phases. It should be remembered that a cell does not reach a phase and stop to be identified; the phases are conveniently identified stages in a continuing process. Because the phases all intergrade, one must depend on certain recognizable features to decide which phase is seen.

The nature of the chromosome complement of a cell should be understood at the outset of a study of cell division. The chromosomes exist in a single nucleus in pairs. Keep in mind that all living organisms produced by **sexual reproduction** start their lives as a single cell. This single cell results from the joining together of two reproductive cells **(gametes)**, an **egg** from the mother and a **sperm** from the father. Each reproductive cell brings an entire set of chromosomes to the new individual. Therefore, **two sets of chromosomes** and their **genes** are found in the cells of the new individual. With some notable exceptions, equal hereditary qualities are contributed by each parent.

Because there are two of each type of chromosome, the number of chromosomes is even. This paired state of chromosomes is called the **diploid condition** and it is maintained in the cells of an organism's body by the process of mitosis. In this process each chromosome originally present in the fertilized egg is faithfully duplicated, producing enough chromosomes for two cells at each division. The fertilized egg divides many times to produce a body which may contain literally millions of cells. Because of mitosis, every one of the cells contains the same paired complement of chromosomes and genes. In human reproduction the sperm cell carries 23 chromosomes and the egg cell carries 23 chromosomes. Thus, when fertilization occurs, the resulting fertilized egg **(zygote)** acquires 46 chromosomes. Through mitosis this complement of 46 chromosomes is maintained for the vast majority of the 100 trillion cells in the mature human body. There are some exceptional tissues, such as liver cells, that vary from the typical 46 chromosomes, but this is statistically unusual and results from duplications that are not associated with divisions.

The beginning of mitosis is recognized by careful examination of the nucleus. Several changes occur which indicate that the division process has begun. Before any of these changes can be seen with a microscope, a nucleus is considered to be in the nondividing phase. This condition is called **interphase** (''between phases''). In reality, it is during interphase that the chromosomes are duplicated. Therefore, although interphase is not considered part of the division process, one of the most important parts of the process is accomplished while the cells appear to be inactive in regard to mitosis.

In early studies of mitosis, workers showed little interest in interphase, since no activities could be discerned within the nucleus. When chemical studies indicated that it was during the prolonged interphase that DNA was replicated, however, considerable attention was focused on this portion of the cell cycle. Interphase has been conveniently divided into categories labeled **G1, S,** and **G2.** S indicates **synthesis of DNA** or the replication phase; G stands for **gaps** in the cycle during which growth and cellular organization occur.

Thus, the cell cycle consists of four major stages: G1, S, G2, and M (mitosis). The process of mitosis usually occupies only a small proportion of the cell cycle (about 5 to 10%), even in rapidly dividing tissues (Fig. 2.8). The time required for a newly formed cell to divide to produce two new cells usually is about 12 to 24 hours, depending on conditions and type of cell.

The cell cycle can be represented as a circle:

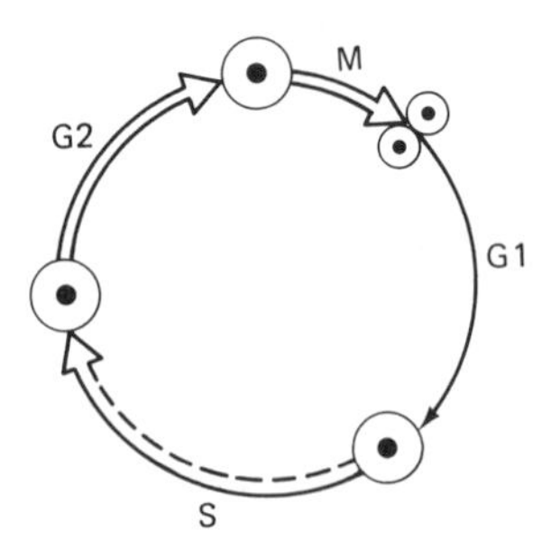

(Single lines indicate the presence of unreplicated chromosomes, while double lines indicate that chromosomes have been replicated.)

The first recognizable phase of mitosis is called **prophase.** During prophase there is a preparation for the separation of the chromosomes. Because the chromosomes are very long, thin, threadlike bodies during interphase, they are tangled into what appears to be a disorganized mass within the nucleus which must be untangled prior to the separation. At the start of and throughout prophase, however, the chromosomes are seen to shorten and thicken. This shortening and thickening allows the chromosomes to be seen as distinct rod-shaped objects of much shorter dimensions. The shortening is due to the fact that the threadlike chromosomes coil up.

While the shortening and thickening of the chromosomes is the most prominent feature of prophase, other important differences may also be seen. The nuclear envelope and the **nucleoli** disappear during prophase and the chromosomes move freely out of the confines of the nucleus. The **spindle** begins to form in the area surrounding the shortening chromosomes.

The *spindle* is so named because of its similarity in shape to the tapered wooden dowel used in an old-fashioned spinning wheel (Fig. 2.9). It consists of a number of **spindle fibers** that are composed of protein molecules organized into tiny tubular structures (**microtubules**). The microtubules run longitudinally through the spindle

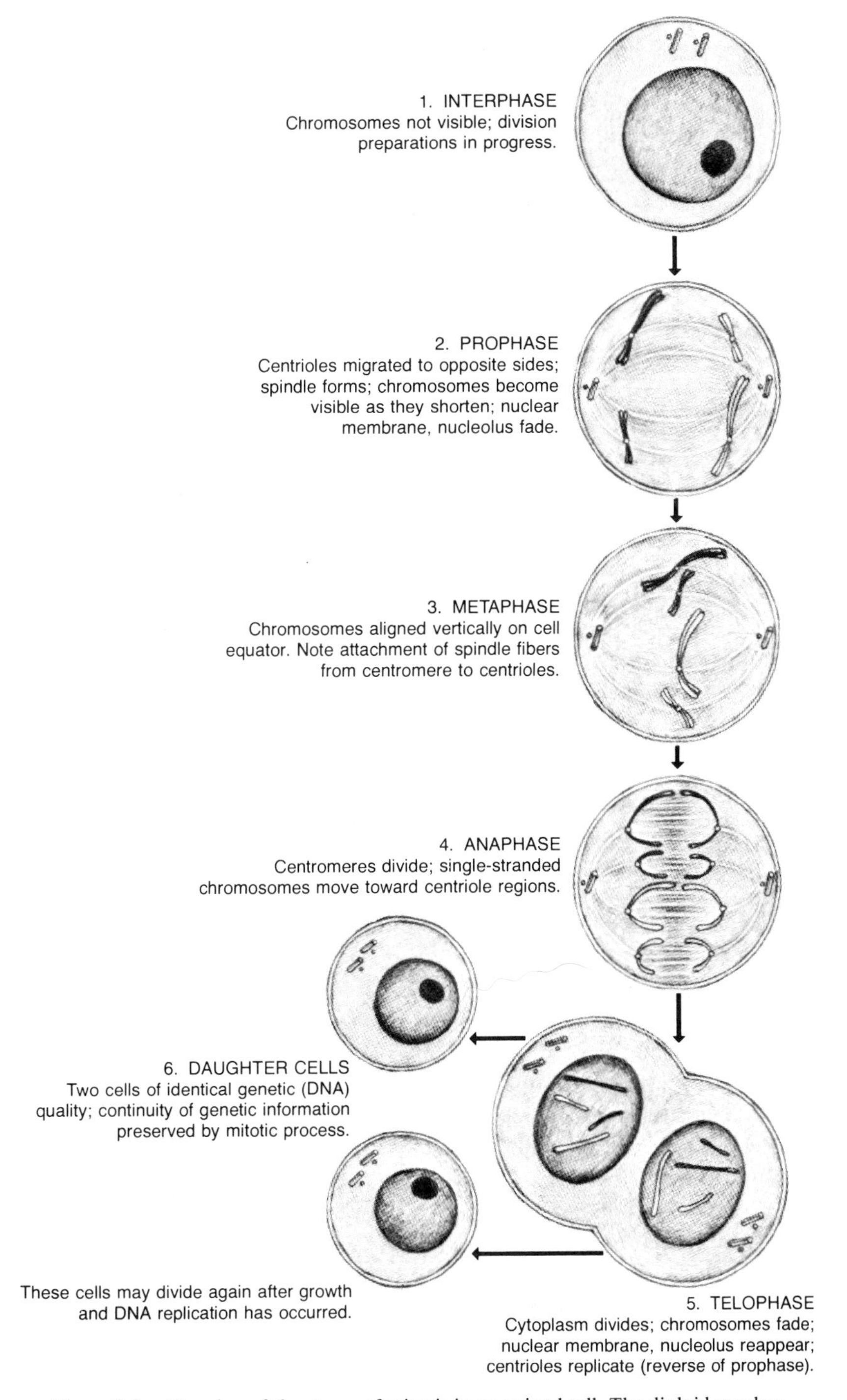

Figure 2.8. Drawing of the stages of mitosis in an animal cell. The diploid number of chromosomes here is four (two pairs). (Reprinted by permission of Scott, Foresman and Company, © 1978.)

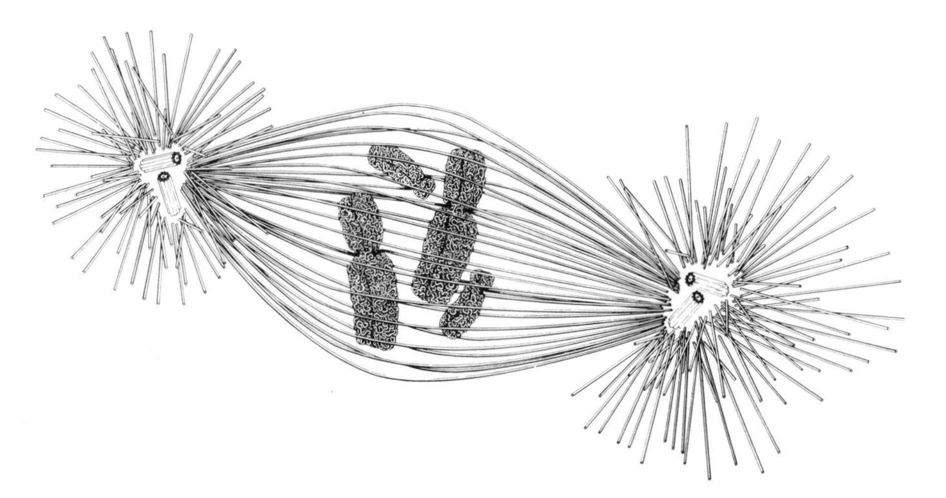

Figure 2.9. Drawing of the spindle apparatus of an animal cell. The spindle and asters are shown to be composed of microtubules. Each pole has two centrioles at its center. Note that the chromosomes are shown in the contracted state as highly twisted and folded strands. Diploid number is four chromosomes (two pairs). (Courtesy E. J. Dupraw and Academic Press, Inc.)

and are all basically parallel to each other. They are somewhat spread apart near the center of the cell and drawn closer together near their ends, or **poles.** Some microtubules are attached to chromosomes at the **centromeres** and are obviously involved in chromosome movements.

The centrosome appears to be associated with spindle formation, because it divides at the start of prophase allowing the separation of the two pairs of centrioles. The original two centrioles in the centrosome are duplicated during interphase at the same time as the chromosomes. These move toward opposite ends of the cell and are found at the poles of the spindle. Radiating out in all directions from the two centrioles are microtubules similar to the spindle microtubules. The two clusters of radiating microtubules at each end of the spindle are called **asters,** because of their star-shaped appearance. Centrioles and asters are found in animal cells, but are not present in higher plants (Fig. 2.10). It has been postulated that they are centers of microtubule formation, but this has been difficult to prove.

Shortening and thickening of the chromosomes and the development of the spindle continues into the next phase of mitosis, **metaphase.** This phase is easily identified, however, since the chromosomes make a distinct organizing movement. Due to some as yet undiscovered force, probably the movement or growth of the spindle microtubules exerting a force on the centromeres, the chromosomes line up within the spindle along a midplane through the center of the cell.

Actually, the ends of the chromosomes may extend toward one pole of the spindle or the other, but the centromere of each chromosome is located directly in the midplane. Centromeres have already been mentioned as spindle microtubule attachment points, and, like the centrioles, some cytologists think these are also centers of microtubule formation.

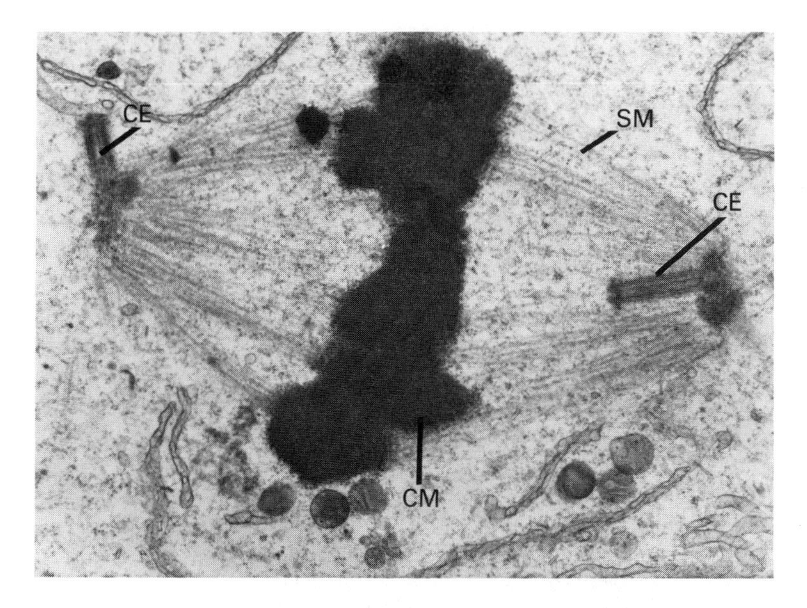

Figure 2.10. An electron photomicrograph of a mitotic spindle in an animal cell (Mag. = 14,000×). The centrioles (ce) and microtubules (sm) can be seen, but the details of the chromosomes (cm)are not clear. Generally, chromosomal structure is difficult to study with an electron microscope because of the torturous, tangled nature of the strand of DNA and protein. (Courtesy J. Richard McIntosh, University of Colorado)

Recall that each chromosome was duplicated during interphase. This duplication occurred in a longitudinal fashion so that two parallel threads were formed. The mechanism which accounts for this exact duplication will be explored more thoroughly in Chapter 4. Each duplicated chromosome is free from its exact replica, except in the centromere region where the two are attached to each other physically. These attached duplicates are called sister **chromatids.** Thus, each chromosome at this stage is represented by two attached chromatids that are identical. Metaphase then is the lining-up of these attached pairs of chromatids along a midline in the cell. The chromosomes are set at a starting point for movement to either end of the cell.

Metaphase ends when the chromatids start to move toward the poles of the spindle. The phase during which this movement occurs is called **anaphase** and the movement begins with the division of the centromeres which hold the duplicate chromatids together (Fig. 2.11). Once the centromeres have divided, the like chromatids (again called chromosomes) are free to move in opposite directions to the ends of the cell. At this point, the mechanism which insures one of each type of chromosome in each new cell can be understood.

The moving force in anaphase chromosomes is not really understood. It appears to be due to a pulling force, and it seems to involve the centromeres, as they lead the way toward the poles. It is important to remember that the spindle microtubules are attached to the centromeres, since it is likely that the motive force is dependent on these microtubules. The ends of the chromosomes trail out toward the midline of the spindle in the center of the cell as the centromeres are drawn to the

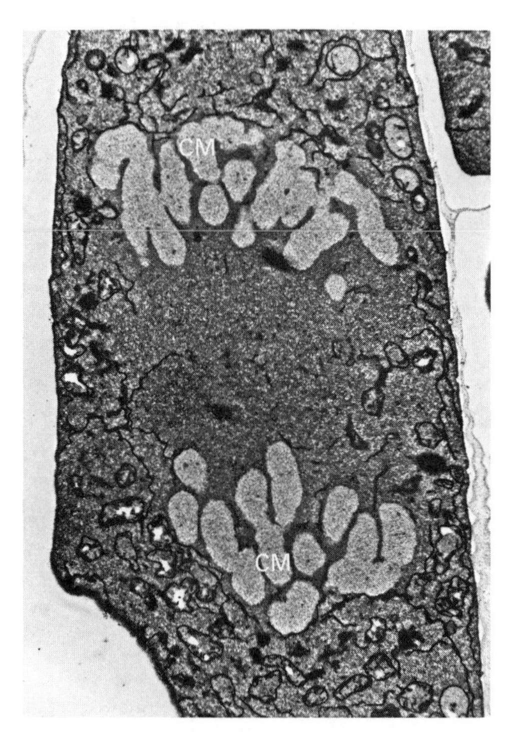

Figure 2.11. An electron photomicrograph of a plant mitotic cell at late anaphase (Mag. = 5500×). Note the groups of chromosomes (cm) at each end of the cell. The spindle is not apparent because the killing agent did not preserve its structure.

poles. It would seem that the microtubules shorten at this stage and thus draw the chromosomes into the two ends of the dividing cell. The cell and spindle lengthen during the separation process, which may also contribute to the poleward movement.

When the chromosomes reach the poles, movement stops and the last phase of mitosis is reached. This last phase is known as **telophase** and involves two groups of chromosomes, one at each end of the cell. These groups represent two new nuclei forming at each end of the cell. In many ways telophase is just the reverse of prophase. During telophase the chromosomes lengthen, become thinner, and form a tangled mass at each end of the cell. The nuclear envelopes reform around each of the two new nuclei and nucleoli reappear in each nucleus. The spindle disintegrates and disappears.

Once nuclear division has been accomplished, **cytoplasmic division** can occur. This does not always occur, and in certain situations cells with several nuclei may result because no cytoplasmic division follows mitosis. Sometimes chromosome duplication occurs but the chromosomes are not separated and organized into new nuclei. Chromosome numbers in excess of the typical complement may result, as in the unusual cells of liver tissue. These are uncommon circumstances, however, and will not be of concern to this book's discussion of normal cell functions.

Cytoplasmic division does not require the precise, orderly duplication, alignment, and separation processes of mitosis. It is simply a matter of splitting the cytoplasm so that each half receives a single nucleus with the other necessary organelles. In animal cells, cytoplasmic division occurs during telophase through **cleavage,** a process in which a furrow or indentation is seen to form around the periphery of the cell. This furrow deepens and progresses inwardly until the two masses of cytoplasm are separated. The weakened plane along which this separation occurs appears to be due to an accumulation of membrane-bound vesicles that have migrated into the equatorial plane of the spindle during telophase. Additionally, there is a band of microfilaments girdling the cell at this point. These contractile structures seem to shorten, causing a constriction that cuts through the cell.

In most plant cells having rigid cellulose walls on their outer surfaces, cleavage cannot occur. A new wall must be built between the two new cells through a process called **cell plate formation.** As the spindle disappears during telophase, it appears to condense toward the midline and the first layer of the cell wall is formed there. This accumulation of material is derived from the migration of vesicles into the central plane of the disintegrating spindle. As the plant cells mature and grow, other layers are added to this original cell plate by cellular activities.

The details of mitosis are important to a thorough understanding of two very important aspects of the hereditary process (Fig. 2.12). First, the repeated duplication of the complement of genes in the chromosomes is essential to provide every cell in an organism with a full set. Second, the orderly separation and movement of the duplicated chromosomes to the two new cells must be accomplished. Both processes are made apparent by carefully analyzing cell division.

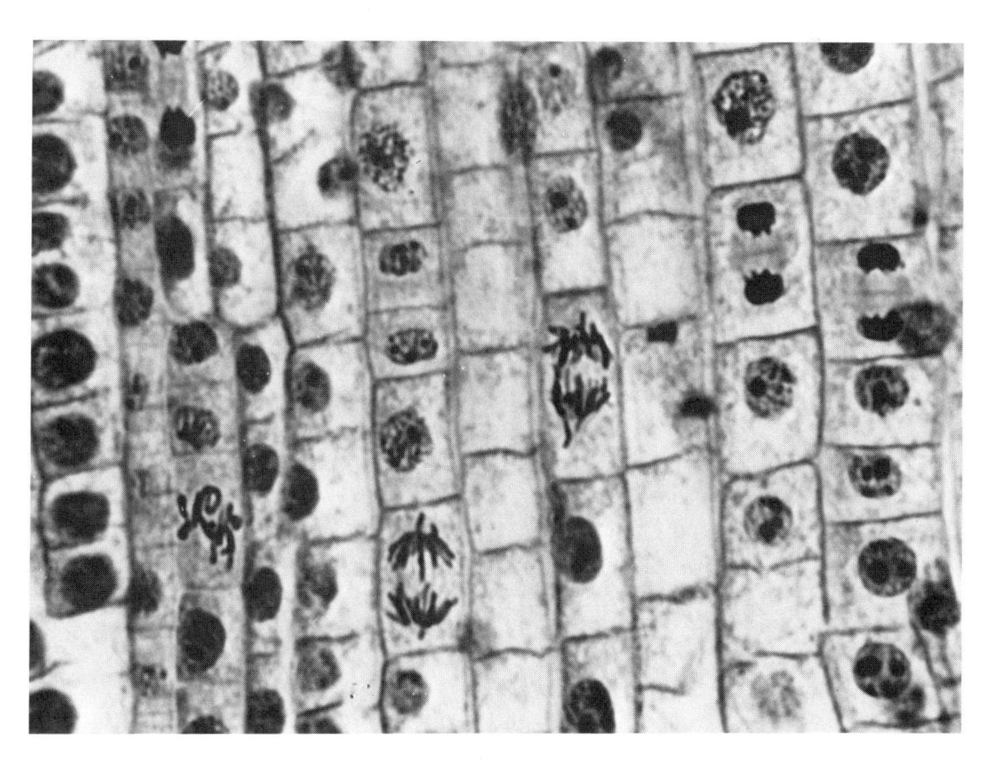

Figure 2.12. A section of a plant root tip which has many cells undergoing mitosis (Mag. = 550×). The section has been stained to show nuclei and chromosomes.

Meiosis

Mitosis is basically the same in all plants and animals and much understanding of the process in humans has been gained from knowledge of the details in other, simpler living things. Cell division, preceded by this precise duplication and reorganization mechanism, permits billions of cells to be developed in an individual's body. Thus, with a few exceptions, every cell in a body has the same hereditary factors in its chromosomes as every other cell in the organism. However, there is a modification in the cell division process that occurs in specialized regions of plant and animal bodies. The modified division process is known as **reduction division** or **meiosis.**

In the late 1800s and early 1900s attention was focused on the cellular details of sexual reproduction. It was recognized that chromosome numbers were constant in a species and that chromosomes were passed on to the next generation through sperm and egg cells. August Weismann predicted in 1887 that a reduction division was essential prior to the fusion of reproductive cells. The process of meiosis was discovered shortly thereafter and studied intensively in several organisms.

To permit the fusion of two reproductive cells (sperm and egg) during sexual reproduction, the paired **(diploid)** nature of the chromosomes must be reduced to a state where only a single chromosome of each type exists in a cell (**haploid**). If this did not happen, the constant number of chromosomes in the cells of each species would not be maintained generation after generation. The process of meiosis accounts for this reduction of chromosome number from diploid to haploid in the production of gametes. In humans meiosis takes place only within the reproductive organs: the **testes** of males and the **ovaries** of females.

Mendel guessed correctly in 1865 that his unit factors or genes were in pairs (diploid). This seemed logical to him, since the cells of organisms came from a fusion of material from two parents. He also recognized that these pairs separated from each other in gamete formation (segregation). This was obvious since a new organism must receive only one gene for each trait from each parent. Mendel hypothesized that every reproductive cell had only a single set of genes (haploid), and that both parents contributed one-half the genes to an offspring. Sutton and Boveri in 1903 saw the parallel between the paired nature of the genes and the chromosomes and realized that the chromosomes were carriers of the genes. They knew that meiosis occurred in gamete formation and that each gamete received only one set of chromosomes, rather than the double set which results from mitosis.

Actually, meiosis is a modified mitosis which consists of two divisions of the nucleus, with only one duplication of the chromosomes. Thus, four cells are produced, each having only half the number of chromosomes of the original dividing cell (Fig. 2.13). Meiosis consists of the same phases as mitosis, but there are two of each phase in the sequence, since there are two divisions: interphase, prophase I, metaphase I, anaphase I, telophase I, an abbreviated interphase followed by prophase II, metaphase II, anaphase II, telophase II, and interphase.

Three major changes in the normal mitotic process occur during meiosis. The first of these occurs in prophase I and is called **synapsis.** In synapsis the chromosome pairs are attracted to each other through some unknown force. There is a precise, point-by-point alignment of each pair of chromosomes. Thus, two chromosomes

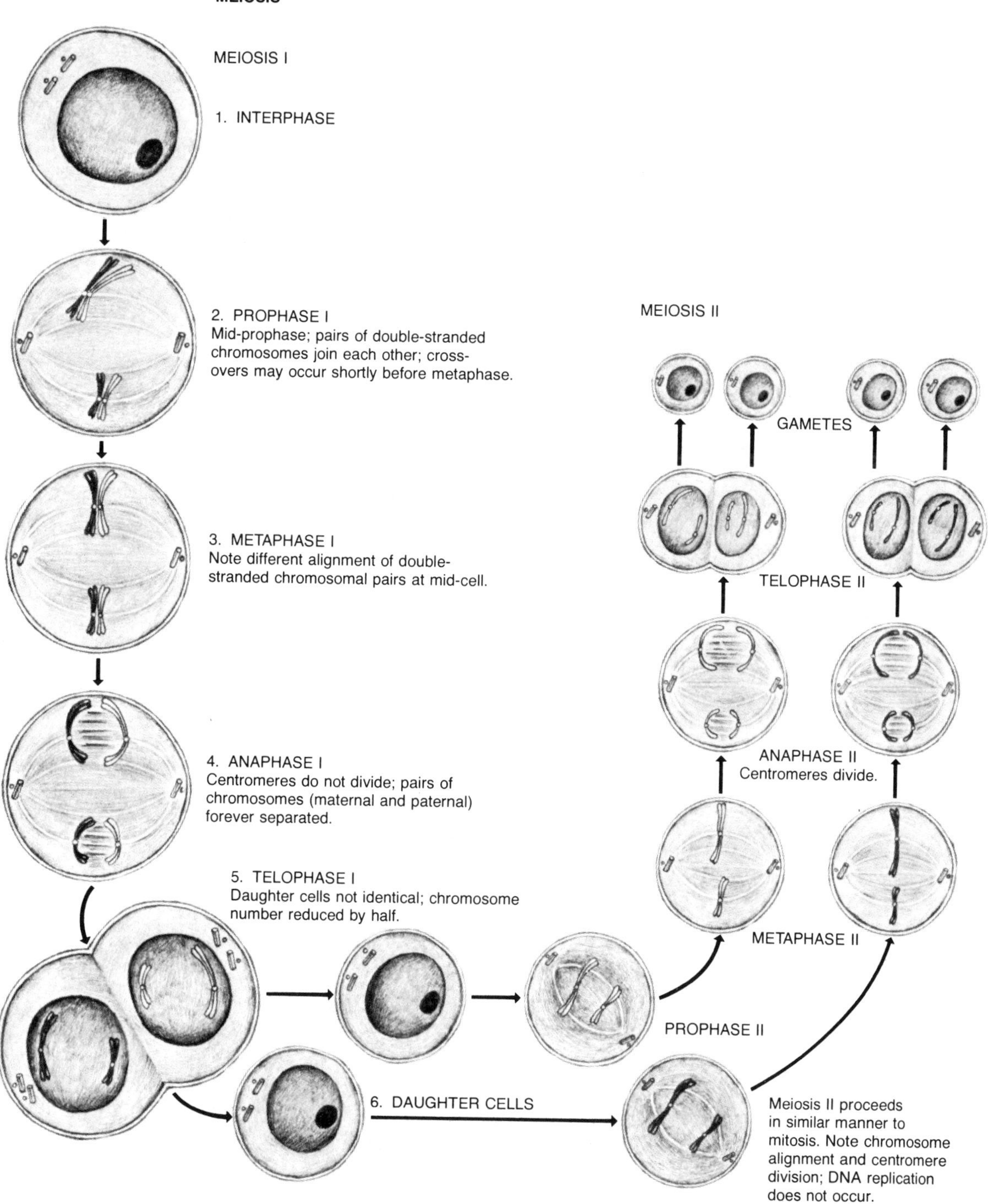

Figure 2.13. Drawing of the stages of meiosis in an animal cell. The diploid number of chromosomes is four (two pairs), while the reduced or haploid number is two (unpaired chromosomes). For simplicity, crossing-over is not indicated, although it commonly occurs during synapsis in prophase I (note Fig. 2.14). (Reprinted by permission of Scott, Foresman and Company, © 1978.)

carrying genes which effect the same traits, but which have been contributed by two different parents (**homologous chromosomes**), are pulled together in a precise parallel arrangement (Fig. 2.14). As in the interphase preceding mitosis, the single chromosomes have been duplicated prior to this movement. In reality there are four threads in very close parallel association. This process is very important to genetics, because it permits a mixing of genes from the two parents through **crossing-over,** the exchange of parts through breakage and joining of chromosomes from different parents.

The second major difference from a normal mitosis is seen when anaphase I begins and the chromosomes start to move to opposite ends of the cell. Normally, in mitosis the centromeres holding the pairs of duplicate chromatids together divide, allowing them to separate and move to opposite ends of the cell. In meiosis, however, the centromeres do not divide during the first sequence of phases. While the mechanism underlying this refusal of the centromeres to divide has not been proved, it is possibly the result of synapsis and the consequent orientation of all spindle microtubules from a single centromere toward one pole. As movement begins, each centromere leads the way toward a pole but it carries with it two chromatids. Keep in mind that there were four potential chromosomes associated in synapsis. As a result of the synaptic association, one centromere moves to one pole, carrying with it two potential chromosomes. Meanwhile, the other centromere moves to the opposite pole carrying with it the other two potential chromosomes. Therefore, as the first series of phases ends, the chromosomes are left in the duplicated condition and each consists of two chromatids attached by a centromere. The two cells produced then enter a shortened interphase.

At this point the third major difference from normal mitosis is realized. The chromosomes are in a duplicated state at the end of the first meiotic series and there

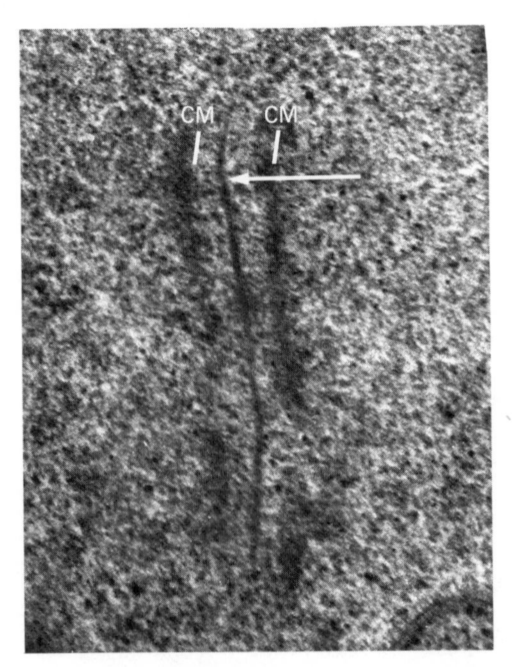

Figure 2.14. Synaptinemal complex in fungus *(Armillaria mellea).* When synapsis occurs during prophase I of meiosis, a structured array of molecules exists between the paired homologous chromosomes. This synaptinemal complex occurs in all cells undergoing meiosis and has a regular form consisting of two lateral elements held tightly to the chromosomes and one central element with a space on either side separating the three elements. Seen at the arrow is a typical central element between two homologous chromosomes (cm—dark, thick, fuzzy lines). Magnification approximately 48,000×. (Courtesy D. C. Peabody, J. J. Motta, and *Can. J. Bot.*)

is no further duplication preceding the second meiotic series of phases. Prophase II therefore begins as a normal mitotic division, except that the chromosomes have been left in a duplicated condition from the first series, not as a result of duplication during interphase. Metaphase II, anaphase II, and telophase II all follow in an essentially normal mitotic sequence in each of the two dividing cells. The centromeres divide normally and individual chromosomes move to the poles in anaphase II.

As a result of meiosis, four cells are produced, each having half the number of chromosomes of the original cell. This reduction is the direct result of a change from the paired condition of chromosomes (diploid) to the unpaired condition (haploid).

With these chromosomal movements in mind, one may understand the physical basis for Mendel's fourth principle, random assortment. Notice that during synapsis each **bivalent,** or set of four strands, consists of a duplicated pair of like chromatids of **maternal origin** aligned with a duplicated pair of like chromatids of **paternal origin.** They are aligned with either the maternal or the paternal pair oriented toward one end or the other of the dividing cell on the basis of chance. Thus, when separating during anaphase, the paternal pair may move to one end, while in another bivalent either the paternal or maternal pair will move in the same direction.

Notice in Figure 2.13 the anaphase I separation follows a pattern in which chromosomes of like origin (same shading) move to the same ends:

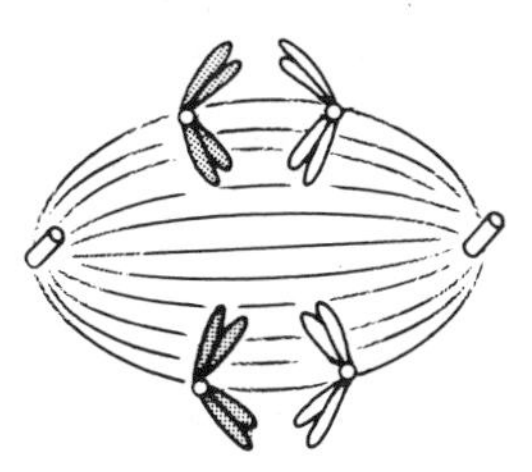

The reverse would yield similar results:

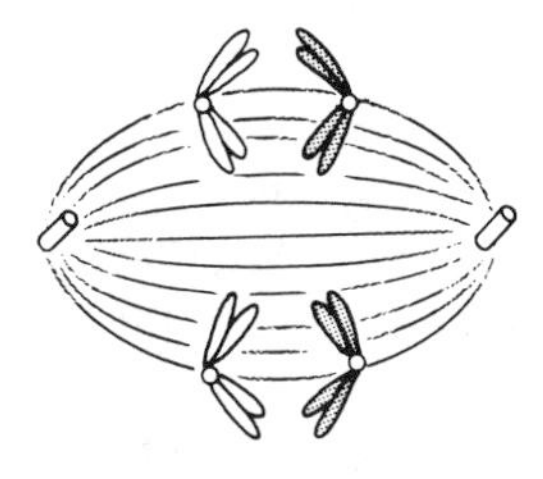

Two other patterns could be expected with equal frequencies, resulting in completely randomized separations:

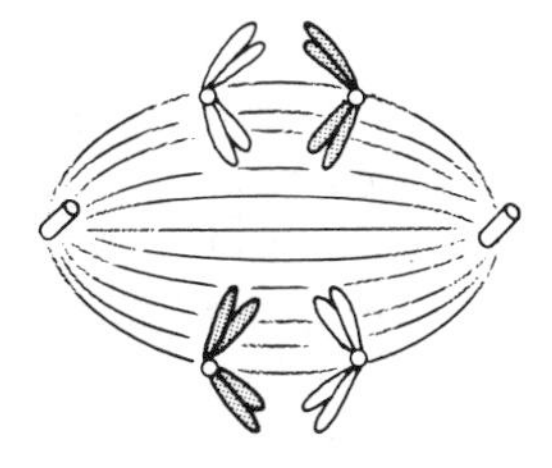

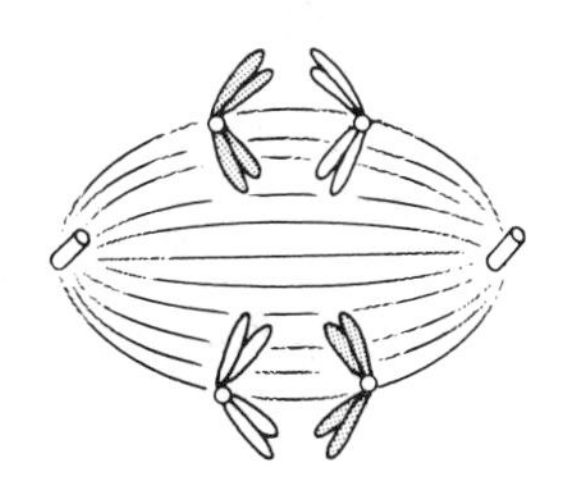

It is apparent that nonhomologous chromosomes are independent in their synaptic orientations and subsequent separation to the new cells. Thus, the block of genes carried on each chromosome is independent of the genes carried on other chromosomes and the result is random assortment.

The randomizing process is further enhanced by crossing-over, which consists of breakage and rejoining of adjacent synapsed chromatids. This phenomenon leads to a strand of genes of maternal origin becoming attached to a strand of paternal origin, while the reverse occurs in the adjacent chromatid. A consideration of the diagram on page 115 will clarify the results of this process.

The essential differences between somatic cell division (mitosis) and reduction division (meiosis) are based on events that take place in prophase I of meiosis. The key feature that accounts for the halving or return to the haploid number of chromosomes is found in the process of synapsis. Because of its importance to the process of meiosis, prophase I was subdivided into five stages by early workers (Fig. 2.15). These substages were based primarily on the actions of the chromosomes as they lined up in a fashion peculiar to the meiotic process. Although more is now known about the process, the substages are still useful.

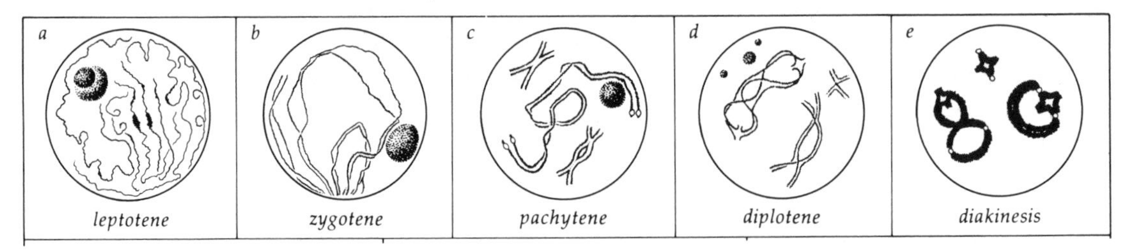

Figure 2.15. The five stages of meiotic prophase I. Drawing shows characteristic features. (Courtesy Bernard John, Australian National University, Canberra)

Careful observation reveals that the chromosomes begin to shorten and thicken in the earliest stage of meiosis, as they do in mitosis. At this point microscopically they appear to be unduplicated. However, chemical analysis has revealed that the DNA has been doubled prior to this time. The workers of the early 1900s thought that chromosome duplication occurred subsequent to this time, and it may be that full-fledged chromosome replication has not yet been completed. At any rate, this stage is called **leptonema** ("slender thread").

When synapsis begins, the homologous chromosomes, representing one each of maternal and paternal origin, pair up in zipperlike fashion starting at various points along their length. This stage is called **zygonema** ("adjoining thread") and the chromosomes still appear to be single strands.

Pachynema ("thick thread") is the next stage and the chromosomes continue to shorten during this stage. Although splitting of each into two chromatids is considered to be complete, the chromosomes cannot be observed clearly as double or replicated threads. Each synaptic figure is considered now to consist of four strands and is referred to as a **bivalent** chromosome (tetrad of chromatids). Significantly, although points cannot be seen clearly where crossing-over has occurred, adjacent nonsister chromatids are considered to have broken and rejoined with their neighbor reciprocally at certain points. In effect, a portion of a chromosome of maternal origin has recombined with a portion of a chromosome of paternal origin. This results in a significant mixing of genes that are linked together within individual chromosomes. This phenomenon will be explored in more depth in Chapter 5. The diagrams on page 115 may be helpful at this time.

At **diplonema** ("double thread") the double nature or two-chromatid structure of each homologous chromosome can be observed microscopically. Furthermore, during this stage the homologous, doubled chromosomes are repulsed from each other and points where crossing-over has occurred between adjacent chromatids can also be observed microscopically. These obvious points are called **chiasmata** (singular = chiasma).

Finally, the end of prophase I is marked by **diakinesis** ("divided across") with the replicated chromosomes in a very shortened and thickened condition. The chiasmata within bivalents can be seen, but they have been moved toward the ends of the chromatids as the homologous pairs show further repulsion. Because of the shortened nature of individual bivalents, this is a good stage for counting the number of chromosomes in a cell.

It is interesting to consider the implications of the general reproductive mechanism as it involves chromosomes and genes. Although an individual serves as a reservoir of genetic information passed on from two parents in equal quantities, only one-half of this information can be passed on to any offspring. The reproducing individual also serves as a mechanism for the mixing in randomized fashion of the two sets of genetic information. A simple mathematical analysis is necessary to describe this process. If only one pair of chromosomes were possessed by an organism, there would be only two possible types of cells produced by meiosis. If there were two pairs, four combinations or mixtures of the parents' chromosomes would be possible (2×2 or 2^2), with three pairs there would be eight ($2 \times 2 \times 2$ or 2^3), and so on. In humans with 23 pairs of chromosomes there would be 2^{23} possible combinations or 8,388,608 different types of cells which could be produced!

When one considers that the large number of genes linked together in a single chromosome are not fixed together in an unbreakable combination because of crossing-over during synapsis, the possibilities for mixing the genes passed along by an individual's parents indeed become astronomical.

To gain a true understanding of genetics in sexually reproducing organisms such as humans, it is most important that meiosis be understood. It is the basis for recombination and variation of parental traits among offspring. Without such a reduction in the number of chromosomes in gametes, the normal Mendelian pattern of inheritance would not exist.

ADDITIONAL READING

AVERS, C. J., *Cell Biology* (2nd ed.). New York: D. Van Nostrand, 1981.

BERNS, M. W., *Cells* (2nd ed.). Philadelphia: Saunders College, 1983.

BRACHET, J., "The Living Cell," *Scientific American* (1961), 205, No. 3, 50–61.

DEROBERTS, E. D. P., and E. M. F. DEROBERTS, Jr., *Essentials of Cell and Molecular Biology* (7th ed.). Philadelphia: Saunders College, 1980.

HOPKINS, C. R., *Structure and Function of Cells.* London: W. B. Saunders, 1978.

HURRY, S. W., *The Microstructure of Cells.* Boston: Houghton Mifflin Co., 1964.

JOHN, B., and K. R. LEWIS, *The Meiotic Mechanism* (Oxford/Carolina Biology Reader No. 65). Burlington, N. C.: Carolina Biological Supply Co., 1973.

________, *Somatic Cell Division* (Oxford/Carolina Biology Reader No. 26, 2nd ed.). Burlington, N. C.: Carolina Biological Supply Co., 1980.

MAZIA, D., "How Cells Divide," *Scientific American* (1961), 205, No. 3, 100–23.

________, "The Cell Cycle," *Scientific American* (1974) 230, No. 1, 54–64.

WOLFE, S. L., *Biology of the Cell* (2nd ed.). Belmont, Calif.: Wadsworth, 1981.

REVIEW QUESTIONS

1. Diagram a cell and indicate by arrows a pathway that might be involved as a ribosome is manufactured in the nucleus, then moves out and becomes attached to the endoplasmic reticulum. Indicate next in similar fashion how a protein might be synthesized, moving through the endoplasmic reticulum to a Golgi body, and then to the outside of the cell.
2. Cite several examples to indicate how the characteristics of an organism might be influenced by genetic control of protein synthesis.
3. What advantages can be seen in the separate compartmentalization of the hereditary material (genes and chromosomes) within the nucleus?
4. What significance can be attributed to the separate, but incomplete, genetic mechanism of the mitochondria?
5. Look carefully at Figures 2.6 and 2.7. Find the points where the three major differences between mitosis and meiosis occur. Copy the pertinent diagrams on a separate paper and indicate with arrows and notations the differences.
6. **(a.)** Using two different colors make a diagram of a diploid cell with two chromosomes. Indicate which chromosome originally was contributed to the cell by the father with one color; use the other color for the maternally contributed chromosome. Using the same color scheme, show what the chromosome complement of the four daughter cells will be if the cell undergoes meiosis. What fraction of the daughter cells carry the paternal chromosome? The maternal chromosome?
   (b.) Do the same with a diploid cell that has two pairs of chromosomes (four total). How many *different* types of daughter cells can be produced?
   (c.) Do the same with a diploid cell that has three pairs of chromosomes (six total). How many *different* results now?
7. Following the same procedure as in Question 6, assume that crossing-over occurs at one place in each synaptic pairing. Remember that crossing-over involves two strands only, so that in the process only the two adjacent chromatids would be affected, not the two outside chromatids. How many different types of meiotic daughter cells would be produced by the diploid cell with one pair of chromosomes? Can you estimate without drawing how many would be produced by the diploid cell with two pairs?
8. Explain how the chance mechanism of independent assortment is explained by an understanding of the process of meiosis.

9. (a.) Using Figure 2.12 on page 35, count all the cells in interphase, prophase, metaphase, anaphase, and telophase. Represent each as a percentage of the total number of cells. Why might not these figures be truly representative of a real situation?
   (b.) If a cell cycle lasts 12 hours (from a newly divided cell to two new daughter cells), how long would each of the five stages be? (Use the percentage in part (a.) to divide the time proportionally.) Do these figures seem reasonable for future predictions?
10. Why does it seem necessary to assume that chromosome replication is complete and that each chromosome consists of two chromatids in pachynema? In answering, look carefully at the fact that chiasmata and crossing-over have been shown to occur in pachynema.

CHAPTER SUMMARY

1. Biological studies of cells utilize microscopes and the history of their development has determined progress in understanding cell structure. The first successful light microscope was constructed about 1600 and cells were observed and named in 1665. Matthias Schleiden and Theodor Schwann recognized in 1838 that living material has cells as the basic unit of life.
2. Basic light microscopes with magnifications of about 1000 times life size reached their technological peak in about 1900 and their continued use has refined the details of many different types of cells. In the 1950s biologists devised techniques that permitted study of cellular fine structure with powerful electron microscopes and details 1000 times smaller were made visible. Fantastic new perspectives of the ultrastructure of cell organelles have resulted from the use of the electron microscope.
3. Tissue culture techniques and centrifugation studies have helped to reveal the details of the functional significance of the myriad of cellular organelles that are now known to cytologists. Important to an understanding of genetics are many cellular components: nuclei, cytoplasm, chromosomes, nucleoli, endoplasmic reticulum, ribosomes, centrosomes, Golgi bodies, and mitochondria.
4. Cells divide or duplicate to form two daughter cells as a part of an organism's growth. The chromosomes must be replicated and apportioned equally to the two cells in this process (mitosis). Based on the condition and placement of the chromosomes, mitosis is recognized as consisting of intergrading stages: prophase, metaphase, anaphase, telophase. Non-mitotic cells are said to be in interphase.
5. A specialized type of cell division found in the production of reproductive cells involves a modified type of mitosis (meiosis). Meiosis involves two full division cycles, but only one duplication of chromosomes. As a result, chromosome number in the reproductive cells is halved. This corresponds to Mendel's process of segregation in which genes are transmitted singly to the offspring.
6. An extremely important part of meiosis is observed during the first division cycle (prophase I) when synapsis occurs. Homologous chromosomes line up in precise two-by-two fashion and crossing-over or recombination of chromosome parts occurs. This contributes greatly to the diversity of offspring as randomized mixing of the various genes is enhanced.

CHAPTER

3

HUMAN REPRODUCTION

Either sex alone is half itself and in true marriage lies
Nor equal, nor unequal.

Tennyson, The Princess (1847)

At the time of conception the full genetic potential for an individual is contained in a single cell. This cell **(zygote)** has received essentially equal amounts of hereditary material from each of two parents. From this simple beginning—the fusion together of two **gametes** or **sex cells**—the complex, well-integrated, and smoothly functioning system of organs and tissues called the human body is developed. About 100 trillion cells, from repeated mitoses in the embryo, are the result of the development of this first cell in the life of an individual. The combination of genes and the resultant proteins they determine, as well as the environmental circumstances surrounding the maturation of the embryo, fetus, child, adolescent, and adult, result in a person who is unique as a member of the species.

The mechanisms involved in the development of a human from a particular single cell in the uterus of a female are complex and varied, but scientists are beginning to understand some of the more important aspects of the reproductive process. Interest in the control of population size and the alleviation of problems of congenital birth defects has been heightened through an increased awareness of the needs of modern society.

Because of the nature of the reproductive process and its intimate association with human culture and society, many misunderstandings have resulted. The complexity of the highly evolved reproductive mechanisms and the difficulty in studying the details of many aspects in human bodies have made advances slow and difficult.

Probably no single person has had more influence on the development of scientific thoughts and ideas than the Greek philosopher Aristotle. His formulation of the study of natural history and use of logic in understanding the laws of nature profoundly influenced formal thinking for 2000 years and more. Unfortunately, many of his ideas were not representative by present standards of sound scientific logic, but his influence was great among his followers and some of his thoughts were accepted as dogma.

Without ever seeing the results of detailed studies based on observations of the anatomical structure of the human reproductive system and without understanding the true nature of sperm and egg production, Aristotle in about 300 B.C. proposed some ideas on the "generation of animals." He especially related his thoughts to humans in regard to the perpetuation of the species.

Another early Greek worker, Hippocrates, sometimes called the "father of medicine," had proposed a theory of inheritance based on the production of tiny particles by each part of the body that were passed on to the offspring (**pangenesis**). Aristotle rejected this idea, but did not propose an alternative to explain how a person acquires the necessary messages for a complete body through the sexual intercourse of his parents. Aristotle's conclusion was that the menstrual fluid of the female contributed the building material of the new individual, while the male supplied a mysterious *dynamis*, an unexplained character or life-giving force.

The idea that all parts of the body contributed particles, seeds, or gemmules to the emissions produced in sexual reproduction was not adequately dispelled by Aristotle. Actually, the hypothesis was only modified by his observations and logic: For a long time educated scientists and philosophers thought that "coagulating semens" from the male and female formed the **fetus**. Even Charles Darwin returned emphatically to the old idea of pangenesis to explain the transmission of hereditary traits for each of the body's organs to the next generation. His second cousin, Francis Galton, in 1871 attempted blood transfusions in rabbits to demonstrate that microscopic particles or **pangenes** freely circulated in the bloodstream. Although, the hypothesis proved to be wrong, Darwin did not reject the idea of pangenesis but simply observed that the blood was not the medium for the collection and transmission of the particles of inheritance. Today one still uses that old cliché, "It's in his blood," when "It's in his genes!" would be more appropriate.

Even before Darwin's return to the erroneous ideas of pangenesis in the middle 1800s, a microscopic observation of one side of the human reproductive story had been made. Anton van Leeuwenhoek reported this observation with his primitive microscopes in 1677. He had seen the tiny swimming **sperm** cells in **semen** from his own body and was puzzled by their possible involvement in the reproductive process.

Van Leeuwenhoek thought that each sperm contained a minute, completely formed human in its head. All that was needed for development was implantation into a woman's body. Other microscopists followed van Leeuwenhoek's idea, even diagramming the preformed **homunculus** in the sperm they observed (Fig. 3.1). There was another group who assumed that a complete being was present in the female gamete or egg.

The idea of preformation was convenient and persisted until more careful studies with better microscopes were done. August Weismann in the 1800s, building on his own observations and those of others, was able to supplant such erroneous concepts with ideas concerning the importance of nuclei in **sperm** and **eggs** to the transmission of traits and the development of new individuals. Once the basic underlying mechanism of the fusion of two **gametes** and their nuclei was discovered, a more careful analysis

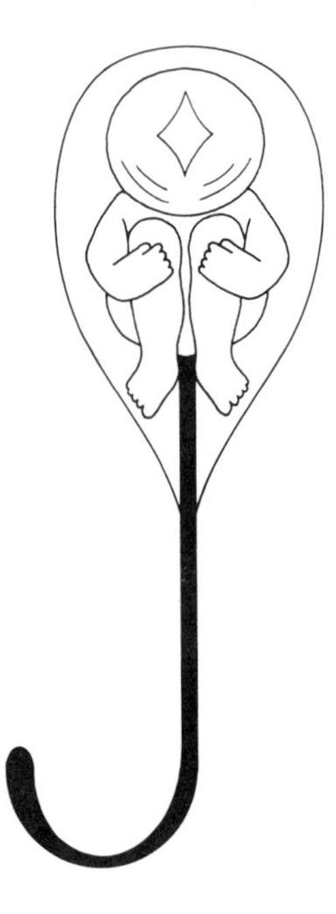

Figure 3.1. Homunculus of the type diagrammed by seventeenth-century microscopists. Early microscopists who apparently subscribed to Aristotle's view—that males contribute substance and females contribute nutrition in reproduction—thought they saw completely preformed humans in sperm cells. These early workers belonged to a group often referred to as "spermatists."

was begun of the features of sex cell production and delivery to the site of **fertilization** and **development.**

The societal taboos surrounding human sexual conduct have hindered the development and dissemination of an understanding of the psychology of sex. Knowledge of physiological aspects has also been slow because humans are not acceptable experimental organisms and the human biological processes are not always the same as those of animals which are more acceptable for experimentation.

In spite of the many handicaps, knowledge of human sexual psychology, physiology, cytology, and embryology has been accumulated in recent years. This text will concentrate primarily on the cellular aspects of the origin of a new life as the hereditary bridge between generations. The delivery of the reproductive cells to the site of **fertilization** and the **development** of the individual will be discussed briefly. Although such topics are important for an understanding of human genetics, as are the roles of hormones in sex (among other psychological and physiological aspects of the process), a detailed discussion is beyond the scope of this text.

SEX DETERMINATION IN THE FETUS

Because equal numbers of chromosomes (23) are supplied in reproduction by both parents, one might expect to find 23 homologous pairs of chromosomes in human body cells. However, there is an important difference between male and female chromosomal complements. In males there are 22 pairs of homologous chromosomes and two essentially unpaired chromosomes of different sizes, carrying different genes. In females there are 23 homologous chromosome pairs. Based on these distinctions, one says that there are 44 **autosomes** and two **sex chromosomes.** The two homologous chromosomes in females are called **X chromosomes;** males are said to carry one **X chromosome** and one **Y chromosome.**

The early embryological differentiation of males and females is apparently based on the presence or absence of the Y chromosome. Evidence for this is found in some unusual combinations of chromosomes that have been observed in humans. Although strictly normal development is not seen in these individuals, it is apparent that the presence of the Y chromosome results in the production of the combination of phenotypic characteristics of the male. Absence of the Y chromosome results in the physical development of a phenotypic female. The control of this developmental sequence by the Y chromosome is not completely understood, but it is apparently initiated very early in life.

Chromosomal Basis

In 1891 H. von Henking, a German, was studying meiosis during sperm formation in an insect. Half of the developing gametes contained a body that stained similarly in his preparations to a chromosome or nucleolus. He referred to the stained object as "a peculiar chromatin element," but did not realize that it was an unpaired chromosome. Because it was an unknown, he used a mathematical notation and called it *X*, but he did not associate it with **sex determination.**

C. E. McClung carried von Henking's observations a step closer to a valid understanding when he showed in 1902 that the nuclei of female grasshoppers have a different chromosome number than males. The X body or chromosome could then be associated with sex determination. Because McClung noted an "accessory" or unpaired chromosome in the males, he erroneously assumed that the X chromosome was a male-determining factor.

Finally in 1905 Nettie Stevens and E. B. Wilson independently recognized that some insects have females with two X chromosomes (XX) and males with only one X (XO). They also demonstrated that other insects have a sex-determination mechanism based on females with two X's (XX) and males with a dissimilar pair which they labeled XY. The **XX = female, XY = male** situation was later found to exist in humans.

Based on the XX female, XY male mechanism, one can predict the ratio of males and females among offspring according to a simple Mendelian type of interpretation. Because homologous chromosomes segregate during meiosis, it follows that the normal disjunction of two X chromosomes during the formation of egg cells by a female would result in all egg cells carrying a single X chromosome.

Although the mechanism is not completely clear, during meiosis in sperm formation the X and Y chromosomes segregate from each other, as do the two X's in egg formation (Fig. 3.2). This probably occurs because the X and Y chromosomes have a short homologous region and show partial synapsis of the largely dissimilar X and Y chromosomes during prophase I. As a result, one-half the sperm produced carry an X chromosome and one-half carry a Y. Thus, the chances for fertilization by either an X-bearing or a Y-bearing sperm are equal and the ratio of females to males at birth should be 1:1 (1/2 females, 1/2 males).

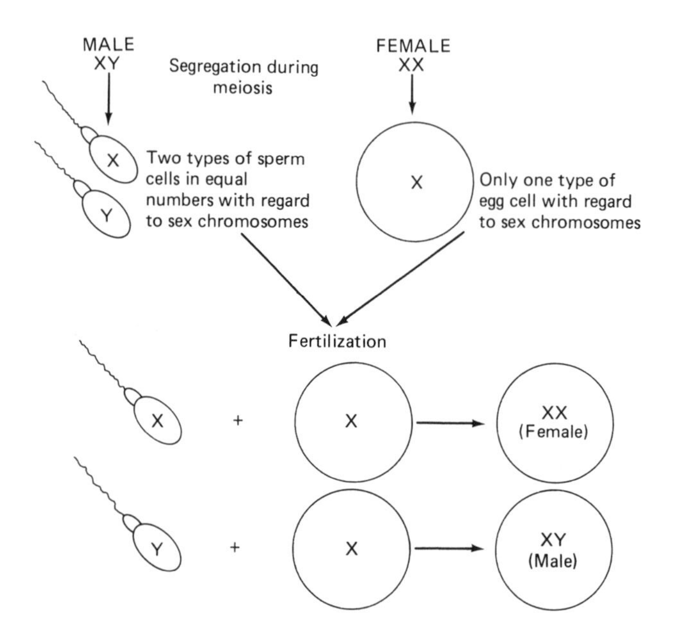

Figure 3.2. Diagram of mechanism of XY segregation and combination in fertilization. Theoretically, because half the sperm cells are X and half are Y, there is a one-half chance for conceptions producing females (XX) and one-half chance for conceptions producing males (XY). In reality, the Y-bearing sperm have an advantage and there are significantly more XY conceptions than XX.

When the ratio of human female to male births is examined, however, there are 106 males for every 100 females. This discrepancy is not understood, but it may be that the Y-bearing sperm have some sort of selective advantage over the X-bearing sperm. Some scientists think it is due to the smaller size of the Y chromosome, allowing the Y-bearing sperm to compete more favorably.

The ratio of 106 to 100 is called the **secondary sex ratio**, since it represents the number of births of the two sexes. The **primary sex ratio**, or the number of zygotes at conception, was once estimated to be considerably higher, possibly in the vicinity of 130 females to 100 males, but such estimates have been discounted in recent years as the number of male and female spontaneous abortuses appears to be similar to the 106 to 100 ratio. By age 21 the ratio of the two sexes is about equal, indicating that males do not survive to reproductive age as effectively as females.

Development of the Gonads

During the first 2 months of development within the mother's **uterus**, it is impossible to distinguish whether the paired gonads which appear during the first 2 or 3 weeks of development will become the **testes** of a male or the **ovaries** of a female. These undifferentiated organs are called **ovotestes**. During the early developmental phases, **primordial germ cells** can be found within the embryonic tissues migrating to the developing gonads. These predecessors of sperm or egg cells move into the undifferentiated gonadal tissue where they increase greatly in number through mitosis. The undifferentiated gonads contain two basic areas: an outer region **(cortex)**, which is amplified and enhanced in its development if the individual becomes a female and the organ becomes an ovary, and an inner region **(medulla)**, which will develop into the sperm-producing sections in the testes of the male. The migrating germ cells located in the cortex may lead to egg development within ovarian chambers while those within the medulla may ultimately give rise to sperm cells produced within tubules of the testes, depending on which type of gonad is formed.

It is possible that the male-determining genes of the Y chromosome act as testis growth promoters which are involved in stimulating the undifferentiated gonads to develop into the testes. In the absence of these male-determining genes at a specific time in the development of the fetus, the gonads will become ovaries. Subsequent gonadal tissue differentiation results in the production of a number of hormones leading to the physical secondary sex characteristics of males or females.

The external genitalia during the first 2 months are also undetermined (Fig. 3.3). As the gonads are potentially male or female, the external sex organs are capable of developing into typical male or female structures. Thus, certain parts of the external genitalia of one sex are homologous to parts of the external genitalia of the other, even though superficially the mature sex organs appear quite different.

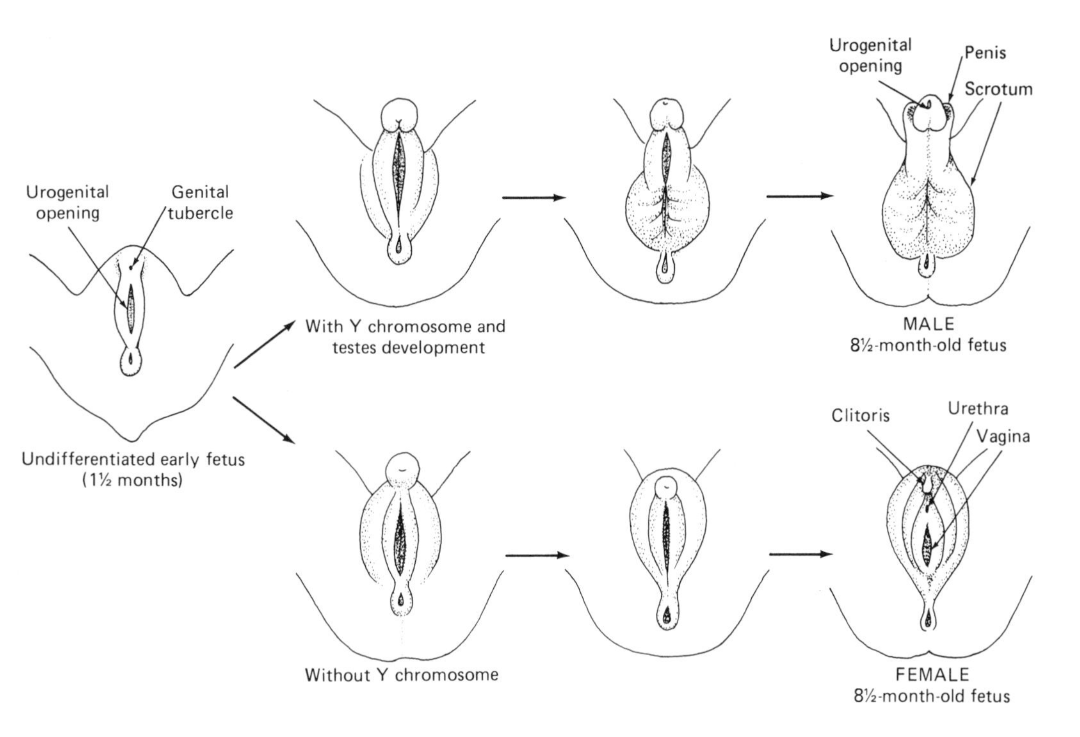

Figure 3.3. Fetal development of external genitalia. Note that in the two sexes there is a common origin for different structures, which are indistinguishable until about the tenth week after conception.

FEMALE REPRODUCTIVE PROCESS

During the second month of fetal life in the developing gonads of a female, there are probably one-half million potential gametes resulting from the migration and many divisions of the primordial germ cells. Rapid mitoses in the developing ovaries give rise to about 7 million potential eggs by the fifth month. The population of germ cells declines due to death and deterioration rapidly after that time and at birth there are about 2 million potential female gametes. By **menopause** very few of these cells remain in the ovaries.

The maturation and release of gametes from the ovaries begins at the age of **puberty** and ceases at menopause, a period of about 30 to 35 years. Generally one gamete is released per month, meaning that the maximum number of germ cells which can mature during a lifetime is about 400. (Of course, egg maturation and release from the ovaries ceases during **pregnancy**.) When one considers that it would not be physically possible for a woman to produce more than 30 to 40 children in a lifetime, the reproductive process seems enormously wasteful. In fact, all biological mechanisms of reproduction are based on this high statistical potential, thus assuring perpetuation of a species.

As the germ cells increase in number in the young ovary by the process of mitosis, they are called **oogonia** and can be distinguished by their size and appearance microscopically from other cells in the organ (Fig. 3.4). At the end of about 5 months, and after a number of divisions, the oogonia cease mitosis, growing in size to become **primary oocytes** (Fig. 3.5). They then enter prophase of the first division of meiosis in the ovaries of the fetus. These primary oocytes continue to grow in size but remain at the prophase I stage until puberty at the earliest, when the first division of meiosis is completed in one cell. This one cell enters the second division of meiosis but again halts its division at metaphase II, where it remains until fertilization or loss from the body. Cytoplasmic division in the first cycle of meiosis in the primary oocyte is highly asymmetrical, since the meiotic spindle is pushed very close to one side of the dividing cell. When the division is accomplished, the result is a tiny, nonfunctional cell called the **polar body** and the much larger **secondary oocyte**. The secondary oocyte is also destined to undergo a highly unequal division and the metaphase II figure can be seen placed very close to the edge of the cell.

Each month during the reproductive life of a female there are around 20 primary oocytes capable of completing meiosis I and entering meiosis II, but only one generally achieves the transition; the others simply disintegrate. The primary oocyte which does function is enclosed in a structure called a **follicle** (Fig. 3.6). The follicle consists of a number of ovarian cells which produce a large drop of accumulated liquid in which the oocyte and its surrounding cells **(corona)** are suspended. As a follicle matures each month, usually in alternately the right and left ovaries, the liquid in the structure increases so that a bulge develops on the outer surface of the ovary. At **ovulation** the follicle bursts and the liquid carries the secondary oocyte in metaphase II with its surroundings corona of follicle cells out into the body cavity of the woman (Fig. 3.7).

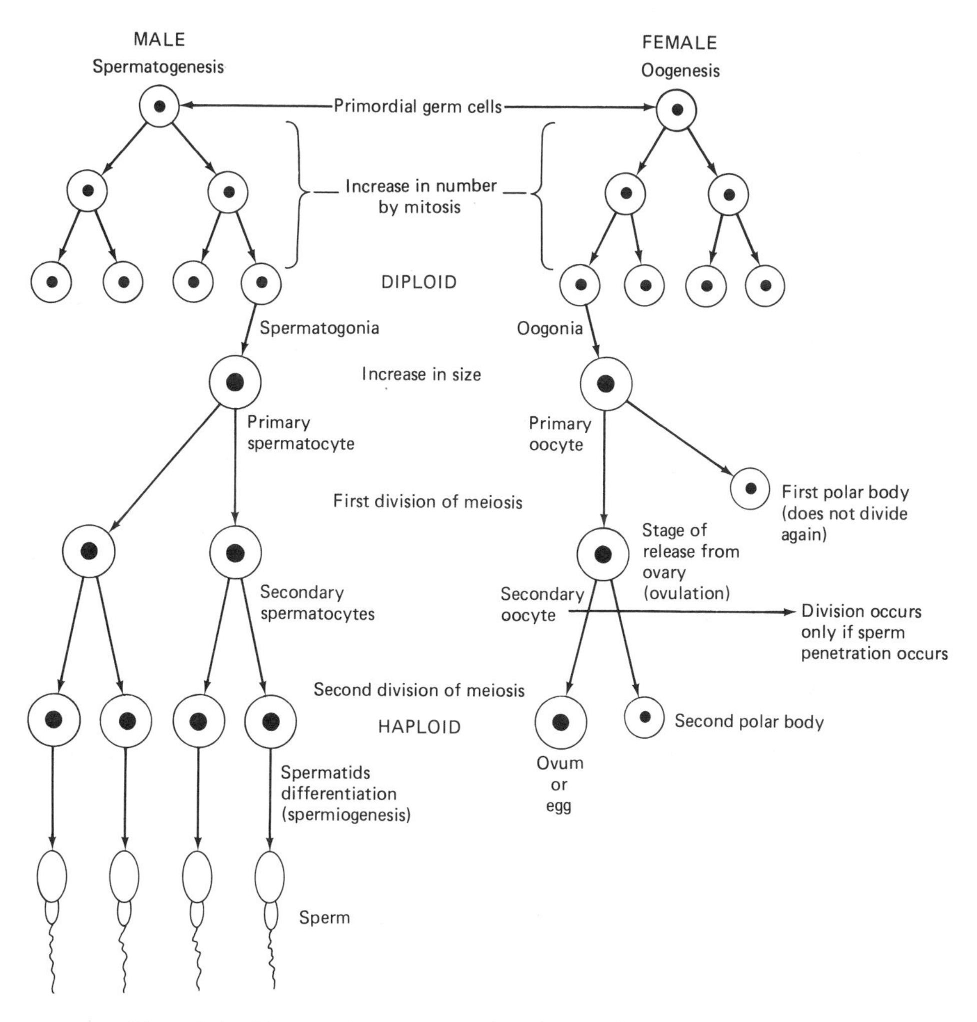

Figure 3.4. Diagram of spermatogenesis and oogenesis. The various stages are shown in diagrammatic form with each type of cell labeled.

To understand the situation following ovulation as fertilization and fetal development occur, it is necessary to refer to the anatomical structure of the female reproductive system (Fig. 3.8). Note that the ovaries and **oviducts** are not physically connected and that the secondary oocyte (the female "gamete" now) and its corona of cells actually passes through the body cavity of the female when going from the ovary to the oviduct. The funnel-shaped opening of the oviduct is lined with motile hair-like processes, whose waving motions create a current that carries the gamete down the oviduct **(Fallopian tube)** toward the uterus. If viable sperm cells are present in the upper third of the oviduct (deposited there within the previous 48 hours) or if they arrive within about 15 hours after ovulation, fertilization is likely to occur. If

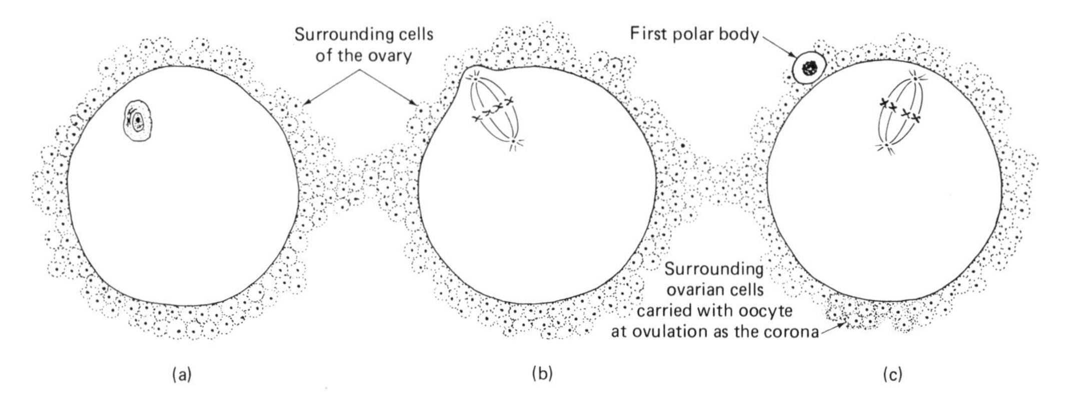

Figure 3.5. Three stages in the development of an oocyte within a follicle of the ovary. (a) Primary oocyte as it is stored in the ovary prior to the beginning of maturation (arrested prophase I). (b) Primary oocyte shortly prior to ovulation. It has entered metaphase I. (c) Secondary oocyte in the condition of release at ovulation (arrested metaphase II). It will not complete meiosis II unless fertilization occurs. Note the first polar body which resulted from meiosis I. Note also the highly asymmetrical placement of division figures (b) and (c).

no sperm cells are present, the gamete passes down through the oviduct to the uterus. It is then carried away in **menstruation** as the lining of the uterus, which has become expanded and filled with blood vessels in preparation for pregnancy, is lost. Thus, meiosis is not completed unless fertilization occurs and without fertilization a typical egg cell does not develop. By strict definition, an egg never develops in the human female, because meiosis is not completed until after the sperm cell has entered the gamete. In fact, if fertilization occurs, there is a reduced (haploid) egg nucleus present after penetration by the sperm and completion of meiosis.

The major features of the female reproductive process are summarized briefly:

1. Primordial germ cells "invade" the paired ovotestes during the first 2 months after fertilization.
2. At about 2 months after fertilization, if no Y chromosome is present, the gonadal tissue differentiates into ovaries, emphasizing the cortical regions of the organs, where oogonia divide repeatedly by mitosis.
3. At about 5 months after fertilization, the oogonia change to primary oocytes, beginning the first division of meiosis. They remain at prophase I until puberty.
4. Beginning at puberty, one primary oocyte per month becomes a secondary oocyte as it completes meiosis I and begins meiosis II, halting again at metaphase II. The follicle containing this secondary oocyte swells until it ruptures at ovulation, releasing the secondary oocyte from the ovary.
5. If fertilization occurs in the oviduct, the resultant embryo may implant in the uterus for 9 months of pregnancy. If fertilization does not occur, the female gamete is lost to the outside of the body in menstruation.

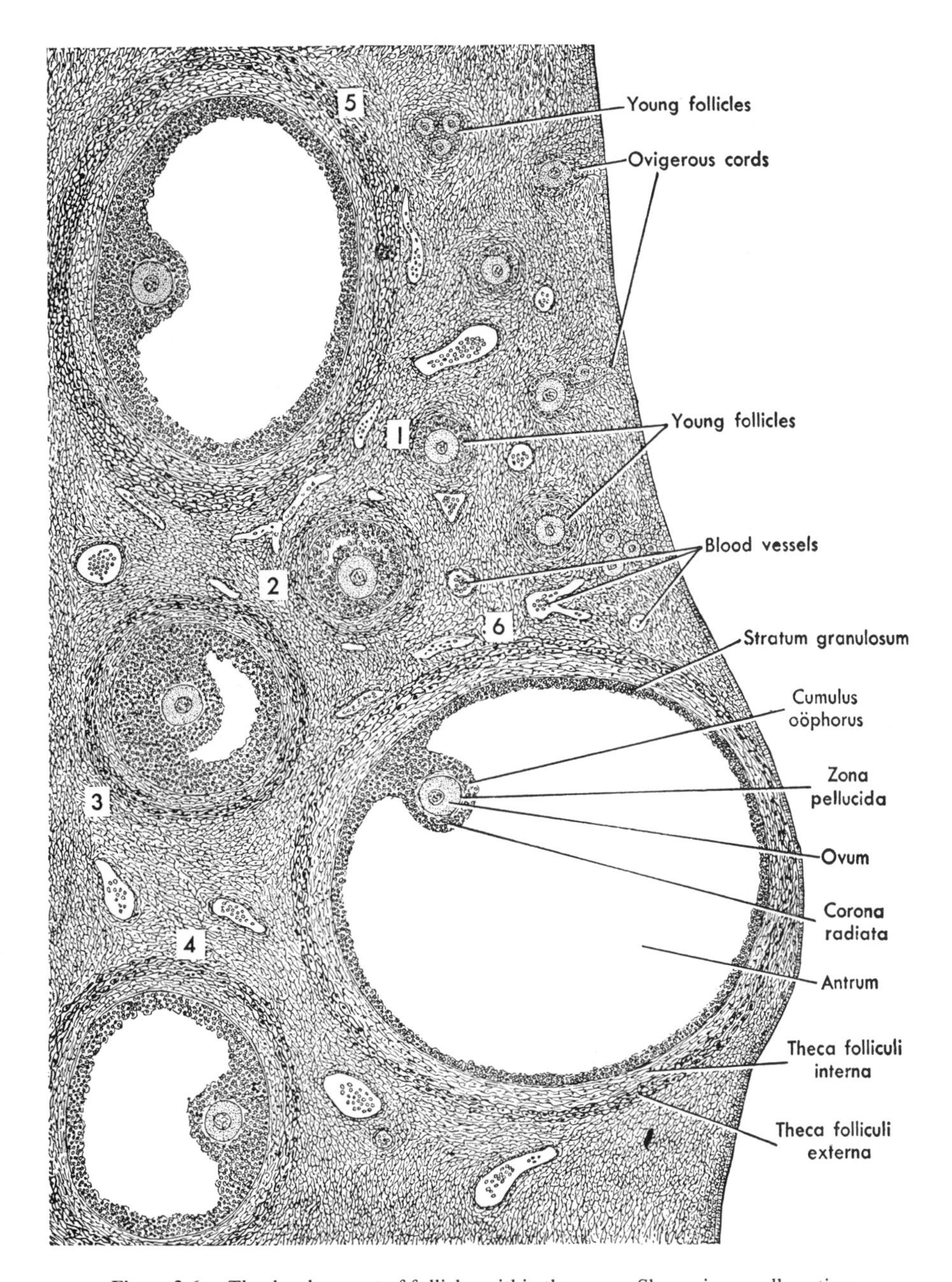

Figure 3.6. The development of follicles within the ovary. Shown is a small portion of the cortex of an ovary. The earliest stages in the developmental series are shown at the upper right with later stages numbered in sequence. Number 6 is about ready to ovulate. (Copyright 1949 by Macmillan Publishing Co., Inc., renewed 1977 by Mary R. Huettner.)

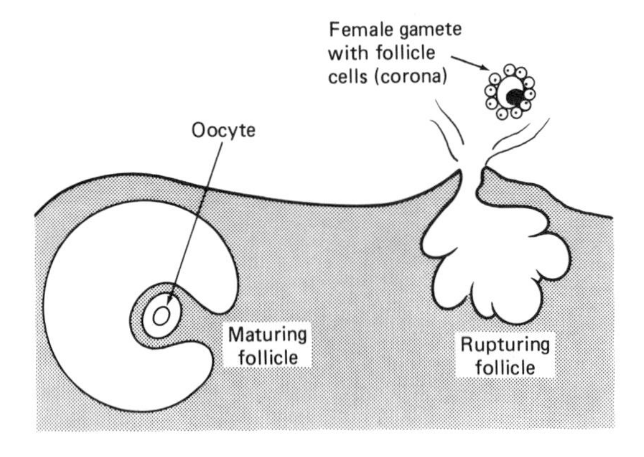

Figure 3.7. Diagram of a section through the cortex of an ovary. One follicle is shown in a developing stage with a primary oocyte, while another follicle has burst, releasing the gamete during ovulation.

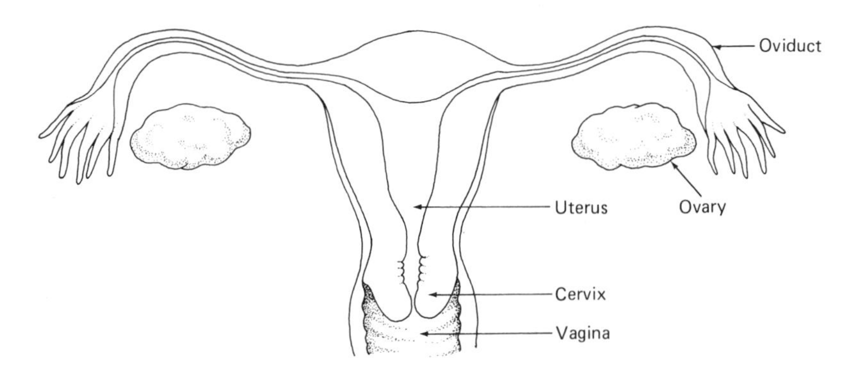

Figure 3.8. Diagram of the human female reproduction system (frontal view).

MALE REPRODUCTIVE PROCESS

If a zygote receives a Y chromosome, the centrally located areas (medullae) of the undifferentiated gonads commence differentiation about the end of the second month of uterine life. The tissues in this area are organized into solid cords containing the germ cells. These cords proliferate during the early development of the individual, elongating and fusing together in one coherent mass of tissue. The germ cells continue to divide, greatly increasing in number among the supporting cells of the cords, distinguishable microscopically by their size and appearance, as in the developing ovaries.

After birth the testes continue to increase in size, but the future sperm-producing tissue consists of a mass of solid cords until puberty. At that time (about 13 to 14 years of age), the cords become hollow tubules (Fig. 3.9). The germ cells are physically located in the cellular walls of these tubules. They are distinguishable as **spermatogonia** and divide by mitosis near the outer periphery of the tubule walls.

An important event in the development of the male usually occurs prior to birth when the immature testes descend to the outside of the body cavity through passages called **inguinal canals.** Here they are retained within a sac of skin and supportive tissue called the **scrotum.** Sperm cells will only develop under conditions of about 3 degrees lower than normal body temperature. Thus, the scrotum provides a lowered temperature environment and acts to regulate thermostatically conditions with muscles that contract in the cold or relax in warm surroundings. Such movements bring the testes close to the warmth of the body or allow them to lower to a position of reduced temperature.

At puberty the spermatogonia continue to divide and begin to push some of their products in toward the hollow centers of the tubules. Some spermatogonia remain at the periphery of the tubules and divide by mitosis throughout the life span of healthy individuals. Although the number of sperm produced may be reduced with age, there is no abrupt cessation of gamete production (there is no male counterpart of menopause).

The spermatogonia that are pushed in toward the center of the tubules begin a sequence of development leading to the formation of sperm cells. First the cells are changed into **primary spermatocytes** and undergo the first division of meiosis. As soon as this is complete, and while forced closer to the center, they become **secondary spermatocytes** undergoing the second division of meiosis.

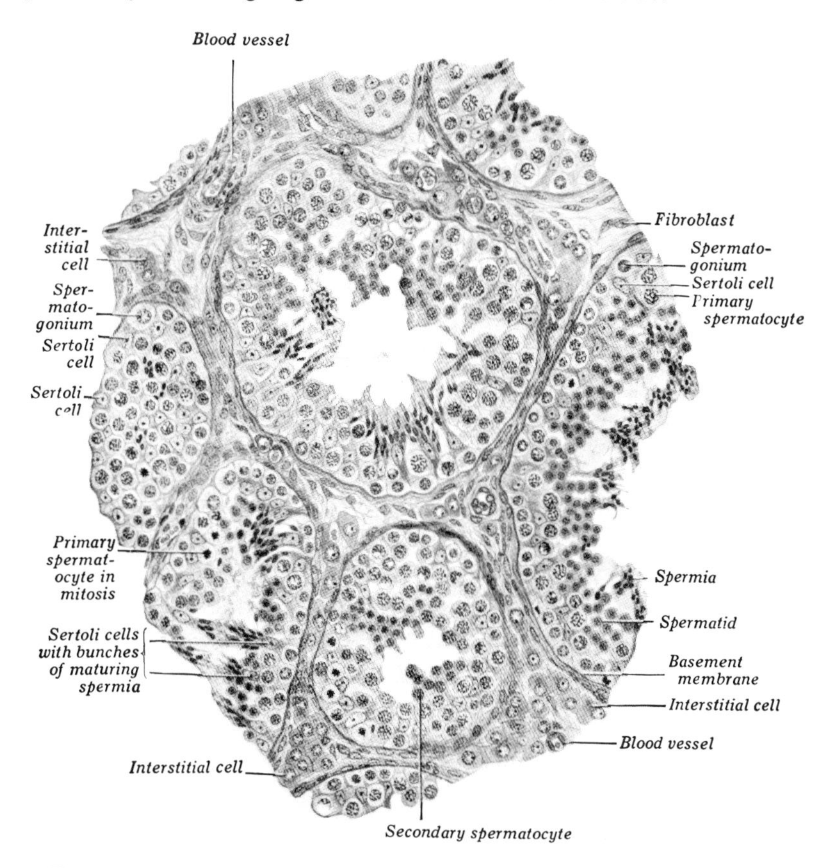

Figure 3.9. Human testis structure. (a) Diagram of section through human testis showing tubular structure. Walls of tubules are made up of cells undergoing various stages in sperm formation. Magnification 170 ×. *(Continued on page 56)*

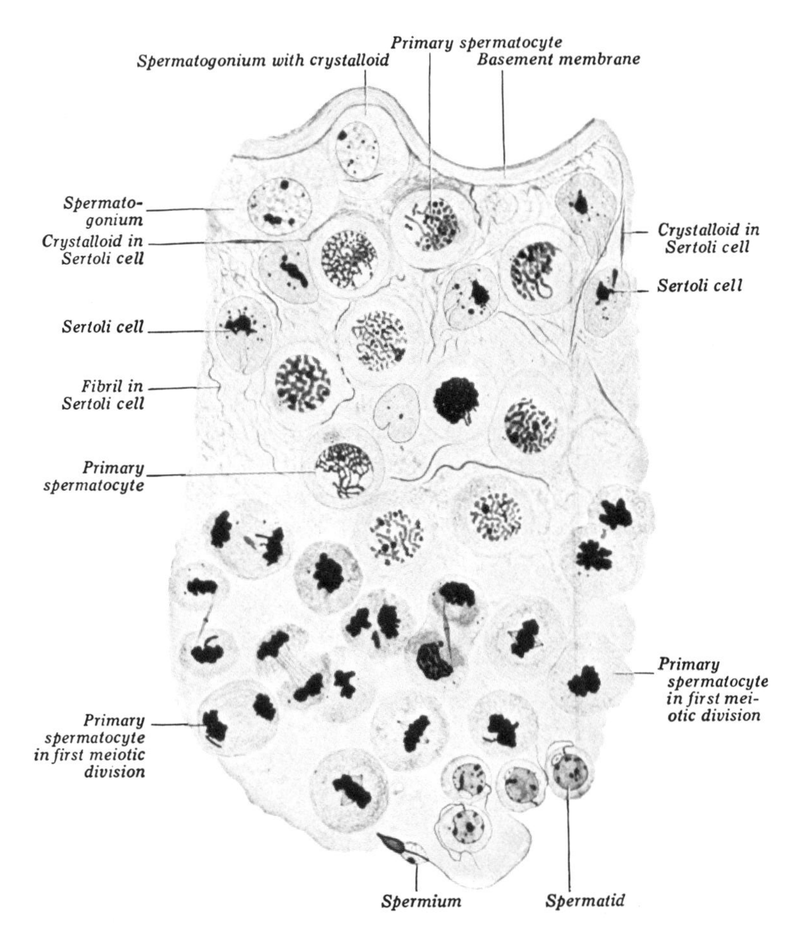

Figure 3.9. *(Contd.)* (b) Diagram of seminiferous tubule from human testis, showing cellular aspects of sperm formation. Magnification 750 ×. (Courtesy A. A. Maximow and Wm. Bloom, W. B. Saunders Co.)

Even when the entire process (**spermatogenesis**) has been accomplished and the cellular products have been moved close to the center of the tubules, the cells are still not classified as sperm. They are called **spermatids** and must undergo a series of developmental steps called **spermiogenesis** to assume the form of sperm cells (Fig. 3.10). The following are major parts of this process: (1) a streamlining operation which removes the excess cytoplasm, leaving primarily nuclear material essential to fertilization; (2) the development of a package of enzymes at the tip called the **acrosome** apparently essential to penetration of the corona and female gamete; and (3) the organization of a swimming apparatus consisting of a long **flagellum** with an energy-providing concentration of mitochondria at its base. The process of sperm formation—spermatogenesis and spermiogenesis–requires about 70 days in human males (Fig. 3.11).

The completely formed, but still nonfunctional, sperm cells are gathered into the center of the tubules and are gradually moved forward by the pressure of form-

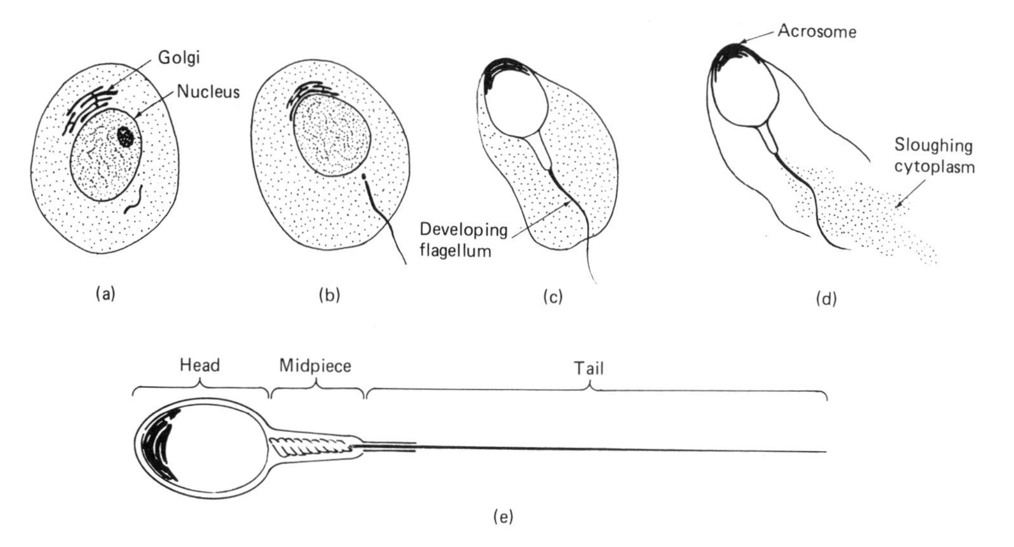

Figure 3.10. Diagram of some stages of spermiogenesis (differentiation of the haploid spermatid into a functional sperm). Diagram (e) represents a mature sperm.

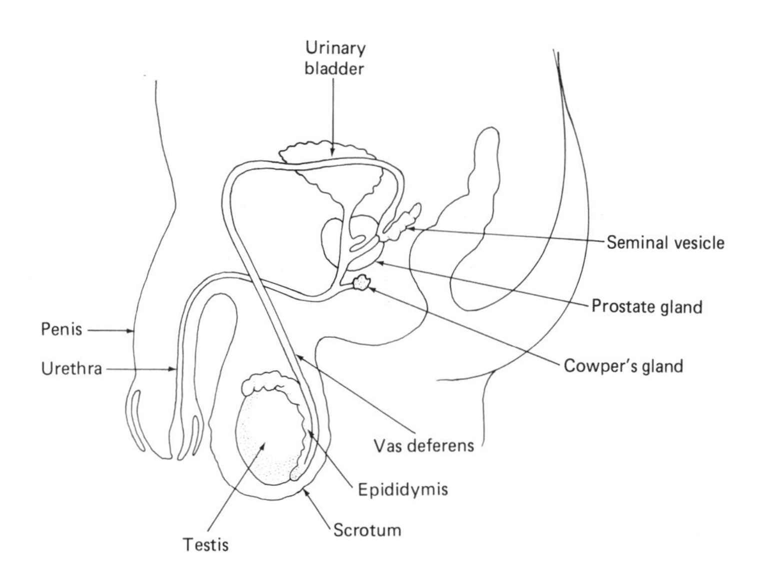

Figure 3.11. Diagram of human male reproductive system (side view).

ing cells to a storage area, the **epididymis**. This is an extension of the tubular system that is convoluted and compressed along the upper surface of the testes. The sperm cells remain here until they are forced to the outside of the body by a build-up of more sperm, or through contractions of the muscles of the reproductive tract during **ejaculation**. If sexual activity is the stimulus for release of sperm cells to the outside of the body, fluid carriers are added to the sperm from the **prostate gland** and the **seminal vesicles**. This mixture of glandular secretions and sperm cells is called **semen** and the tubules where sperm cells are formed are called **seminiferous tubules**. However, it is apparent that sperm are incapable of effective fertilization until they have spent a period of time (thought to be around 7 hours) in the female reproductive tract. The process is not understood but it has been labeled **capacitation**. Recent studies have revealed techniques of inducing capacitation of sperm outside the female body, thus permitting studies of fertilization and early embryology in the laboratory, as well as implantation of a **zygote** in the uterus of woman who is sterile.

DELIVERY OF GAMETES TO THE SITE OF FERTILIZATION

It has been mentioned that female gametes (called **ova** but really secondary oocytes with a surrounding corona of cells) are picked up as they are released from the surface of the ovary and moved down the oviduct toward the uterus. The normal site of fertilization is known to be the broad region in the upper third of one or the other of the two oviducts. It generally takes about 3 days for the egg to be moved down an oviduct to the uterus. If sperm are not encountered within approximately the first 15 hours, the egg deteriorates and is not fertilizable. However, sperm can remain viable in the female reproductive tract for about 48 hours and they might be available for fertilization if **copulation** occurred 2 days prior to ovulation.

If fertilization (**conception**) does occur, the fertilized gamete completes the second division of meiosis II after sperm entrance, again pinching off a tiny, nonfunctional polar body. The result is a zygote containing a haploid egg nucleus and a haploid sperm nucleus within a single cell. Both mother and father have contributed a single set of chromosomes carrying half of their genes to the next generation. Again, the female gamete is never an egg in the true sense of the word since meiosis is completed only after sperm penetration, but the result is the same as in organisms that produce haploid gametes prior to sperm entrance.

The male half of the reproductive cell delivery system is more elaborate in some ways than ovulation since it involves a sequence of events that produces gametes independent in their movements and functions.

The actual deposition of these independent cells in the reproductive tract of the female by the male involves a series of sexual activities variously referred to as **sexual intercourse**, **coitus**, or **copulation**. Generally such activities are preceded by an amount of foreplay, during which the reproductive systems are readied for the sex act. During preparation for coitus the **penis** of the male enlarges as it becomes hard and erect which results from the entrapment of blood within the spongy tissues of the normally soft and flexible organ. At the same time, the **Cowper's gland** secretes a liquid which lubricates the male's reproductive tract in preparation for the passage of sperm to the outside. This exudate is alkaline and helps to neutralize the acidic urinary products in the tract.

Glands at the opening of the female **vagina** during sexual stimulation provide secretions that lubricate the area and facilitate entrance of the penis, while relaxation of muscles in her pelvic region allow rapid movement of the male organ within the muscular walls of the vagina. The friction created by this movement, along with psychological factors, combine to cause rhythmic contractions within the muscles surrounding the male's reproductive system (**orgasm**) causing the ejaculation of semen into the vagina.

The sperm cells are then moved up the oviducts at a rate too fast to be accounted for by the swimming of the sperm cells alone. Several factors could account for such movement. The pistonlike action of the penis in the vagina may force semen through the **cervix** into the uterus and to the oviducts. Some substances in semen may chemically stimulate contractions of the female reproductive tract, which aid in the transport of sperm through the cervix and uterus. Females also experience orgasms involving similar muscular contractions, which may help manuever the semen farther into the reproductive tract. Regardless of the factors involved, once in the oviducts the sperm swim and orient themselves in such a manner that they contact the female gamete head first.

A single ejaculation during coitus of 3 milliliters of semen is likely to result in the deposition in the vagina of over 300 million sperm. Only one of these sperm will be directly involved in the penetration of the female gamete. Although the human reproductive process appears to be exceedingly wasteful, this is simply another example of the evolution of a system that provides a statistical guarantee for the perpetuation of the species. As a general rule, there is an extreme overabundance of male gametes in living organisms. The overwhelming majority of the sperm cells are removed from the female's reproductive tract by vaginal secretions and phagocytic activity of cells lining the tract.

Only a small percentage of the sperm deposited in the vagina are able to reach the area of the oviducts where fertilization occurs. Because it is probably a matter of chance that the sperm contacts the female gamete, the number of sperm arriving in the broad upper one-third of the oviducts is not excessively large. Furthermore, there may be a requirement for an accumulation of sperm in the vicinity of the gamete as a means of providing the acrosomal enzymes necessary to allow sperm to pass through the corona or covering of follicle cells. Whatever the reasons, a male is considered incapable of causing fertilization of a female gamete naturally if less than 60 million sperm cells are released in a single ejaculation.

Upon fertilization, the 23 chromosomes from the father and the 23 chromosomes from the mother are contained within the same cell. At this point the potential for an entirely new individual is realized, with a combination of genes and chromosomes capable of determining the characteristics of an offspring. The zygote quickly begins the first mitotic division in the oviduct and in about 1 1/2 days there are two cells. The young **embryo** continues dividing without much increase in size along its 3-day journey to the uterus. A solid sphere of about 16 cells is formed on the third day as it enters the uterus. Some 8 days after ovulation the embryo has grown in size and is probably made up of 100 or so cells that have the form of a hollow sphere. It becomes attached to the vascularized wall of the uterus, forms a **placenta**, and remains attached parasitically within the mother's body for about 9 months. During pregnancy it is called an embryo for the first 8 weeks and a **fetus** for the remainder of the period. An amazing developmental sequence occurs under the direction of its genes and the environment of the mother's uterus.

For convenience, the 9-month period of **gestation** is said to consist of three **trimesters**. During the first 3 months, all major organs are formed and the organism is about 3 inches long. At the start of the fourth month, its weight is about 1/2 ounce and its appearance is definitely human. However, the capability to survive outside the mother's body is lacking.

The most danger exists for damage to the developing structures during the first trimester. Aberrant chromosomes and defective genes, as well as foreign chemicals and radiation, are particularly effective in causing serious developmental defects. Between about the third and eighth weeks, major features of the nervous, respiratory, digestive, cardiovascular, reproductive, and urinary systems are formed. It is interesting to note that a primitive single tubular heart is beating at the end of the first month. By the sixth week, the heart has the four-chambered form and blood is circulating to the other body parts.

The second trimester is a period of a great growth. Important developments occur in the brain and the eyes become light sensitive. However the hearing apparatus remains nonfunctional. Fingernails, fingertips, and hair develop. Pronounced movement is noticeable to the mother. An important feature of development to medical genetics is that fetal skin cells begin shedding in the second trimester in amounts large enough to recover for tissue culture by extracting a small amount of the fluid in which the fetus is suspended. The cultured cells can be examined for various gene and chromosomal defects, allowing the prenatal identification of serious problems. The process, called **amniocentesis**, will be significant in later topics.

During the third trimester, the fetus grows in size with a few new developmental features. The fetus also becomes capable of survival outside the uterus. Births at 7 months have about a 10% chance of survival; this increases to about 70% in the eighth month and 90% in the ninth.

THE BRIDGE BETWEEN GENERATIONS

As the sperm contacts the surface of the female gamete, a cone of cytoplasm appears to form under the sperm head as the acrosome releases its package of enzymes (Fig. 3.12). The cell membranes surrounding the two gametes fuse together, forming a continuous channel through which the sperm nucleus enters the cytoplasm of the female gamete. The surface of the female gamete changes immediately upon penetration by a single sperm cell and no other sperm cells are able to attach to or penetrate the membranes of the egg. Microscopic observations seem to indicate that a fertilization membrane is raised around the egg at the moment of a single sperm's entrance. During the first mitotic division of the young embryo's life, the hereditary contributions of mother and father constitute the original diploid set of chromosomes. Repeatedly and faithfully copied chromosomes are passed into each pair of cells that is produced by mitosis as the single cell multiplies over a period of years to form a many-celled mature individual. Capable of reproducing in the fashion of its parents, this member of the next generation serves as a repository of hereditary factors passed on from the two parents. The sexual reproductive system of humans provides the chance for characteristic-determining genes to be assorted, mixed, and passed on in the same

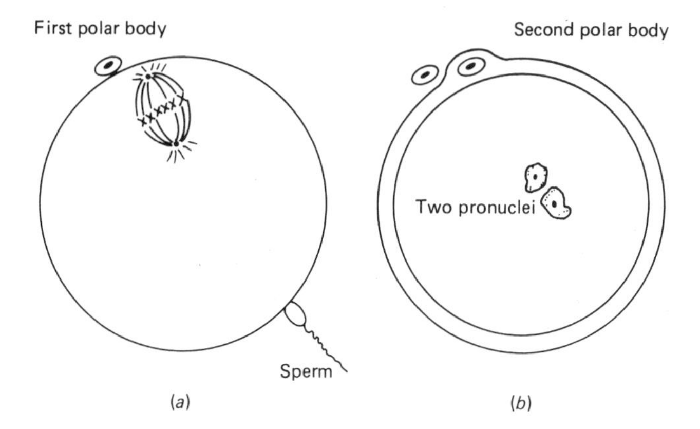

Figure 3.12. Diagram of the process of fertilization. (a) Ovum has completed the first division of meiosis (note first polar body) and has now reached metaphase II. Note sperm entering head first at lower right. (Only one sperm is shown, although hundreds may actually be present on the surface. For clarity no corona of surrounding ovarian cells is shown.) (b) Second meiotic division has been completed (note second polar body) and a male pronucleus and female pronucleus are fusing as the zygote is formed.

manner to future generations. Thus, a thread of DNA connects all members of humanity; this fragile thread can be best observed in the union of gametes from two parents.

Familiarity with the sexual reproductive processes of humans is a prerequisite for thorough understanding of human heredity and its many ramifications. The process that account for characteristics of individuals based on a set of genes from each of two parents are dependent on the ability to produce and deliver fertile reproductive cells by persons of all genetic constitutions.

Only through adequate knowledge of the sexual reproductive process will members of society be able to consider properly the facts necessary to make decisions concerning birth control, prenatal determination of disabling genes, and abortion. The usefulness and risk of procedures designed to reduce the effectiveness of the sexual process and options to prevent the birth of humans carrying defects based on hereditary factors cannot be debated unless grounded on a factual understanding. Most importantly, through a study of human reproduction as it relates to genetics, one can learn to know oneself in relation to the broad spectrum of human deversity and inheritance.

ADDITIONAL READING

AUSTIN, C. R., *Ultrastructure of Fertilization*. New York: Holt, Rinehart & Winston, 1968.

AUSTIN, C. R., and R. V. SHORT (Ed.), *Reproduction in Mammals: 1. Germ Cells and Fertilization*. Cambridge, England: Cambridge University Press, 1972.

______,*Mammals: 2. Embryonic and Fetal Development*. Cambridge, England: Cambridge University Press, 1972.

EDWARDS, R. G., "Mammalian Eggs in the Laboratory," *Scientific American* (1966), 215, No. 2, 72–81.

EDWARDS, R. G., and R. E. FOWLER, "Human Embryos in the Laboratory," *Scientific American*, (1970) 223, No. 6, 44–54.

EPEL, D., "The Program of Fertilization," *Scientific American*, (1977) 237, No. 5, 128–38.

GOLANTY, E., *Human Reproduction*. New York: Holt, Rinehart & Winston, 1975.

GROBSTEIN, C., "External Human Fertilization," *Scientific American*, (1979) 240, No. 6, 57–67.

KAPLAN, R., *Aspects of Human Sexuality*. Dubuque, Iowa: Wm. C. Brown, 1973.

REVIEW QUESTIONS

1. What sort of evidence can be cited to disprove the ideas of Aristotle and Hippocrates regarding human reproduction?
2. One of the problems in understanding the cellular aspects of human reproduction is the difficulty of recovering ova produced by females. What sort of procedure (including sophisticated surgery) can you suggest as a possible means of obtaining a human female gamete to be used in research studies or for in vitro (outside the body) fertilization? Describe briefly any specialized or unique types of equipment or techniques that would be required.
3. Based on a thorough knowledge of human sexual reproduction, it should be possible to identify points in the process that would be easily blocked or interfered with to prevent conception. List and briefly describe several mechanical or chemical manipulations that might be made as means of birth control. In each case tell whether the step taken would lead to a temporary or permanent means of contraception, and if any damage to the health of the individual might be likely.
4. Review the process of meiosis as described in Chapter 2 and explain the differences that are found in a comparison of the basic process in human males and females.
5. What significance to the survival of the species can you attribute to (1) the vast number of male gametes produced and (2) the large amount of cytoplasm retained by the female gamete?
6. If a scientist wanted to study fertilization of an ovum by a sperm and the early stages of embryo development in a test tube, what sort of problems might be encountered? Review both male and female natural processes to determine where difficulties might be encountered in an artificial environment.
7. **(a)** What might be the result if one ovum were released from each ovary simultaneously during ovulation, and then both were fertilized in the oviducts and implanted in the uterus?

   (b) Suppose that only one ovum was released and fertilized. Prior to implantation, the embryo divided into two viable halves which became implanted in the uterus. How would the result differ from that of part (a)?
8. Explain how the condition or stage of development of reproductive cells in males and females differs at the age of puberty.
9. In what ways does the monthly cycle of gamete production and preparation of the uterus for embyro implanatation relate to the menstrual cycle in the human female?

10. (a) Trace the pathway of sperm cells from the seminiferous tubules to the site of fertilization during and following copulation. Specify any substances that are added to the sperm along the way and list their functions.

(b) Trace the pathway of female gametes from follicle to site of fertilization. Are any substances added to these cells by the female reproductive tract before or after fertilization?

CHAPTER SUMMARY

1. Historically a knowledge of the cellular aspects of human sexual reproduction has been clouded by misunderstandings and undocumented speculation. A solid foundation of information now exists that is fundamental for a knowledge of human genetics and peripheral topics, such as birth control, sex determination, and artificial control of gene transmission.
2. Human sex determination is based on the inheritance of the sex chromosomes (X and Y). Females are XX and males are XY. Because offspring are produced from the mating of XX to XY, it can be predicted that there is one-half chance for male or female offspring. The actual birth ratio of males to females is 106:100.
3. Genes on the human Y chromosome determine that testes will develop in a young fetus. In the absence of the Y chromosome, other genes determine that ovaries will develop.
4. External genitalia and the ovotestes are indeterminate until about the second month of fetal life, when the presence of the Y chromosome causes testes development, followed by other male characteristics. In the absence of a Y chromosome, ovaries develop and secondary female sex characteristics appear.
5. In the ovaries the germ cells form oogonia which divide by mitosis until about the fifth month of fetal life when 7 million potential gametes exist. Although many die and disintegrate, some begin meiosis as primary oocytes and reach an arrested state at prophase I.
6. Upon reaching puberty, each month one primary oocyte completes meiosis I and enters the second cycle of meiosis as a secondary ooctye, where it is arrested at metaphase II. The second meiotic division is completed only after ovulation and fertilization by a sperm.
7. In the fetal testes the germ cells reside in the developing tubular system and increase their number greatly by the process of mitosis. After puberty is reached in males, spermatogonia continue to divide by mitosis, but some are pushed in toward the center of the tubules and begin the first division of meiosis as primary spermatocytes. They then undergo the second division of meiosis, pushing haploid undifferentiated spermatids into the ducts of the testes where final conversion to sperm cells occurs.
8. Sperm cells are released in overwhelmingly large numbers during ejaculation. They are deposited in the vagina and eventually are carried into the oviducts where an ovum may be present.
9. Fertilization occurs in the oviduct and the embyro is implanted in the uterus, where development of the fetus takes place. An infant is born approximately 9 months after fertilization.

C H A P T E R

4

THE CHEMICAL MECHANISM OF THE GENE

If there were no internal propensity to unite, even at a prodigiously rudimentary level—indeed in the molecule itself—it would be physically impossible for love to appear higher up.

Pierre Teilhard de Chardin,
*The Phenomenon of Man (1955)**

One of the constant features of science appears to be the ability of scientists to discover the mode of action of some natural law before its basic underlying mechanism is understood. Such a discovery was seen in Mendel's formulation of the principles of heredity. Many years before anything was known about the chromosomes or their manner of action, Mendel, and later Correns, de Vries, and von Tschermak were able to fathom the laws governing their actions. In the historical development of genetics it was logical to work out these principles first, for little was known about chromosome chemistry or structure. Even so, one should consider the molecular nature of a gene before becoming involved in more detailed studies of heredity. To gain an understanding of the traditional principles involved, the underlying molecular mechanism should be kept in mind.

CHROMOSOME STRUCTURE

Early workers did not know the precise chemicals involved in chromosome structure and they simply called the chromosomal material **chromatin.** Chromosomes of higher organisms are now known to be composed of a few particular types of **proteins** and a characteristic chemical called **deoxyribonucleic acid (DNA).** Until the mid-1970s there was no clear, verifiable concept of the structure of chromosomes or the manner of association of the chemicals therein. Several significant pieces of research indicated that DNA was responsible for hereditary transmission of traits and was the stuff of which genes were made. However, it was particularly confusing to compare the results of genetic studies to what could be seen microscopically in chromosomes.

*Courtesy of Harper & Row Publishers, Inc. English Translation by Bernard Wall & Introduction by Julian Huxley. Copyright © 1959 by William Collins Sons & Co., Ltd., London and Harper & Row, Publishers, Inc.

The genetic material of living organisms consists of nucleic acids, but there are two cellular "packaging" arrangements. In fact, biologists divide all life into two major categories on this basis. Organisms without organized nuclei are called **prokaryotes** (evolutionarily, "prior to nuclei"). Organisms whose nuclear material is clearly separated from cytoplasm by membranes are called **eukaryotes** ("true nuclei").

Prokaryotic chromosomes are made up of DNA alone, but eukaryotic chromosomes, such as those in humans, are composed partially of protein along with the DNA. The DNA is tightly bound to the protein molecules, categorized as **histones.** The histones are basic proteins that possess a globular three-dimensional form and exhibit characteristically similar chemical properties.

Because the chromosomes of prokaryotes are much easier to utilize in detailed molecular studies than the chromosomes of higher organisms, present understanding of gene functioning is based significantly on such organisms. Their simple structure has been exploited in many studies, especially with the colon bacterium (*Escherichia coli*), to produce a huge amount of information about gene action. Sometimes the information cannot be directly related to the mechanism in eukaryotes, but such studies have provided insight into valid concepts at all levels of organization.

Research using genetic information consistently indicates the presence of a single, uninterrupted strand of DNA extending from one end of a eukaryotic chromosome to the other. On the other hand, it is clear that histones are bound to the DNA and contribute to a chromosome's structure. They may also be important in regulating the availability of chromosomal DNA for various chemical activities. Until the 1970s microscopic techniques were not adequate to reveal the minute molecular structure. Their structure will be discussed in more detail in Chapter 5.

CHROMOSOMES AS CARRIERS OF HEREDITARY FACTORS

It was observed that two cytologists working independently, William Sutton in America and Theodor Boveri in Germany in 1902, discovered that the chromosomes are the carriers of the hereditary factors or genes. They recognized a parallel in the modes of action of the unseen genes and the observable chromosomes of a nucleus.

The evidence which led Sutton and Boveri to postulate that the genes were carried by the chromosomes was based on their knowledge of the action of chromosomes during the production of reproductive cells. They were aware that chromosomes occur in normal **somatic cells** (body cells) in pairs. Almost 40 years earlier Mendel had demonstrated that his hereditary factors occurred in pairs. Cytologists knew that the chromosomes separated from each other during meiosis. This fact paralleled Mendel's discovery that the paired nature of his factors was lost during the production of reproductive cells. Finally, it was apparent from the new studies of meiosis that the paired chromosomes separated from each other in an independent fashion. The movement of one chromosome from a single pair toward one pole had no bearing on which member of another pair moved in the same direction. This was the physical basis for the observation of Mendel and his followers that the paired genes were assorted among the reproductive cells in an independent fashion.

It occurred to Sutton and other cytologists that there must be many more genes than pairs of chromosomes. Thus, it became obvious that more than one gene would be found on each chromosome. Naturally such genes would not be assorted independently of each other since they would be physically a part of the same piece of material. Such genes are said to be **linked;** the phenomena of **linkage** and **crossing-over** will be considered later. To simplify, one gene at a time, or select genes that are not physically located on the same chromosome, will be discussed.

Once it became clear that the chromosomes were the carriers of the genes, it was possible to speculate concerning the chemical and physical nature of the gene, even though the precise organization of molecules in the chromosomes was still not understood. Geneticists and cytologists engaged in analyzing the many variations in genetic activity and chromosome dynamics, but biochemical analyses were not sophisticated enough to permit a thorough and accurate evaluation of the chemical nature of chromosomes. It was not until the late 1940s that it became apparent that DNA was the hereditary material of the chromosomes.

CHEMICAL IDENTITY OF THE GENE

Over the years the evidence in favor of DNA as the hereditary material built up to an undeniable level. DNA was demonstrated to be duplicated precisely in a living cell, a crucial requirement for the material of the gene. No other chemical of life possesses this inherent replicating property. Early workers recognized that DNA is found in constant and equal amounts in the cells of each species of living organisms. This is a feature that might well be anticipated by anyone familiar with the genetic mechanism. Through chemical analysis it was found that DNA occurred in reproductive cells in only half the amount of the somatic cells, paralleling Mendel's mathematical demonstration of segregation and the fact of reduction in meiosis.

More modern molecular experimentation provided two pieces of evidence that left little doubt that DNA was the chemical responsible for the determination of hereditary traits. In 1928 Frederick Griffith in London observed a change of genetic characteristics in a living bacterium when it was cultured with dead ones of the same species. Oswald Avery with his coworkers in New York proved in 1944 that the chemical from the dead bacteria which caused the change was DNA. It was found that DNA could be extracted from bacterial cells and purified. This pure DNA could then be added to a tube in which other bacteria were growing. Characteristics possessed by the first bacterium could be transferred to the cells growing in the presence of the purified DNA. It was obvious that only DNA was responsible for transferring these characteristics. The changed cells were said to be transformed and the process was called **transformation.**

Shortly thereafter another means of transferring pure DNA from one bacterial cell to another was found. Viruses which infect bacterial cells were discovered in 1915 and came to be known as **bacteriophages.** Careful analysis of these viruses revealed that they consisted of a protein cover on a DNA core. An important observation leading to their experimental use was that during infection only the DNA is injected into a bacterium. In 1951 Alfred Hershey and Martha Chase at Cold Spring Harbor, New

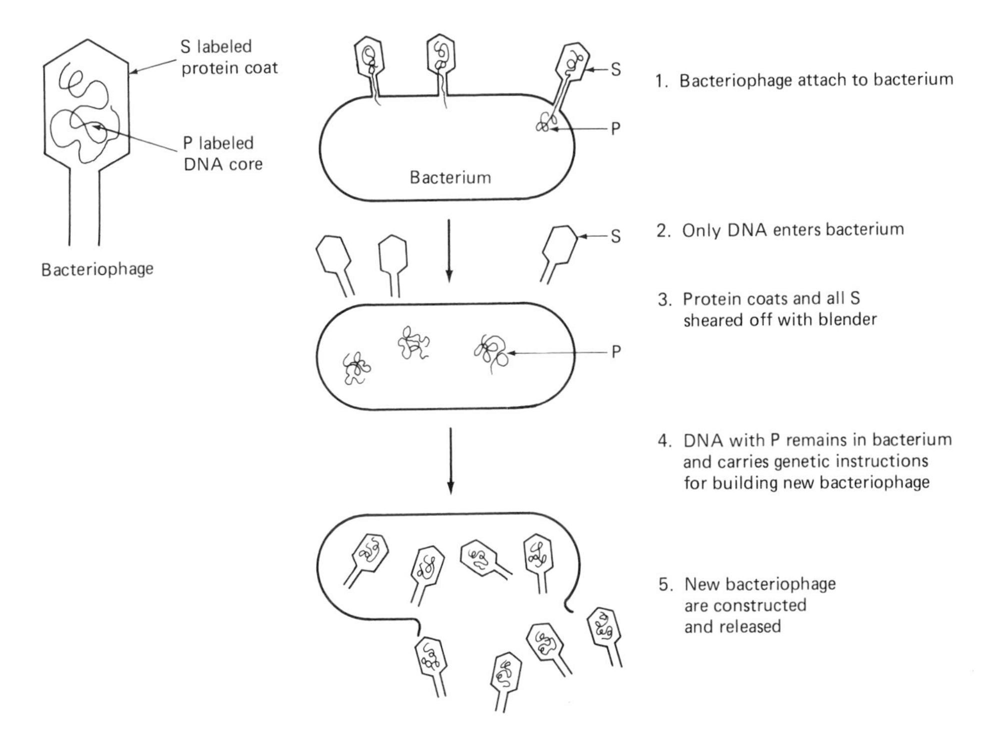

Figure 4.1. The Hershey-Chase experiments. Some viruses that infect bacteria (*bacteriophage*) have a protein coat and a DNA core. In experiments summarized on this diagram, Hershey and Chase showed that only the DNA enters the bacterial cells. Bacteriophage protein coats were "labeled" with radioactive sulfur (S) in one experiment and DNA cores were "labeled" with radioactive phosphorus (P) in another. Components could thus be "traced" during infections. The experiments provide convincing evidence that DNA is the hereditary material of the cell.

York, used radioactive labeling techniques to prove this most significant point (Fig. 4.1). It was later shown that hereditary characteristics can be transferred from one bacterium to another by bacteriophage. Because only DNA is placed in the cell by the infecting virus, it is apparent that the hereditary change of bacteria is controlled by DNA. The process of altering bacteria by infecting with bacteriophage that have acquired a specific type of DNA from other bacteria is called **transduction.** The process serves as an additional proof for the assignment of the basic genetic function to DNA.

MOLECULAR STRUCTURE OF DNA

The molecular structure of DNA was not understood until 1953 when James Watson of the United States and Francis Crick of England working as a team proposed a model for DNA structure. They were assisted by Maurice Wilkins of England whose laboratory provided the **X-ray diffraction** investigations of the chemical. In 1962 these three men shared the Nobel Prize in physiology or medicine for their discoveries. Rosalind Franklin, a name that is often not mentioned in recapitulations of this significant piece of work, did the X-ray analyses which revealed the true structural picture to Watson and Crick.

It had been known for many years that large molecules having orderly and repeating arrangements of submolecules will cause a diffraction (bending) of X-rays which reveals their placement in the material. Thus, Watson, Crick, and Wilkins could make accurate predictions concerning spatial arrangements of the subunits in the large molecule of DNA. The constituents of DNA were known prior to this time, but their arrangement in a self-replicating molecule with genetic properties was not understood. The work of these scientists started a long series of studies which ultimately led to an understanding of how DNA controls the activities and appearance of living organisms.

DNA is composed of only three types of submolecules: a sugar **(deoxyribose)**, an inorganic phosphate unit **(phosphoric acid)**, and four organic nitrogen bases **(adenine, guanine, cytosine,** or **thymine)**. None of these chemicals is really unusual or uncommon and one might expect to find them on the shelf of a well-stocked laboratory. It is their combination into the DNA molecule in all living cells that accounts for the amazing properties of the hereditary material.

The sugar molecules (deoxyribose) are attached to each other through phosphate units, so that a long chain is formed consisting of deoxyribose alternating with phosphate. One of the four nitrogen bases is attached to each sugar molecule in the chain. Actually, when DNA is being assembled in a cell, the basic building block is a **nucleotide** (Fig. 4.2). A nucleotide consists of a phosphate unit, a deoxyribose

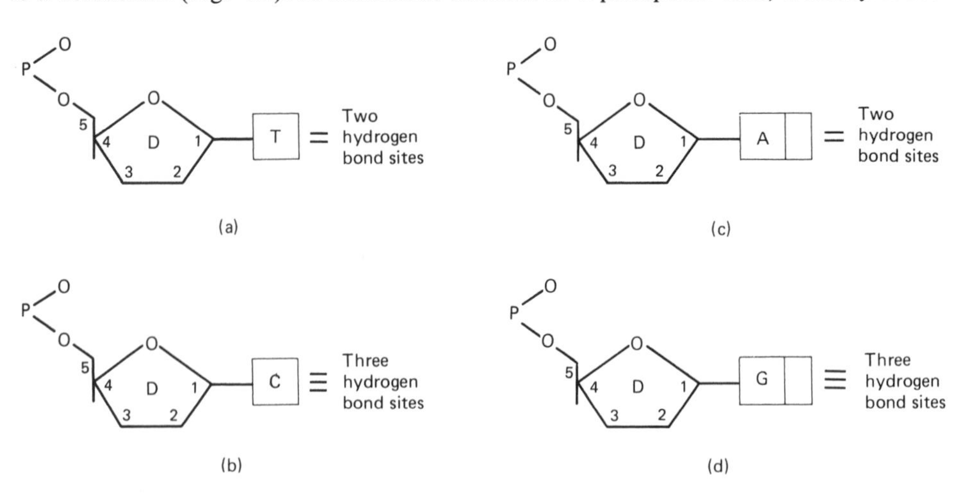

Figure 4.2. The four nucleotides of DNA. The building blocks of DNA are nucleotides, shown diagrammatically: (a) thymidine phosphate; (b) cytidine phosphate; (c) adenosine phosphate; (d) guanosine phosphate. Each is composed of an inorganic phosphate unit (P), a sugar submolecule (D—deoxyribose), and one of four nitrogen bases (T, C, A, G). Two nitrogen bases are single-ring pyrimidines (a) and (b) and two are double-ring purines (c) and (d). Thymine (T) and adenine (A) are capable of joining with each other, since they form two hydrogen bonds, while cytosine (C) and guanine (G) can pair up and join through three hydrogen bonds. The numbers shown in the sugar portion (D) indicate the conventional sequence of numbering the carbon atoms in the molecule (clockwise as shown). Note that the phosphate is attached to the number 5 carbon atom and the nitrogen base is attached to number 1.

sugar unit, and one of the four nitrogen bases. Thus the long strand is built up by connecting the phosphate submolecule of one nucleotide to the deoxyribose submolecule of the next nucleotide, leaving the nitrogen bases projecting out to the side (Fig. 4.3).

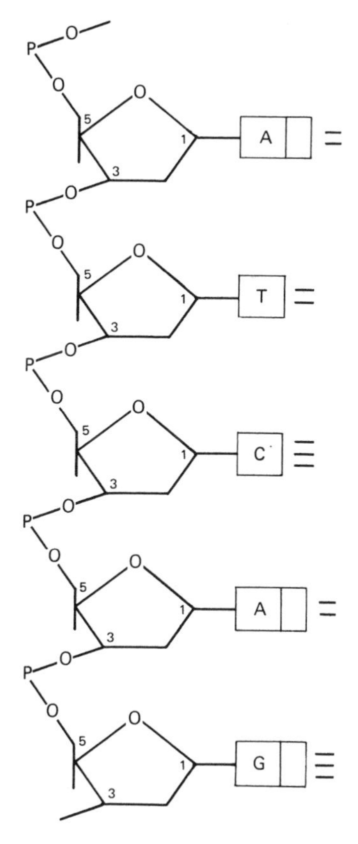

Figure 4.3. Single strand of nucleotides. Very long chains of deoxyribonucleotides are formed by the attachment of phosphates and sugars. The backbone of DNA structure is this alternating strand of phosphates and sugar molecules with nitrogen bases projecting out laterally. Note that the phosphate groups are attached to the sugar molecules at the number 3 and number 5 carbon atoms in the deoxyribose. This gives a reference point for directionality in the chain. By considering the deoxyribose submolecules only, the direction can be expressed as 3→5, reading from bottom to top.

Because the four nitrogen bases can be arranged in a random fashion along the sugar-phosphate chain, the great amount of variability required of a genetic molecule is achieved. It is as if a four-letter alphabet were used to spell hereditary messages. The four nitrogen bases represent the four letters and there are many possible combinations. For example, if a chain consisting of only 10 nucleotides were considered, there would be over 1 million (4^{10} or 1,048,476) different combinations possible for the four different types of bases! DNA molecules in life consist of thousands of nucleotides in each chain. Is it any wonder that life is so variable, when the basic control molecule can vary so much and yet be composed of so few units?

Watson and Crick, with the help of Wilkins, using Franklin's analyses, constructed molecular models which showed that DNA was really a double-stranded molecule (Fig. 4.4). This double nature is due to the fact that the nitrogen bases have the ability to attach to other nitrogen bases by the sharing of hydrogen atoms. To form a double chain, it is possible only to have attachments between adenine and

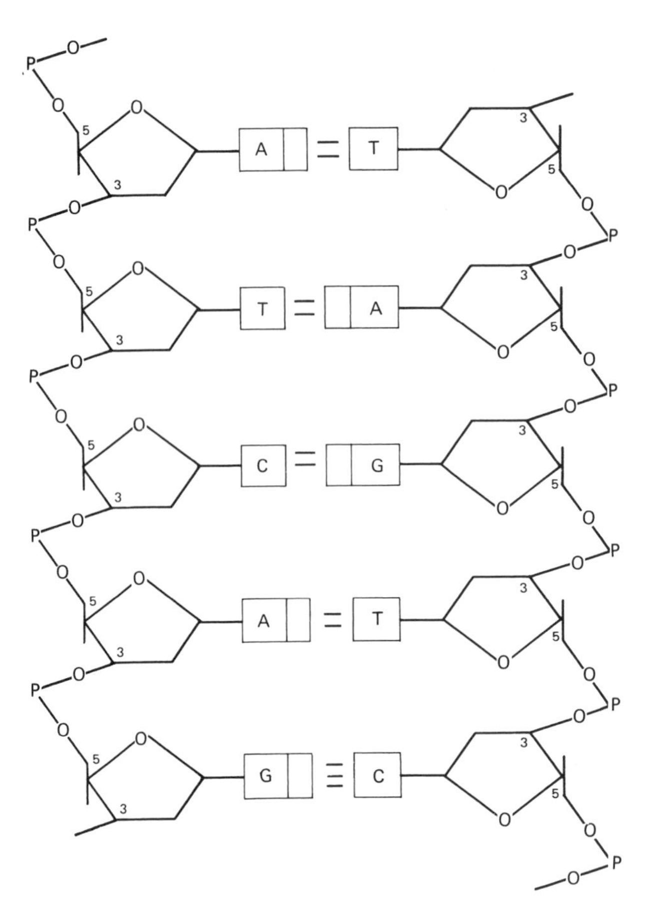

Figure 4.4. Double-chain structure of DNA. Two complementary chains are held together by hydrogen bonds between an adjacent purine and pyridimine (two between adenine and thymine and three between guanine and cytosine). Note that this complementarity across the DNA molecule is specific for pairing (A with T and C with G). Note that the two chains are "antiparallel." That is, the directionality indicated by the 3 carbon→5 carbon sequence in one chain is reversed to 5 carbon→3 carbon in the opposite chain.

thymine and between cytosine and guanine. This is due in part because the **purine** molecules (adenine and guanine) are larger, double-ring molecules and the **pyrimidines** (thymine and cytosine) are smaller, single-ring molecules. Spatially, it is obligatory that a purine be paired with a pyrimidine. Beyond this, the hydrogen bonding pattern is specific and the A—T combination shares two hydrogen bonds and the C—G combination shares three. Any other combination of paired bases is not possible in the chain. Thus adenine and thymine must always be opposite each other in the chain, as must guanine and cytosine. Prior to this discovery, chemical analysis had already shown that adenine and thymine occur in approximately equal amounts in DNA, and that cytosine and guanine are also present in equal quantities.

Another important characteristic of DNA now becomes more understandable: its amazing ability to be replicated precisely (Fig. 4.5). If the hydrogen bonds between nitrogen bases are broken and the two half-chains allowed to separate, it is easily seen that each half-chain might act as a model for the formation of a new double-stranded molecule. Through the selection of nucleotides from the surrounding cell material, according to the specific selection of adenine by thymine, thymine by adenine, cytosine by guanine, and guanine by cytosine, two new double-stranded DNA molecules can be constructed. Each half-chain will build a new half-chain exactly

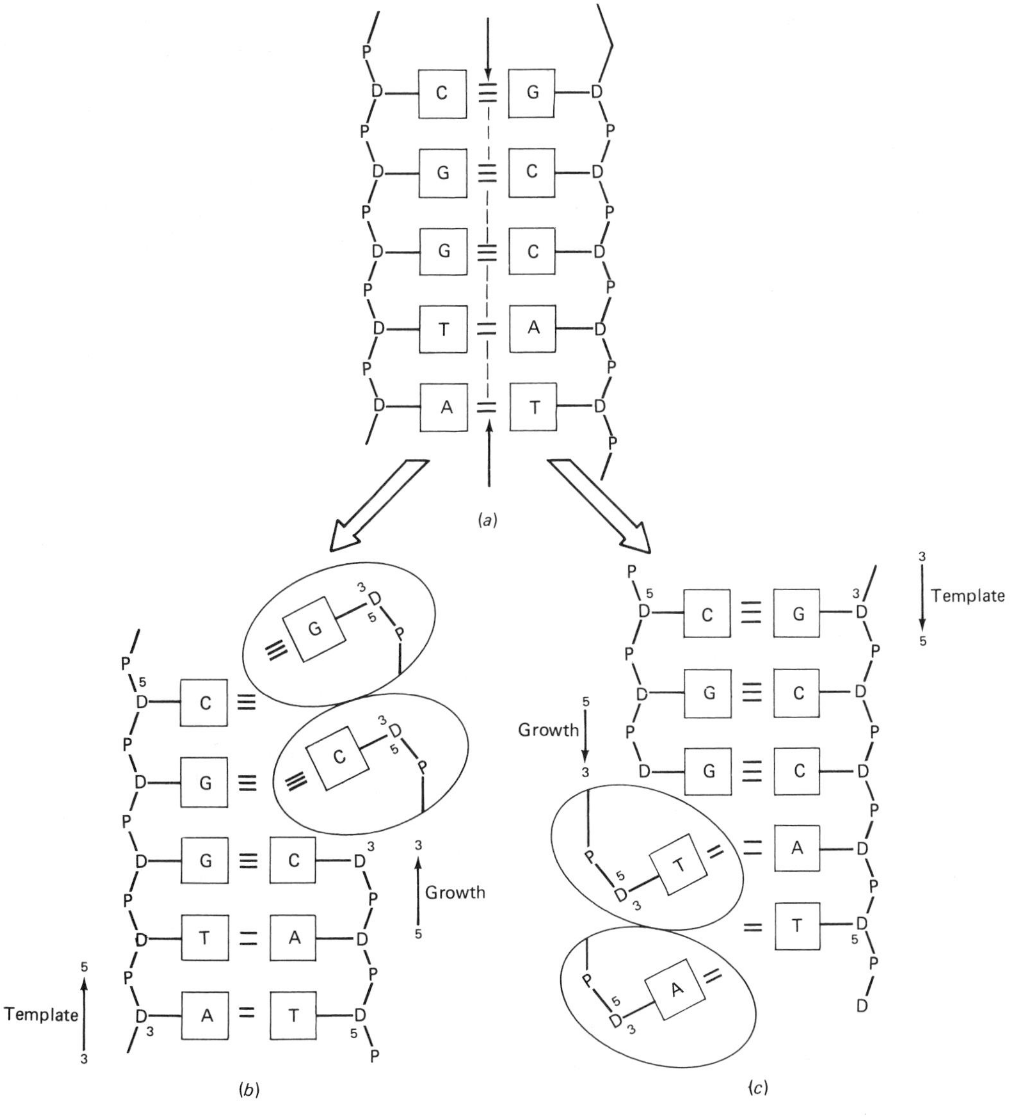

Figure 4.5. Replication of DNA. The double strands of the DNA molecule separate from each other (a) and serve as templates for the formation of new half-chains (b) and (c). Note that because of A—T, G—C complementarity, the nucleotides selected for each position follow a specific sequence dictated by the old half-chain (''incoming'' nucleotides circled). Two new double chains are thus produced that are exactly like the original double-stranded molecule. Note that the nucleotides utilized in the assembly process attach to the forming chain through the phosphate (P) on the number 5 carbon of the sugar (D). Thus, it is said that the assembly is in the 5→3 direction, using the 3→5 chain as a template.

like the one just removed. Both new double chains will be precisely like the one originally present! Note from Figures 4.2, 4.3, 4.4, and 4.5 that the chains are manufactured in the 5→ 3 direction. This is because the enzyme involved recognizes only the 3→ 5 chain as a template to be copied. Imagine a chain of chemical units, thousands of molecules long, being duplicated in a living system without a single mistake. Only modern, smoothly functioning computers can come close to accomplishing such a feat.

Finally, the double-stranded DNA molecule was recognized to be arranged in the form of a helix (Fig. 4.6). This helical structure accounts for the patterned diffraction of X-rays because of the stacked, repetitive nature of the molecule (Fig. 4.7). To envision the structure of the molecule, think of a flexible ladder. The sides of the ladder are represented by the deoxyribose-phosphate chain and the rungs repre-

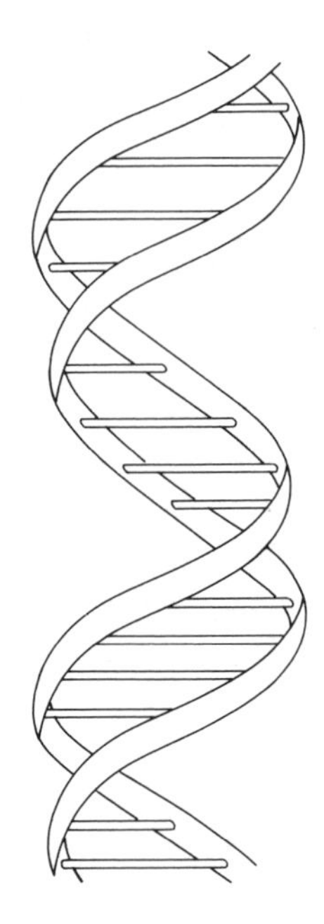

Figure 4.6. Helical structure of the DNA double chain. Intertwined "ribbons" represent deoxyribose-phosphate chain, while connecting lines represent paired nitrogen bases.

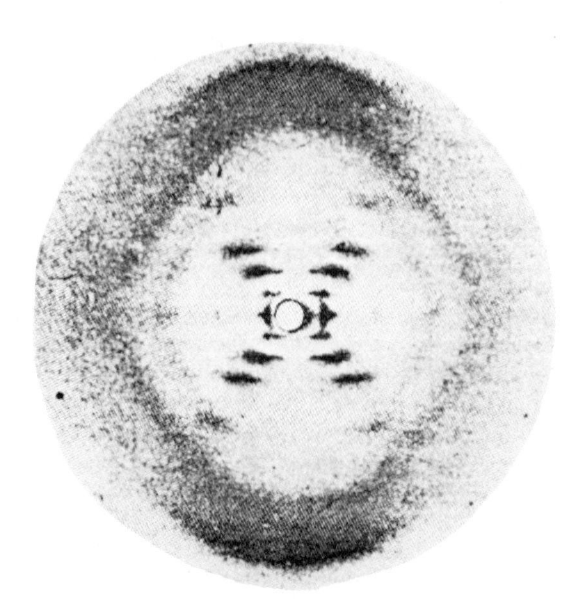

Figure 4.7. X-ray crystallograph of DNA. The highly repetitive nature of the dark areas was evidence to Watson and Crick in 1953 that DNA has a helical form. (Reprinted by permission from *Nature,* Vol. 171, p. 740. Copyright (c) 1953 Macmillan Journals Limited.)

sented by the nitrogen base pairs joined by the hydrogen bonds. If the two ends of the ladder were held and then twisted in opposite directions, a double helix would be formed. The two sides of the structure would form two intertwined chains twisted about each other in opposite directions, and connected by nitrogen bases and hydrogen bonds.

TRANSCRIPTION OF DNA TO RNA

DNA may also serve as a model for the manufacture of a compound called **ribonucleic acid** (RNA). Although not every detail of RNA manufacture is known, RNA is under the direct control of DNA. Again the DNA serves as a kind of template in the assembly of a number of component submolecules which make up RNA. It is important to note that there are specific enzymes required for each chemical step in both the **replication** of DNA and its **transcription** to RNA. Thus there is genetic control over these fundamental genetic processes through the production and control of the essential enzymes. The enzyme, as in DNA replication, is so specific that the RNA chain is built in the 5 → 3 direction, using the 3 → 5 chain as template.

DNA replication and transcription to RNA assume the aspect of routine cellular processes in this context because they are fundamental procedures in all living cells. Obviously, they are essential to life and therefore they are often accepted as elementary processes. In fact they are very complex chemical schemes requiring many factors for proper completion. Because the goal of this text is simply an understanding of the chemical basis for the hereditary mechanism, the descriptions are oversimplifications that have left out many significant features such as essential enzymes and other cofactors.

The RNA molecule is similar to DNA in several ways. Its basic chain is composed of phosphate submolecules placed alternately with molecules of a five-carbon sugar (ribose). The ribose is exactly like deoxyribose, except that it contains one more oxygen in its structure. Attached to each of the ribose molecules in the chain is a nitrogen base. The nitrogen bases involved are the same as those in DNA, except that uracil is utilized instead of thymine.

Because the final product of RNA manufacture usually remains as a single-chain molecule, it does not consist of the neat double-helical structure observed in DNA. The nitrogen bases of RNA have the ability to form hydrogen bonds as in DNA and in some regions along the length of an RNA molecule they may fold back and adjacent nitrogen bases may form hydrogen bonds. The folded and twisted short region on the RNA molecule might be likened to a twisted hairpin in three-dimensional shape.

The physical structure of RNA in the cell has been more difficult to analyze than DNA. Because of its basic single-stranded nature, it does not produce X-ray patterns as readily as DNA. Based on its function in the cell, the physical structure can be classified into three categories: (1) **ribosomal RNA;** (2) **transfer RNA;** and (3) **messenger RNA.** The ribosomal RNA (rRNA) is a structural component with proteins, of the ribosomes. Its function in the activity of the ribosomes is not completely understood, but it appears to serve as a sort of skeletal substance on which the ribosomal proteins are arranged. Fully formed ribosomes containing RNA and protein molecules are clearly essential to protein manufacture in the cell. Likewise, transfer RNA and messenger RNA are necessary for the synthesis of proteins. The building blocks of proteins are amino acids which are assembled into large functional molecules only in the presence of the three species of RNA.

Transfer or soluble RNA (tRNA) is composed of chains approximately 80 bases in length which are folded back on themselves with adjacent complementary nitrogen bases joined through hydrogen bonds (Fig. 4.8). These small units of RNA have some unpaired nitrogen bases projecting from the molecule at various points. Within a living system containing specific enzymes and energy sources, the tRNA may be joined with amino acids. This is a highly specific process, with certain structurally distinct tRNA units capable of joining only with certain amino acids. The real significance of this process becomes apparent when one recognizes that proteins are composed

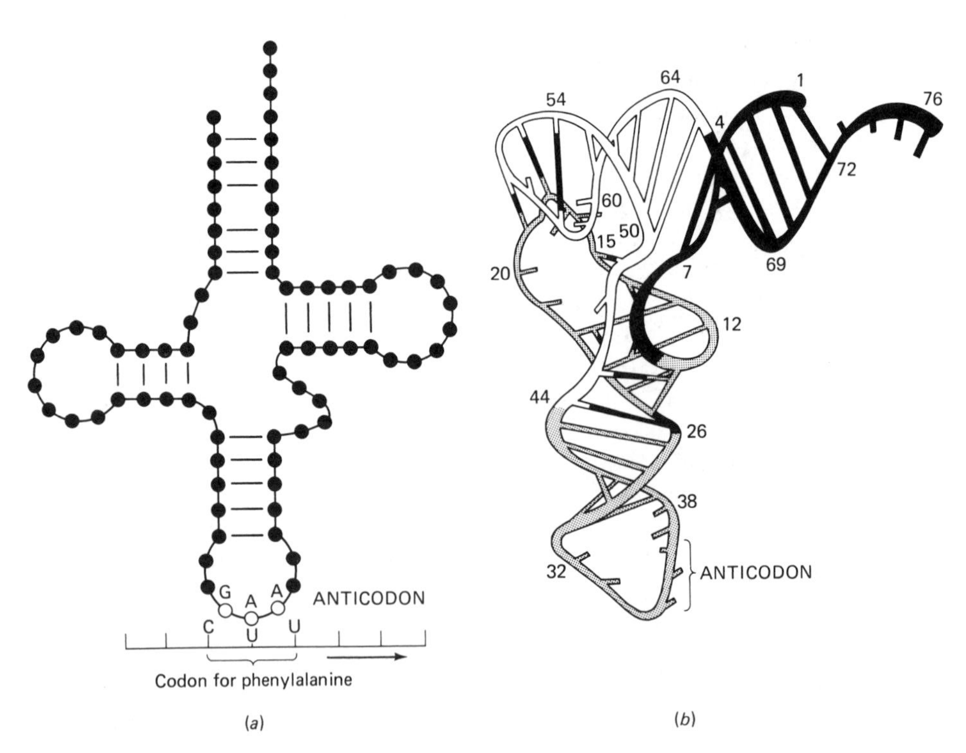

Figure 4.8. Diagrams of transfer RNA molecule. Shown is the tRNA molecule from yeast cells which transports the amino acid phenylalanine. One mRNA codon for phenylalanine is UUC (shown on a strand representing mRNA, read right to left). Note that the anticodon, or complementary sequence of bases, is AAG, exposed as unpaired bases in one of the loops. Note also the presence of some paired bases (dots connected by lines in (a), ladderlike rungs in (b). The pairing is due to hydrogen bonding, similar to that in DNA, resulting in helically twisted segments and an overall twisted "cloverleaf" configuration. Flattened, idealized structure is shown in (a). Three-dimensional form is shown in (b). (From "The three-dimensional structure of transfer RNA," by Alexander Rich and Sung Hou Kim. Copyright (c) January 1978 by Scientific American, Inc. All rights reserved.)

of amino acids. This ability to join with amino acids allows tRNA to serve as a transporter of these molecules to the site of protein synthesis.

Messenger RNA (mRNA) molecules are long, single-stranded molecules which serve as a mechanism for carrying the specific instructions for protein manufacture from the DNA of the chromosomes out into the cytoplasm. Once in the cytoplasm, the precisely copied arrangement of nucleotides in mRNA becomes a means of directing cell activities and structure through the synthesis of specific enzymatic and structural proteins.

TRANSLATION OF THE DNA MESSAGE INTO PROTEINS

It has been stressed previously that proteins are the chemical basis for all activities in the cell. All the structural components of a cell—membranes, plastids, mitochondria, ribosomes, chromosomes—are partially composed of proteins. In addition all the chemical activities of the living cell are under the control of enzymes. The basic molecular portion of every enzyme is a specific protein. So important are enzymes that they are recognized as the essence of life. Even the structural arrangement and composition of cells is based on enzymatically controlled cell activities. Anything capable of directing the manufacture of proteins, as DNA, is capable of directing all the life activities and characteristics of an organism.

Protein and Amino Acid Structure

To understand the hereditary mechanism, it is necessary to understand the the structure of proteins. In brief, the proteins are very long chains of amino acids. There are 20 amino acids essential to the structure of all proteins of living cells. All these compounds contain the same chemical arrangement in one part of their molecules. Attached to a single carbon is an organic acid group (—COOH) and an amine group (—NH_2) thus, the derivation of the name amino acid. This fundamental part of the molecule may be diagrammed structurally:

$$\begin{array}{c} \text{H} \\ | \\ \text{H}_2\text{N}—\ \text{C}\ —\text{COOH} \\ | \\ (\text{R}) \end{array}$$

Note that an undefined structure remains at the position indicated by (R). Various types of chemical side chains are attached at this position. Different molecular arrangements of carbon, hydrogen, oxygen, and sometimes sulfur atoms at this point account for the different amino acids. As in all organic molecules, the common structural characteristic is a carbon chain with various other atoms bonded to the carbons. The structures of the 20 essential amino acids are shown in Table 4.1.

TABLE 4.1. The 20 essential amino acids and the genetic codes.

Name	*Abbreviation*	*Formula*	*DNA Triplets*	*Messenger RNA Codons*
Alanine	ALA	NH_2—CH—COOH | CH_3	CGA, CGG, CGT, CGC	GCU, GCC, GCA, GCG
Arginine	ARG	NH_2—CH—COOH | C_3H_6 | NH | CN_2H_3	GCA, GCG, GCT, GCC TCT, TCC	CGU, CGC, CGA, CGG, AGA, AGG
Asparagine	ASN	NH_2—CH—COOH | CH_2 | CO | NH_2	TTA, TTG	AAU,AAC
Aspartic acid	ASP	NH_2—CH—COOH | CH_2 | COOH	CTA, CTG	GAU, GAC
Cysteine	CYS	NH_2—CH—COOH | CH_2 | SH	ACA, ACG	UGU, UGC
Glutamic acid	GLU	NH_2—CH—COOH | C_2H_4 | COOH	CTT, CTC	GAA, GAG
Glutamine	GLN	NH_2—CH—COOH | C_2H_4 | CO | NH_2	GTT,GTC	CAA,CAG
Glycine	GLY	NH_2—CH—COOH | H	CCA, CCG CCT, CCC	GGU, GGC GGA, GGG
Histidine	HIS	NH_2—CH—COOH | CH_2 | (imidazole ring: N, NH) | H	GTA, GTG	CAU, CAC

TABLE 4.1. (continued) The 20 essential amino acids and the genetic codes.

Name	*Abbreviation*	*Formula*	*DNA Triplets*	*Messenger RNA Codons*
Isoleucine	ILE	NH_2—CH—COOH, | $CHCH_3$ | C_2H_5	TAA, TAG TAT	AUU, AUC AUA
Leucine	LEU	NH_2—CH—COOH, | CH_2 | CH / \ CH_3 CH_3	AAT, AAC GAA, GAG GAT, GAC	UUA, UUG, CUU, CUC, CUA, CUG
Lysine	LYS	NH_2—CH—COOH, | C_4H_8 | NH_2	TTT, TTC	AAA, AAG
Methionine	MET	NH_2—CH—COOH, | C_2 H_4 | S | C-H_3	TAC	AUG
Phenylalanine	PHE	NH_2—CH—COOH, | CH_2 | (benzene ring with H, H, H, H, H)	AAA, AAG	UUU, UUC
Proline	PRO	NH—CH—COOH, H_2 H_2 H_2 (ring)	GGA, GGG, GGT, GGC	CCU, CCC CCA, CCG
Serine	SER	NH_2—CH—COOH, | CH_2 | OH	AGA, AGG, AGT, AGC TCA, TCG	UCU, UCC, UCA, UCG AGU, AGC
Threonine	THR	NH_2—CH—COOH, | CH_2O | CH_3	TGA, TGG, TGT, TGC	ACU, ACC, ACA, ACG

Table 4.1. (Continued) The 20 essential amino acids and the genetic codes.

Name	*Abbreviation*	*Formula*	*DNA Triplets*	*Messenger RNA Codons*
Tryptophan	TYP	NH_2—CH—COOH, side chain CH_2 attached to indole ring (ring H, NH, H, H, H, H)	ACC	UGG
Tyrosine	TYR	NH_2—CH—COOH, side chain CH_2 attached to benzene ring (H, H, H, H) with OH	ATA, ATG	UAU, UAC
Valine	VAL	NH_2—CH—COOH, side chain CH with CH_3 CH_3	CAA, CAG, CAT, CAC	GUU, GUC, GUA, GUG
Initiation (start protein signal)			TAC	AUG (at beginning of chain only)
Termination (end protein signal)			ATC, ATT, ACT	UAG, UAA, UGA

RNA Dictation of Amino Acid Sequence

The mechanism involved in selecting the exact linear order of amino acids to be linked together in the structure of a protein molecule is found in the messenger RNA. Keep in mind that this structure consists of long chains of ribose nucleotides, the order of which is specified by the order of deoxyribose nucleotides in the DNA of the chromosomes. The DNA serves to direct the manufacture of RNA in a very specific manner. The mRNA moves out into the cytoplasm carrying a message from the DNA; this sequence of nitrogen bases specifies the linear arrangement of amino acids in the protein. Thus mRNA carries a message for protein manufacture from the nucleus to the cytoplasm.

It is possible to determine how many nitrogen bases are required for the specification of the 20 different amino acids. If each base served as a message for one amino acid, only four different amino acids could be coded for (one each for A, T, G, and C)—in effect, a language in which each word consisted of only one letter. If two nitrogen bases represented each word, 16 amino acids could be specified (four things taken two at a time equals 4^2). At least three nitrogen bases per word would therefore be required for 20 amino acids. In the jargon of the molecular biologist, each of these words or sequences of three nitrogen bases is called a **codon**. There are 64 possible codons, based on four things taken three at a time (4^3). This clearly provides more than enough combinations to specify 20 amino acids. In fact, the DNA—RNA—protein code is redundant, with some amino acids having as many as six different codons. Futhermore, some triplet codons serve as ''punctuation marks'' or messages to terminate a protein or start a new one.

When the DNA triplet which specifies an amino acid in the sequence is known, it is automatically known what RNA codon will appear in the messenger RNA. It is the complement of the DNA sequence, based on A, G, C, and T of DNA requiring selection of their complements U, C, G, and A for the messenger RNA.

In 1961 Marshall Nirenberg and coworkers at the National Institutes of Health (N.I.H.) prepared the first synthetic messenger RNA and determined that the triplet UUU (three uracil molecules in sequence) was the codon specifying the amino acid phenylalanine in protein synthesis. By 1966 all possible combinations utilized in specifying amino acids had been deciphered. Although a number of workers contributed to this intense effort, the two leaders in the research, Nirenberg at N.I.H. and A. Gobind Khorana at the University of Wisconsin, shared the Nobel Prize in physiology or medicine in 1968.

Table 4.2 shows the mRNA codons specifying the amino acids in protein synthesis. The letters represent the four RNA nitrogen bases (U = uracil, C = cytosine, A = adenine, G = guanine). The bases are arranged in checkerboard fashion according to first nitrogen base (vertically), second nitrogen base (horizontally), and third nitrogen base (vertically within each group). The abbreviations for the names of the amino acids are the same as those provided in Table 4.1. Note that one codon serves as the start signal for a protein when at the beginning of a message (AUG, the codon for methionine) and three codons serve as stop or termination signals in translation to protein (UAA, UAG, and UGA). Table 4.1 lists both the RNA codons and their complementary DNA codons.

Mechanism of Protein Synthesis

Messenger RNA alone is incapable of manufacturing proteins; ribosomes and transfer RNA are also required for the process. Based on sophisticated molecular studies, two sizes of ribosomal particles have been found and it is known that one of the large size and one of the smaller size must be joined together to form a ribosome that is active in protein synthesis. In a cell the two ribosomal subunits join together in a single structure as they bind to the mRNA.

Once attached to the messenger RNA molecule, the ribosome moves along the single strand of mRNA (or the mRNA moves past the ribosome with the same effect) (Fig. 4.9). The mRNA strand is physically situated in the groove located at the juncture of the large and small ribosomal particles. As the ribosome lines up with triplets of nucleotides on the mRNA, receptive points are formed. At these points, three nitrogen bases on a transfer RNA molecule can momentarily be bonded by pairing with three nitrogen bases on the messenger RNA. This pairing process is based on the adenine-to-uracil and guanine-to-cytosine hydrogen bonding noted earlier.

Table 4.2. The mRNA codons specifying the amino acids selected in protein synthesis.

1st Base	*2nd Base* U		C		A		G		*3rd Base*
U	UUU	Phe	UCU	Ser	UAU	Tyr	UGU	Cys	U
	UUC	Phe	UCC	Ser	UAC	Tyr	UGC	Cys	C
	UUA	Leu	UCA	Ser	UAA	STOP	UGA	STOP	A
	UUG	Leu	UCG	Ser	UAG	STOP	UGG	Typ	G
C	CUU	Leu	CCU	Pro	CAU	His	CGU	Arg	U
	CUC	Leu	CCC	Pro	CAC	His	CGC	Arg	C
	CUA	Leu	CCA	Pro	CAA	Gln	CGA	Arg	A
	CUG	Leu	CCG	Pro	CAG	Gln	CGG	Arg	G
A	AUU	Ile	ACU	Thr	AAU	Asn	AGU	Ser	U
	AUC	Ile	ACC	Thr	AAC	Asn	AGC	Ser	C
	AUA	Ile	ACA	Thr	AAA	Lys	AGA	Arg	A
	AUG	Met and START	ACG	Thr	AAG	Lys	AGG	Arg	G
G	GUU	Val	GCU	Ala	GAU	Asp	GGU	Gly	U
	GUC	Val	GCC	Ala	GAC	Asp	GGC	Gly	C
	GUA	Val	GCA	Ala	GAA	Glu	GGA	Gly	A
	GUG	Val	GCG	Ala	GAG	Glu	GGG	Gly	G

Remember that the transfer RNA molecules are carrying amino acid molecules picked up in the cytoplasm.

It can be postulated that there is a **receptive site** on the functioning ribosome on which a tRNA is first attached to the ribosome-mRNA complex and a **bonding site,** to which the tRNA is then moved, allowing another tRNA to move into the vacant receptive site. Except for initiation and termination steps, there will be two tRNA's attached to the ribosome while a protein is forming. The amino acids carried by the two tRNA's are thereby brought into close proximity and a chemical linkage forms between them. Each time the ribosome moves along the mRNA, one tRNA falls away, leaving its amino acid as part of the lengthening chain, and another tRNA moves into the receptive site carrying another amino acid. Of course, the used tRNA's can return to the cytoplasm and pick up other amino acids for transfer.

As stated, this is a highly specific process. A tRNA molecule cannot be selected for a specific site of attachment on the mRNA unless it has a specific triplet of unpaired nitrogen bases. These must be the exact complement of the triplet in mRNA that is associated with the functional ribosome. The tRNA molecules will join with and transfer only certain specific amino acids, due to the structure of the two molecules and the enzyme which activates their joining.

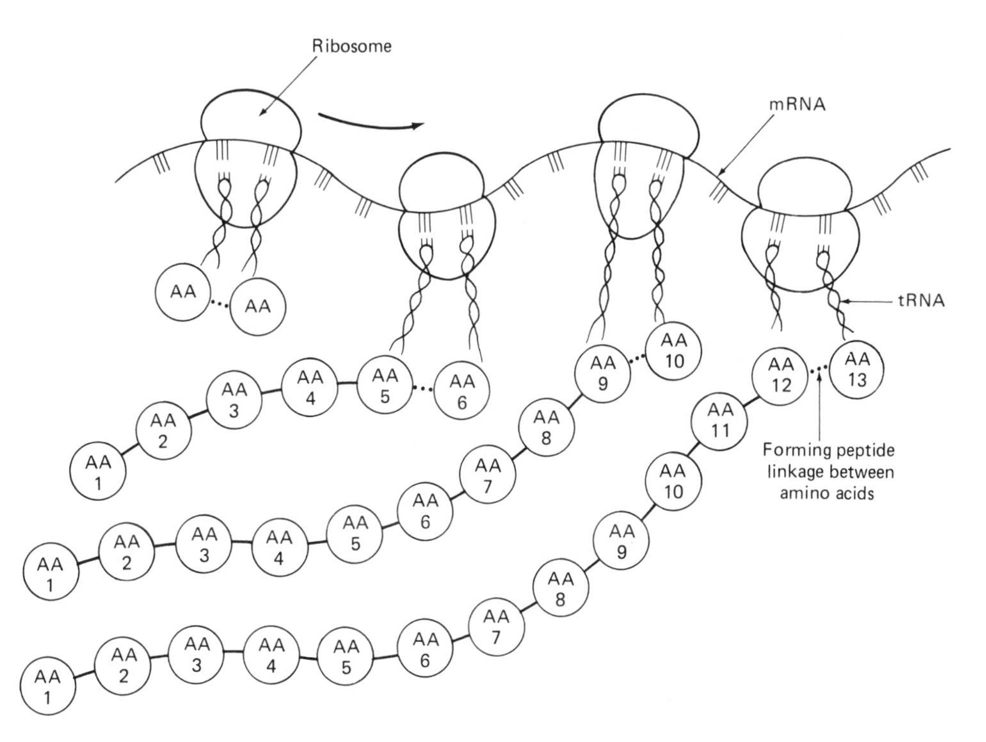

Figure 4.9. Diagrammatic representation of ribosomes moving along messenger RNA molecule. Transfer RNA molecules whose anticodons match triplet codons of nitrogen bases on mRNA bring specific amino acids into location on the ribosomes. The amino acids then form a peptide linkage with the adjacent amino acid on the ribosome and ultimately form a long primary strand (the forming protein).

If two amino acids are brought together on a ribosome in the cell in the presence of the proper enzymes, the two amino acids will be linked together as a water molecule is removed. This is done by removing an H from the amine group of one acid and an OH from the acid group of the other amino acid. The two amino acids are thus bound together, the water goes off into the cytoplasm, and two potential bonding sites remain, one on each end, where other amino acids may be attached in the same fashion. Such bonds hold the adjacent amino acids together firmly and are called **peptide linkages.** Immense protein molecules may be constructed by the formation of these peptide linkages among many amino acids. The process is shown graphically in Figure 4.10 using chemical symbols.

Although the linear sequence of amino acids constitutes the **primary structure** of protein molecules, several other organizational levels also exist. These other levels are based essentially on the primary structure and are in effect dependent on the arrangement of the various amino acids. Thus the chain is most often coiled into a **secondary** helical structure and held in such form by hydrogen bonds with nearby amino acids. This helical chain is bent, folded, twisted, and bonded together in a three-dimensional shape or **tertiary** structure. Some proteins may be composed of several such globular subunits bonded together in a large single molecule (quaternary structure). The three-dimensional shape of a protein molecule is specific for that

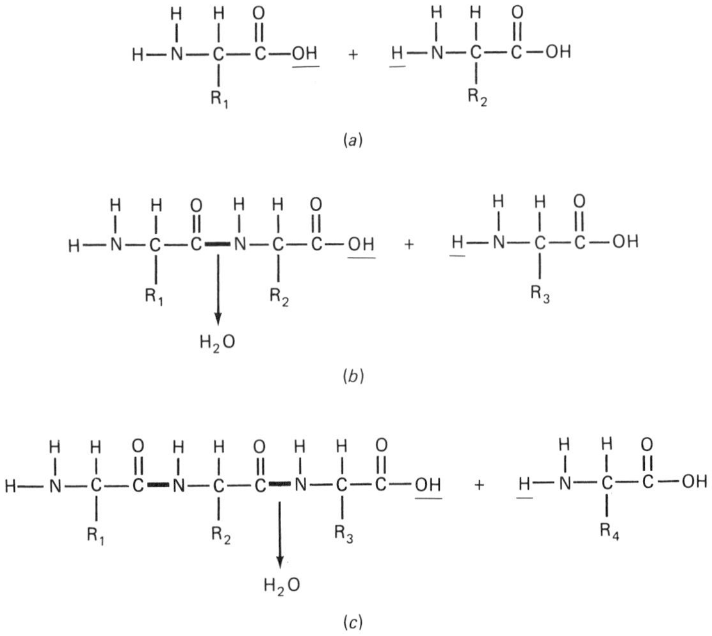

(a)

H_2O

(b)

H_2O

(c)

Figure 4.10. The synthesis of protein molecules. A peptide linkage forms between two amino acids as an acid group (—C(=O)—OH) and an amine group (H—N(H)—) are brought adjacent to each other on a ribosome. A water molecule (H_2O) is formed as each peptide linkage forms using the —OH from the acid and the H— from the amine. Several hundred such linkages may form in a single protein. (a) Two amino acids; (b) One peptide linkage (thick line between C and N) and a third amino acid; (c) Two peptide linkages (thick lines) and a fourth amino acid. R's represent variable side chains of atoms which account for the differences among amino acids.

particular combination of amino acids; the specific function that a protein can perform in a cell is partially dependent on this shape.

All enzyme proteins and those performing other functions where three-dimensional shape is important are globular molecules composed of folded and twisted alpha helices (Fig. 4.11). Some proteins in the body, however, are assembled into strong supporting structures. The primary fibrous structure of a protein may be aligned parallel to similar fibrous molecules and be joined to form **pleated sheets** instead of the **alpha helix.** In such tissues as skin, bone, and tendon, a third structural arrangement is found where three primary strands are wound around each other to form a very strong proteinaceous fiber that is well-suited to its supportive role.

A review of the process is worthwhile. Only four different symbols or letters are involved in the "alphabet" (A, C, G, and T or U). They are arranged in triplets or codons which specify each of the 20 amino acids composing the proteins of a living system. There are 64 possible different combinations of the four nitrogen bases taken three at a time. This represents more than three times the number of codons required for the 20 amino acids. The arrangement of codons in DNA can vary in an infinite number of ways; any conceivable linear arrangement of amino acids in a protein is possible.

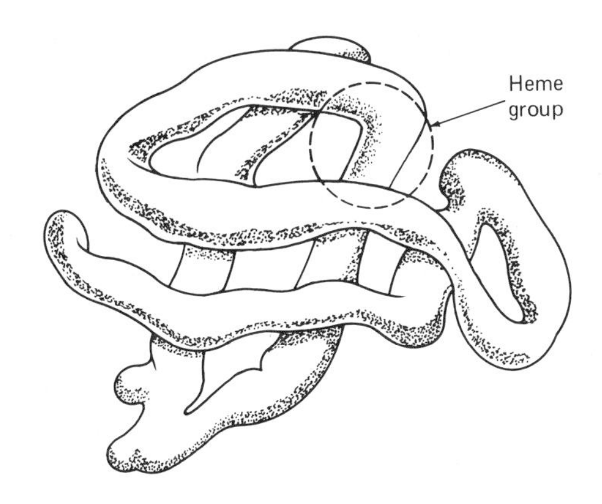

Figure 4.11. Diagrammatic representation of the myoglobin molecule showing three-dimensional shape. The tubular strands constituting the molecule represent alpha helices (secondary structure) of the chain of amino acids (primary structure). The bent and folded form of the molecule (tertiary structure) is specific for the molecule.

Given a mechanism which is capable of such extreme variability, and yet one which is precisely self-replicating, living cells have been able to assemble a vast array of proteins through billions of years of experimentation. These proteins are the characteristic chemicals of life. They consitute part of the structural material of living cells and they serve as enzymes, often simultaneously in a cell. Not only do they constitute part of the structural system, they are also responsible for the chemical reactions characteristic of life. Everything any living organism is capable of doing is dependent on proteins and the mechanism which manufactures them. Human cells are capable of producing thousands of different proteins for the multiple functions of life.

As the study of genetics has advanced, the operation of this mechanism of cellular control involving DNA, RNA, and proteins has become more obvious. Direction of activities at the cellular level is reflected in the entire organism, if only superficially.

MUTATIONS: HERITABLE CHANGES IN GENE STRUCTURE

It should come as no surprise that the precise arrangement and mechanism of replication of DNA is not completely free from the possibility for change or alteration. No component of a living system is exempt from deviations from a so-called normal situation and a small proportion of deviant forms, even at the molecular level, should always be anticipated. A significant characteristic of any changes that do occur in DNA is that these alterations, or errors, will be copied in the replication process and passed on to future generations as faithfully as the original DNA pattern. Once a deviant form becomes a part of the functioning system, it follows the rules and patterns of normal forms.

Almost since the rediscovery of Mendel's work the idea has existed that the hereditary mechanism was not an unalterably stable mechanism. In fact, the first person to formalize the theory was one of the original three rediscoverers of Mendelian genetics, Hugo de Vries. About 1903, using his studies of plants as a basis, de Vries proposed that sudden, unexpected new traits sometimes appear. He referred to this process of abrupt change as **mutation** and to the new forms as **mutants.**

To provide experimental evidence for the process of mutation, Thomas Hunt Morgan at Columbia University began the study of innumerable fruit flies in the early 1900s. His first mutant was a white-eyed male that appeared suddenly in 1910 in a culture of normally red-eyed flies.

Hermann J. Muller, an undergraduate who worked in Morgan's laboratory, went on to study mutation in greater depth. In 1927 Muller provided evidence that mutations could be induced artificially. Using heavy doses of **X-rays,** he increased mutation rates in the fruit flies by about 1500 times the normal low-level spontaneous rate.

Spontaneous mutations are those that seem to occur naturally through the action of no known external stimulus. In truth it is difficult to be sure that there has been no external stimulus. Certain factors in the environment are known to increase the rate of mutations: for example, ionizing radiations, such as X-rays, isotopes, and cosmic rays, or extreme temperatures and chemicals of various types. Of course, there are always various chemicals in a cell's environment, as well as natural types of radiation. Therefore, mutation-inducing factors are always present to exert an environmental effect on the stability of the molecular basis of genes.

Mutation Frequency

It is very difficult to recover most mutations and thereby come up with an accurate figure to describe the frequency of their occurrence. A safe but very large range in which to estimate the mutation rate for a single gene is probably in the area of one mutation per 10,000 to 1,000,000 gametes. In other words, out of every 10,000 to 1,000,000 gametes produced by an individual, only one would show a mutated gene at any given locus.

The figures for most human genes are not very reliable, but if a figure of 1/100,000 were selected as the mutation rate for any given gene, one out of 100,000 gametes might carry a different form of the gene than the parental type. It is estimated that a single ejaculation from a male during copulation produces 360,000,000 sperm. In actual numbers, about 3600 sperm then would carry a mutated gene at any given locus. This is a startlingly high figure in absolute numbers, but still very small proportionally.

Again using the estimate of 100,000 genes per human cell, statistical estimates under these conditions predict that about 63% of the gametes (approximately 277 million in a single ejaculate) would carry one or more new mutations at some locus, while only 37% would be free of new mutations. Possibly more surprising is the estimate that each human carries five to eight lethal mutations as a reservoir or heritage of mutations in past generations. These are masked because they are recessive and may be thought of as latently detrimental to future generations.

Types of Mutations

Because mutations can be of various sorts and often involve a single base pair in the DNA molecule, the term **point mutation** evolved. In general, three types of point mutations are recognized: (1) the **substitution** of one base pair for another; (2) a **dele-**

tion of a base pair from the chain; (3) the **insertion** of an additional base pair in the DNA molecule. The first type results in an altered codon and different amino acid at one point in the ultimate protein product. While it could be significant, often the protein is not altered enough to interfere with its function. The latter two types would cause a **frameshift** in the translation of the gene to protein, resulting in more serious consequences than the simple substitution of a single amino acid in a protein molecule.

The insertion or removal of a single base from the codon would obviously throw off the sequence of codons from that point on. The triplets would be all entirely new, since the frame of reading would be shifted laterally by one unit:

Normal mRNA base sequence: AUA GGA CUA GAU

Base insertion mutation: AUA UGG ACU AGA U

Base deletion mutation: AUA GAC UAG AU-

A very large change in the array of amino acids selected could result and the protein might be totally different from the one normally produced. In fact, it would probably be a "nonsense" product and fulfill no functional role in the cells.

In all cases the real significance is that the new sequence of bases is copied faithfully after the mutation. It is this regularity of replication that makes mutations so important. A one-time change in a small number of cells might be insignificant, but a change that becomes a permanent replicating form (unless a back mutation occurs) cannot be discounted, especially when it occurs in reproductive cells and becomes heritable.

Equally important is that, when the DNA message is transcribed to RNA, the bases selected for the mRNA strand will correspond to the new message and produce an altered protein. Of course the new form of protein might be more effective at accomplishing its task in the cell, but it is not very likely. An established, functioning protein or enzyme is not likely to be improved by substituting amino acids in the molecule.

Causes of Mutations

The molecular reasons for alterations in the base pairs of DNA can be understood only by delving deeply into the chemical structure of the genetic material. Although such a study is beyond the scope of this text, such alterations may result from several causes. Spontaneous changes of nitrogen bases that occur with a low frequency may cause pairing other than the normal A—T, G—C pattern, or defective DNA manufacturing enzymes may be responsible. In some cases the presence of chemicals in the cell environment may modify the nitrogen bases or even substitute for them in the molecules and cause distortions.

Physical factors in a cell's environment are often more effective **mutagens** than chemicals. Some forms of irradiation, such as ultraviolet light, act directly on the atoms of the DNA molecule to cause changes. Other more effective forms of irradiation, such as X-rays or rays from radioactive materials, cause either direct alterations or changes in chemicals around the DNA (ionizations). Even elevated temperatures are mutagenic and it has been estimated that a 3° C rise in human testis temperature may double the mutation rate for sperm cells.

Much of the so-called spontaneous mutation rate in humans is due to environmental factors. More than half the radiation in the environment is natural—from the sun (cosmic rays) or radioactive materials. Medical and dental X-rays account for about two-thirds as much as the natural sources. Hopefully, the medical value of such radiation outweighs the detrimental effects. It is possible with these artificial sources to shield and protect to some extent the gonadal regions of the body.

Although a lively debate has taken place concerning the mutagenic properties of nuclear weapons fallout and nuclear power plants, these sources are very low in terms of total human radiation exposure. They account for less than 10% of the medically based sources and probably will continue to be minor sources.

MUTATIONS: THE BASIS OF ALLELIC GENES

One might postulate that mutations act exactly as the recessive genes Mendel discovered. When mutations occur in a cell with a normal gene on the homologous chromosome, the normal gene continues to produce its protein, which performs its usual function in the cell. As noted, the mutated gene form produces a nonsense protein which often performs no function at all. Only when two mutated genes occur in the same cell is the function lacking. It is probably true that the organism with one normal and one mutated gene produces less gene product than the one with two normals, but the difference usually goes unnoticed.

Many so-called recessive genes probably arose in this fashion. Naturally, even the dominant genes must have once arisen by mutation. The great diversity of living things is due to the large variety of proteins produced; this variety is due to modifications in DNA that occurred over millions of years.

Mutations should not be considered "bad" in terms of the development and change of living things (although to a well-adapted organism that term applies). Without mutations there would be no diversity and no reason to investigate the hereditary process.

Although mutations are rare, they can be carried along as recessives by individuals in a population without apparent detriment to the individuals. However, if an individual receives a mutated gene form for a certain key protein from each parent and no normal gene, the situation may be lethal. Thus, there is a selective disadvantage for individuals having the mutation in the population, as individuals who receive two mutated genes do not survive.

Some mutations are not selected against in this fashion: They may actually confer an advantage on the person carrying them. Such mutant forms become established in the human gene pool and are maintained at a frequency considerably higher than the mutation rate, since such persons may be better equipped to pass on their genes. Conceptually, the diverse human characteristics known to be gene controlled must represent a great number of mutations over millions of years of evolution that have conferred various advantages on distant ancestors.

Sickle-Cell Hemoglobin

A good model for an understanding of the mutation process is the genetic disease or deficiency in human blood called **sickle-cell anemia.** It occurs when an abnormal type of hemoglobin is produced by the body. Under periods of oxygen deficiency, such as in a high altitude or during physical exercise, the abnormal hemoglobin is found not to conduct oxygen effectively. It precipitates into elongate crystals and the red blood cells that contain it are distorted into sickle-shaped crescents (Fig. 4.12).

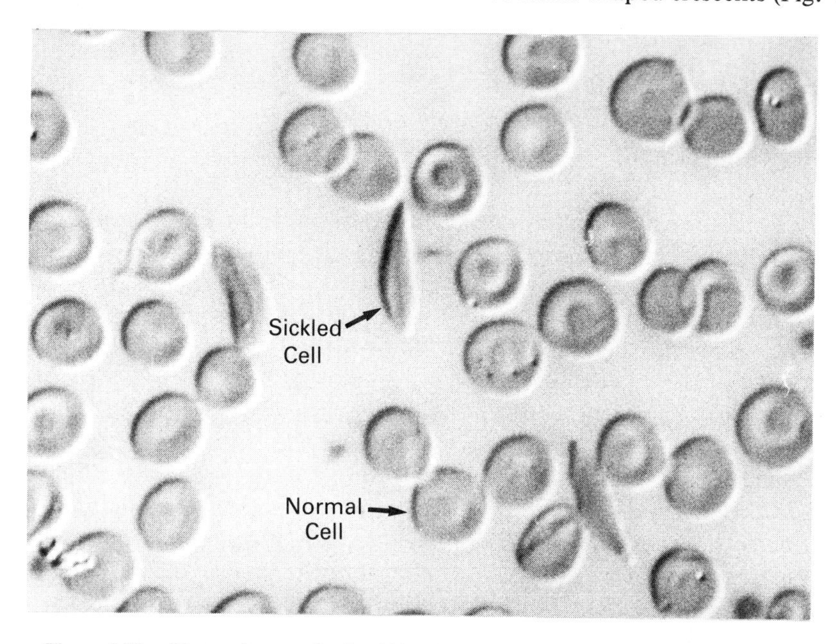

Figure 4.12. Photomicrograph of red blood cells showing normal cells and cells that have sickled under lowered oxygen conditions.

Because of the inability of the abnormal **sickle hemoglobin (HbS)** to transport sufficient oxygen when the hemoglobin has crystallized, the disease is often fatal unless a highly protected life style is maintained. Furthermore, the misshaped red blood cells have a tendency to lodge in small capillaries, shutting off blood supply in certain areas and causing considerable pain. Another condition involving HbS, **sickle-cell trait,** occurs when red blood cells possess both **normal hemoglobin (HbA)** and the sickle hemoglobin (HbS) (Fig. 4.13). There is usually about 60% HbA and 40% HbS, due to the lower production rate of HbS. In this mixed state a much greater oxygen deficiency is required to cause sickling; the condition has rarely been fatal, except under very rigorous conditions. This condition is due to a combination of one HbA allele and one HbS allele for the production of both types of hemoglobin simultaneously. Persons with sickle-cell trait are essentially normal.

Sickle-cell anemia, however, is due to the presence of two alleles for HbS. Because of blockage of blood supply to the heart, liver, or other vital organs, the disease has sometimes been fatal.

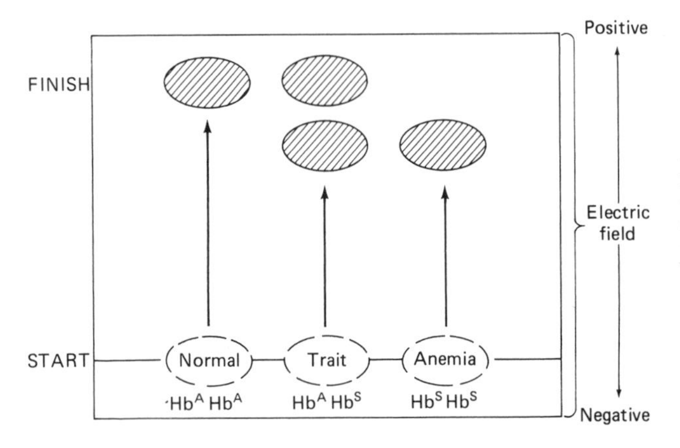

Figure 4.13. Gel electrophoresis of hemoglobin. When hemoglobin is spotted on a sheet of gel at the "start" line and an electrical current is applied, there is a migration of the molecules. Hemoglobin S migrates at a different rate to a different final position than hemoglobin A. Thus, electrophoresis of hemoglobin will distinguish persons of three different genotypes.

Considerable interest has focused on the technique of **screening** various populations to identify those with sickle-cell anemia or sickle-cell trait. Once identified, certain precautions and medical attention may be warranted. In addition the person identified as carrying the HbS allele may want to consider carefully the genetic possibilities for passing the trait on to future offspring. A simple and relatively inexpensive test can be conducted with a high degree of accuracy using a very small blood sample. The test is based on the insolubility of sickle hemoglobin when oxygen is removed in a salt solution (**sickledex test**). If the test is positive, a more thorough and definitive test is necessary to distinguish the two types of persons.

There has been a recent effort within some black communities to have potential parents opt for an analysis to discover carriers of the trait. Unfortunately, once recognized only negative options can be suggested to prevent the birth of affected children.

Several states that have progressive and fairly thorough screening programs for newborn infants check for the HbS gene. If an infant is identified as having the sickle trait, it is then known that one or both parents are sickle trait persons as well. **Counseling** is warranted in such cases to explain the situation to the family and to make suggestions for the care of the child.

It has been found that the hemoglobin S allele confers a very effective resistance to malarial infection in persons with sickle cell trait. Because of this, the HbS gene is found at a comparatively high frequency in high malaria risk areas. The trait is most often associated with blacks of African ancestry, although it is found at relatively high frequencies among other populations from areas traditionally high in malaria (Fig. 4.14). Hemoglobin S occurs among 20% of some African populations. Among American blacks, because of the lack of the malaria parasite in the environment and intermarriages with other racial groups, it has declined to 8% of the popualtion. These levels are well in excess of the rate for mutation to the HbS form, a situation that has resulted from the selective advantage where malaria is prevalent.

A series of therapeutic chemicals have been tested for their ability to prevent sickling of red cells containing hemoglobin S. Although the sickling process may be reversed or prevented, the presence of the antisickling agents in the person's body has not been compatible with good health. The search for a harmless, effective chemotherapeutic continues with some hope for success.

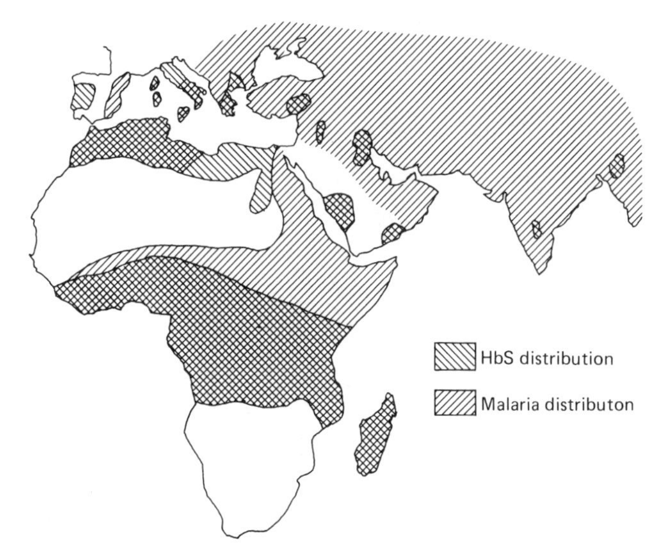

Figure 4.14. Map of Africa and Mediterranean region. Shaded areas indicate approximate regions of high malaria incidence and above average frequency of sickle hemoglobin. The cross-hatched areas illustrate the rough correspondence of the two phenomena.

By chemical standards, hemoglobin is a very large molecule, consisting of over 8000 linked atoms. Careful analysis has revealed that this gigantic molecule is composed of heme, the oxygen-transporting portion, and globin, a protein consisting of 574 individual amino acid molecules of 19 different types. An involved and complicated series of experiments by Vernon Ingram in 1957 with the abnormal hemoglobin of sickle cell anemia proved that one single amino acid, valine, is substituted for another which is normally present at a specific point in the molecule, glutamic acid (Fig. 4.15). This change of a single amino acid in the hemoglobin molecule results in a potentially fatal disease!

The protein portion of the hemoglobin molecule is actually composed of four separate amino acid chains (polypeptides) linked together. There are two alpha (α) chains of 141 amino acids each and two beta (β) chains of 146 amino acids each. Each one of the four chains is folded and twisted into a globular unit very similar to the myoglobin molecule depicted in Figure 4.11 on page 83 (see Fig. 4.15). It is within the β chain that the substitution of valine for glutamic acid occurs in S hemoglobin. Thus, it is within the specific DNA segment which codes for the β chain, that a mutation occurs.

It is known that the change of the normal glutamic acid for valine in S hemoglobin is due to a change in one nitrogen base in the RNA message for hemoglobin formation. One codon specifying glutamic acid in mRNA is guanine-adenine-guanine (GAG). A codon specifying valine in mRNA is guanine-uracil-guanine (GUG). It is incredible that this profoundly important human deficiency could result from the substitution of uracil for adenine at one point in an mRNA molecule containing about 1600 nitrogen bases.

Theoretically, the β-globin gene would need only 438 nitrogen base pairs to specify the amino acid sequence (3×146). However, for various reasons, there are many more (about 1600). In a series of experiments in 1977, eukaryotic genes were transferred into bacterial cells and maintained there in a functional state. This type of research has lately come to be known as genetic engineering. With mammalian

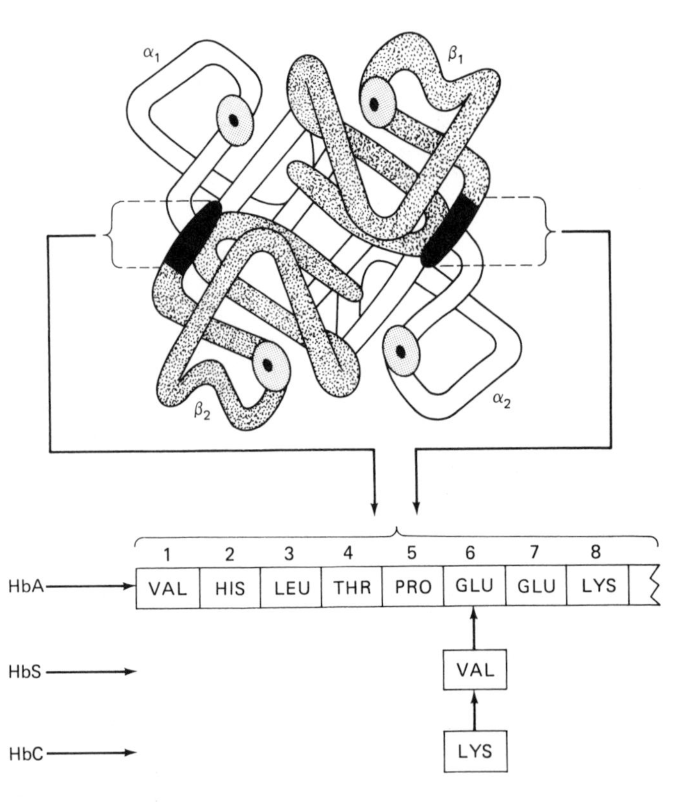

Figure 4.15. Diagram of the hemoglobin molecule. Each of the four submolecules contains a heme group (shown as disc-shaped structures). These heme groups contain four porphyrin rings and an iron atom. The terminal regions of the two globin molecules are indicated and the sequence of amino acids is given. Note that when amino acid number six is changed from glutamic acid to valine, sickle hemoglobin is the result (HbS). If this same position contains lysine, hemoglobin type C is the result (HbC).

genes present in simply grown and easily manipulated cells, amazing experiments could be done. One of these involved the annealing of a gene product (mRNA) back to the chromosome fragment containing the original DNA template.

Molecular biologists were surprised to find that the mRNA was considerably shorter than the original DNA segment. This meant that there were intervening sequences of nucleotides that were not utilized in producing the final protein molecule. These sequences became known as introns, while the functionally important sequences were called exons.

It was subsequently shown that the full DNA sequence is transcribed to mRNA first; the mRNA is then prepared for service as a messenger through a mechanism called splicing. Enzymes cut out precisely the intronic nucleotides (excision) and then rejoin the cut ends (ligation). All mammalian genes, with the notable exception of histone genes, appear to contain introns.

The gene (DNA template) for β globin has two introns (dividing the gene into three exons). One of these introns includes 120 nucleotide pairs and the other has

550 nucleotide pairs. Thus, over one-third of the 1600 nucleotides are excised prior to the translation process.

Other Hemoglobin Mutants

Once the basis of mutations like that for HbS is recognized, it should be simple to predict the large potential for many other mutant forms of hemoglobins. A large protein molecule like hemoglobin presents an easy target for the modification of amino acid content while still preserving the essential functional structure of the molecule. Although hemoglobin has 200 mutant gene forms in the human gene pool, most are rare, representing random mutations. Theoretically, there could be a number of amino acid substitutions for every one of the 287 units in the α-and β-hemoglobin chains, but some probably result in nonfunctional or lethal changes. Many random mutants probably exist with no detrimental effect on persons carrying them.

Two other fairly common types of hemoglobin mutants are HbC (found in some West African populations at a rate as high as 27%) and HbE (most common in Southeast Asia at a rate as high as 30%). Both represent single amino acid substitutions in the β chain. HbC has a lysine unit substituted for glutamic acid at exactly the same position as the substitution in HbS. HbE has a lysine substituted for glumatic acid also, but at a point 20 amino acids farther down the β chain. The HbS, HbC, and HbE mutants are good examples of functional mutants whose presence is only diagnosed because of special situations that arise as a result of their involvement in anemia. All three exist in certain areas at a rate well above the mutation rate, due to their antimalarial action.

Thalassemia

The blood disease **thalassemia** (Greek: sea) was first described for persons whose ancestry was traced to the Mediterranean. This inherited defect is found with a frequency of 10 to 15% in some Italian communities.

Functionally, the problem stems from the insufficient quantity of hemoglobin manufactured in the body. Because hemoglobin contains four submolecules—two α-chains and two β chains—a pronounced lack of β-chain submolecules would result in a reduced amount of hemoglobin. Actually, there is a series of different thalassemias which results in reduced hemoglobin amounts and anemia. The type most commonly known is **Cooley's anemia,** which affects the β chain. Interestingly, total β-chain production is not stopped—the quantity is just strongly reduced. Furthermore, the β chains produced are normal in structure. The situation presented a paradox until the techniques of genetic engineering permitted detailed analysis of the exact base sequence of the gene and its product. Transplantation of the gene to human tissue culture cells (HeLa) allowed workers to follow the production of mRNA from the mutated β-globin gene.

The mutation causing Cooley's anemia involves a tricky mechanism (see Fig. 4.16). The mutation consists of the substitution of an A-T base pair where C-G is normally located in the first intron of the gene. The resultant sequence closely resembles a signal to cut the mRNA and excise the intron. Because the message is

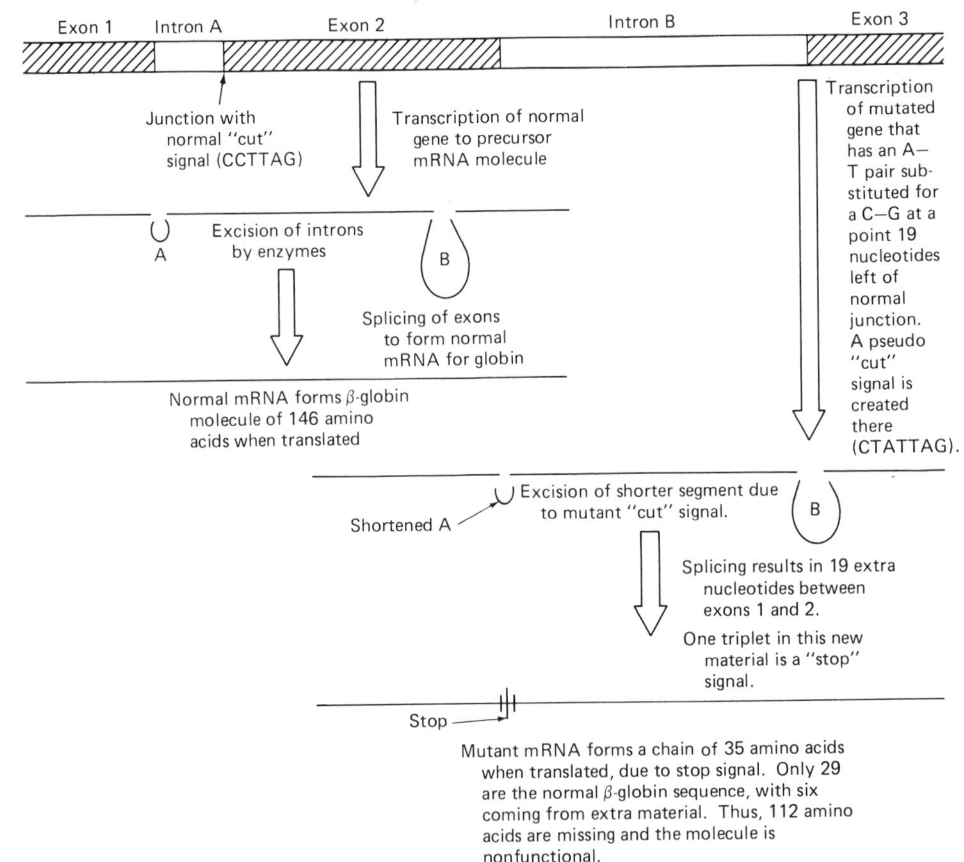

Figure 4.16. Diagram of the β-globin gene (DNA) and its transcription to messenger RNA. At the left, a precursor mRNA molecule is formed that then has the introns excised; the mRNA fragments are annealed to produce the functional mRNA for the β-globin molecule. To the right is shown the result of a mutation that is the basis for β thalassemia or Cooley's anemia. A false "cut" signal is produced by the mutation which causes the excision of a smaller intron (A). The extra nucleotides in the functional mRNA include a termination signal that is responsible for a much shortened and nonfunctional protein molecule.

inside the intron, all of the intron is not removed from the final mRNA copy. The mRNA with the supernumerary nitrogen bases (19 extra) then moves into the cytoplasm and is translated into protein. Six incorrect amino acids are added and then a codon in the intronic sequence is reached that is a chain termination codon. Translation ends at that point, far short of the full β-globin chain (only 35 are present and six are wrong).

The fact that some normal β globin is produced by the mutant is based on the presence of the normal signal for intron excision. Apparently the proper signal is sometimes recognized and correct mRNA is produced. Unfortunately for persons with the mutant, the correct signal operates only about 10% of the time.

The situation indicates that a single functional gene product, such as hemoglobin, may require the action of more than one DNA segment (one for the α chain and one for the β chain). In addition it provides an example of the importance of intronic gene substance in gene action.

Persons who show the severe anemic symptoms possess two alleles for thalassemia and require frequent transfusions from early childhood. Most often they die prior to reproductive age. Because normal production of adult hemoglobins requires that the β chain not develop until about two years of age, the disease is not a problem in infants. **Fetal hemoglobin** effectively performs normal blood functions in such situations.

Thalassemia follows the pattern of a recessive gene, except for the production of a small amount of normal hemoglobin. Persons who carry the normal gene with the mutant form show a milder form of anemia that is not generally fatal and does not require transfusions. Sometimes under periods of stress the mild anemia may cause problems, but it usually is not detrimental. This milder form is called **thalassemia minor** and the severe homozygous form, **thalassemia major.**

Research has indicated that the presence of the thalassemia mutant may confer a partial resistance to malaria, as other hemoglobin mutants. It is likely that this resistance accounts for its high incidence in malaria-risk areas around the Mediterranean Sea. Again, the partial protection against malaria would account for a positive selective advantage for those carrying a mutant form and a normal gene, even though it is fatal where two mutant genes are possessed.

Because it is fairly simple to identify persons with thalassemia minor, and because two parents showing the trait are required for a child to have thalassemia major, matings may be identified in which such births can occur. As in other cases where risk of severe defects is recognized, counseling of potential parents is important.

Although this text will continue to refer to normal, functional hemoglobin as HbA, there are three types of functional, oxygen-carrying hemoglobin molecules in adult human blood. Recall that the hemoglobin molecule is composed of two **alpha (α) chains** of 141 amino acids and two **beta (β) chains** of 146 amino acids. Thus, **HbA** (or more specifically **HbA_1**) may be given the formula $\alpha_2\beta_2$. It normally constitutes about 97% of the hemoglobin in a person's blood. Two to 3% of normal hemoglobin is **HbA_2.** It contains two α chains and two **delta (δ) chains** ($\alpha_2\delta_2$). Delta globin also has 146 amino acids, but 10 are different from those found in β globin.

A third hemoglobin is found in adult blood at a level that is usually considerably less than 1%. **Fetal hemoglobin (HbF)** constitutes the bulk of the hemoglobin (about 85%) in fetal blood. It contains two α chains and two **gamma (γ) chains** ($\alpha_2\gamma_2$). Gamma globin, similar to β globin, has 146 amino acids having 39 differences in the sequence.

To produce these three types of hemoglobin, four separate genes are necessary: one each for α globin, β globin, δ globin, and γ globin. Obviously, the α and β genes are the preponderantly active genes in normal adults.

It is apparent that the γ globin gene is "switched off" or operating at a very slow rate in the normal adult. This fact is important in the treatment of thalassemia, where there is a very low amount of β globin produced, or possibly in sickle cell anemia, where there is an aberrant form of β globin.

A recent attempt to find a means of therapy has been to replace HbA and β globin with HbF and γ globin by "turning on" the gene for γ globin. A common anticancer drug, **5-azacytidine**, is known to activate repressed genes and has been used to activate the essentially dormant γ globin gene and thus raise the amount of HbF in the blood of thalassemic individuals. Its administration for seven days caused a seven-fold increase in γ globin and at least transiently normalized the globin synthesis. The 5-azacytidine may not be specific for only the γ globin gene and it is not know what unwanted gene products may be stimulated. Nevertheless, it is an interesting idea that has shown promise in the early stages and may well suggest a means of therapy for the future.

POLYMORPHISMS

The study of different forms of genes that develop as a result of mutations, or gene **polymorphisms**, is most important to modern genetics. The word *polymorphic* means having many forms (poly = many, morph = form or shape). As used in genetics, it defines a gene that occurs in two or more forms. Because it is often used to refer to the gene product or phenotype as well, there may be polymorphic *traits*.

As seen, a gene may be changed at the level of single nucleotides without any real alteration in its function. Thus, a gene could be highly polymorphic with all its different forms unnoticed at the level of the phenotype. In other cases the different forms of the gene could result in considerable diversity in the structure or function of the organism. To make the concept realistic, the term is restricted to genes whose alleles occur at a frequency higher than would be expected from random mutations. The different hemoglobins and the various alleles controlling them are a good example of polymorphism. It is significant that some polymorphic forms occur in specific populations, as African malaria areas, where the HbS, HbC, and HbE alleles have been maintained at frequencies well above the mutation rate. Table 4.3 lists some of the more than 150 known abnormal forms of hemoglobin. Some forms are due to mutations in the α chain and others to mutations in the β chain. For each abnormal type, the table shows the position (numerical sequence) in the chain where a change has occurred and the amino acid which has been substituted. The normal amino acid at each position in question is given for normal hemoglobin (HbA). Abnormal hemoglobins were originally labeled alphabetically in order of discovery (although HbS was the first discovered). They are presently named for the place of discovery.

Histocompatibility Genes

An important system of at least four closely linked genes (loci A, B, C, D) in chromosome 6 is often cited as an example of a cluster of highly polymorphic chromosome loci. This complex of genes presents one of the most complicated systems known in humans; the basis can be understood by analyzing how two of the loci (A and B) operate. The products of the genes control the production of antigens, and thus

Table 4.3. Abnormal Hemoglobin Types and Their Substituted Amino Acids.

Substitutions in the 141 amino acid α chain:

Amino acid sequence number	16	30	57	58	68
HbA	Lysine	Glutamic Acid	Glycine	Histidine	Asparagine
HbI	Aspartic acid	a	a	a	a
Hb Norfolk	a	a	Aspartic acid	a	a
HbM Boston	a	a	a	Tyrosine	a
HbG Philadelphia	a	a	a	a	Lysine
HbG Honolulu	a	Glutamine	a	a	a

Substitutions in the 146 amino acid β chain:

Amino acid sequence Number	6	7	26
HbA	Glutamic acid	Glutamic acid	Glutamic acid
HbS	Valine	a	a
HbC	Lysine	a	a
HbE	a	a	Lysine
HbG San Jose	a	Glycine	a

[a]Indicates that amino acid in this position is the same as HbA.

the acceptance or rejection of tissue transplants between two different persons. Therefore, they are referred to as **histocompatibility genes** (histo = tissue).

The products of the alleles in the system are complex proteins embedded in the membranes of cells in an individual's body. These molecules may be foreign to another's body and are referred to as **antigens.** The foreign molecules of viruses or bacteria that invade the body during diseases are also antigens. A major feature of an antigen is the **immune reaction** that it stimulates. In response to the invasion of the antigen, the body manufactures specific **antibodies** that attack and destroy the antigen.

Because the histocompatibility gene products or antigens occur on the membranes of **white blood cells (leukocytes)** and are easily studied with these cells, they are known as **human leukocyte antigens (HLA).** They are also present in other types of cells and have a profound effect on skin grafts and organ transplants. The foreign antigens stimulate antibody formation and these in turn cause rejection of the transplanted tissue.

It is difficult to accept the idea that such a complicated system of cell surface antigens has developed in humans with the sole function of preventing the transplantation of foreign tissues in the body. However, no other function has yet been associated with the HLA cell surface antigens. They may simply be cell surface structural features, highly tolerant of diverse polymorphisms: When mutations occur there is no selection against the new form.

Regardless of their functional role, they serve as a very effective means of identifying individuals. Close attention must be given to them in organ and tissue transplants. The only acceptable match that can be made is between identical twins. However, if HLA matches can be made between siblings, transplant surgery for such organs as kidneys is often very effective.

The same principle applies to bone marrow transplants: Because white blood cells originate in the bone marrow, the HLA antigens are in this region. Additional care is required, since the transplanted bone marrow cells can effectively begin to grow and produce antibodies against the normal cells and HLA antigens of the individual's body.

The two loci of major importance (HLA-A and HLA-B) are so close together in chromosome 6 that crossing-over rarely is observed (Fig. 4.17). Therefore, they usually remain linked in their passage from parent to offspring. The complexity exists in their high degrees of polymorphism. Researchers continue to discover more alleles and add them to those already known. It appears that there are at least 20 different alleles for HLA-A and 40 for HLA-B. There are thus about 800 different combinations of these 60 antigens alone (20 $\times$ 40). These linked combinations of A and B genes are called **haplotypes.** Because each person has two number 6 chromosomes and therefore two separately linked A and B pairs, the number of combinations of haplotypes is calculated to be over 320,000 (800 $\times$ 800 = 640,000 divided by 2 because of duplicated haplotypes in the chromosome pairs). The difficulty in

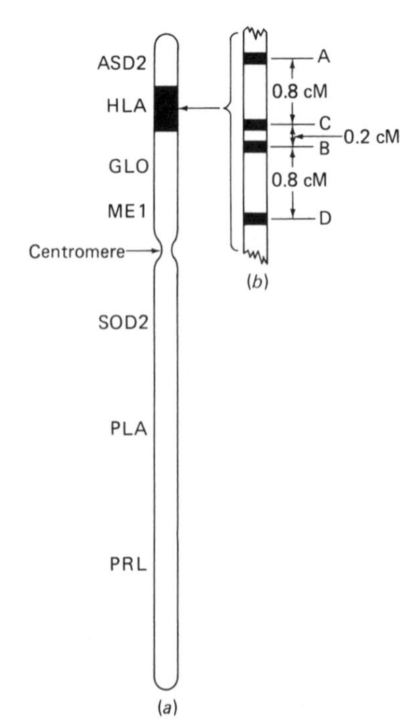

Figure 4.17. Chromosome map indicating the approximate location of the HLA loci on chromosome 6. (a) Chromosome 6 showing some of the genes in the linkage group with the approximate position of the HLA gene complex shown in the short arm. (b) Enlarged HLA region showing the order and distances between the four major human leukocyte antigen gene loci. Distances given are in map units called centimorgans (described in Chapter 5). All loci are closely linked and the A and B loci are approximately 1 map unit apart. The other gene loci are indicated: ASD2 = heart defect, GLO = glyoxylase enzyme, ME1 = malic enzyme, SOD2 = mitochondrial superoxide dismutase enzyme, PLA = plasminogen activator (blood clotting), PRL = prolactin enzyme.

finding a compatible donor for organs or tissues becomes apparent in view of such numbers.

When the other histocompatibility loci are considered (HLA-C and HLA-D,) the number of combinations is in the millions. Probably no two individuals have the same combination, except identical twins. The system has even been used to identify the father in disputed parentage cases. In such situations there is virtual certainty in the identification process.

Immunoglobulins

While the HLA system is complicated because of the high degree of polymorphism and the many combinations of closely linked alleles, the body's **immune system** presents complications for other reasons. The production of antibodies to counteract the invasion of foreign cells or substances (antigens) is obviously of utmost importance to the survival of humans. Although this text will not deal with the molecular details involved in antigen-antibody reactions, the genetic basis of the antibody response requires a brief look. As seen, HLA antigens stimulate a response in the tissues of the body; antigens on red blood cells also call into action the antibody mechanism.

A superficial understanding of the immune reaction is necessary to make studies of topics like HLA and blood types more meaningful. The genetic control of antibody production, however, presents in and of itself some exceptions to classical Mendelian genetics. The detailed molecular aspects of gene control of immune substances also do not seem to follow typical patterns of gene action.

Some reasons for deviations from typical patterns can be sensed intuitively by considering the massive functional role that is demanded of a system that must defend the body against a huge array of potential invaders. The body can manufacture large amounts of a specific antibody against virtually any foreign antigen within days, or a few weeks at most, after exposure. Probably more than 1 million different specific antibodies can be made in this way.

The task of investigating such an array of compounds and the genetic mechanism that controls their manufacture seemed impossible at the outset, but researchers found some novel means of studying the process. The first clue came in 1847, but little could be done with the molecular details until the 1950s and 1960s.

While studying the rare bone marrow cancer **multiple myeloma** in 1847, Henry Bence-Jones in England found that a large amount of a particular globular protein was excreted by patients. The **Bence-Jones protein** became a good means of diagnosing the disease, but its significance went further. After more than 100 years as a curiosity, researchers refocused on the protein as an obvious antibody produced by the cancerous marrow cells. In 1972 Gerald Edelman at Rockefeller University and Rodney Porter at Oxford University received the Nobel Prize for their efforts in determining the structure of antibodies. Their initial problem was to obtain enough of one specific antibody to allow its careful analysis.

With literally as many as a million different antibodies produced at different times and in response to different environments, there was a real problem. Each person's body contains many different antibodies simultaneously, all so closely related that chemical separation is unusually difficult. When it was found that myeloma cells could be cultured in mice and that a particular antibody-producing cell could be fused with a myeloma cell, it became possible to produce large amounts of specific antibodies. These **monoclonal antibodies** are so named because they come from a clone or line of cells derived from a single parent cell. The hybrid cells resulting from the fusion (**hybridoma cells**) produced the single monoclonal antibody of the selected cell and relatively large amounts were available for study.

The results of careful assays of the antibodies showed a unique and interesting structure. The antibodies consist of globular protein units called **immunoglobulins** (Fig. 4.18). Each molecule consists of four chains of amino acids linked together. Two of the chains are large chains ("heavy") made up of four units ("domains") of about 110 amino acids each. The other two chains are shorter ("light"), containing two units or domains of about 110 amino acids each. A terminal domain in each chain is called "variable" because its sequence of amino acids varies with the specific antigen attacked by the antibody. The other eight domains (three in each long chain and one in each short chain) are "constant," with the same sequence of amino acids in each antibody. The Bence-Jones protein was found to be made up of the two **light chains** attached to each other and excreted in abundance in the patients' urine.

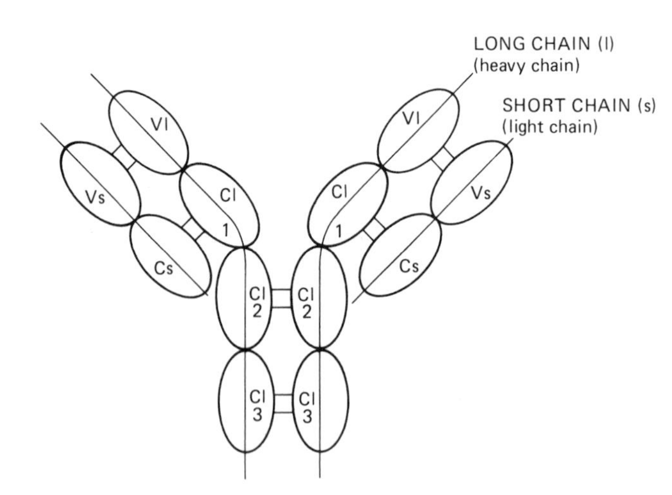

Figure 4.18. Diagram of immunoglobulin molecule. The immunoglobulin molecule is composed of 12 domains, each a sector of about 110 amino acids. They are arranged in chains, two long (l) and two short (s). The two long chains are identical, consisting of four domains (l), and the two short are identical, consisting of two domains (s). Each of the chains has one variable domain (V) that differs with regard to the antigen attacked. The other domains are constant (C) in amino acid structure and do not react with specific antigens.

The genetic mechanism for controlling antibody production is not exceptional if only the **constant domain** is considered for the light and heavy chains. A gene for the light chain would require the controlled selection of about 110 amino acids and the **heavy chain** would necessitate codons for the selection of about 330 amino acids. What about the two variable domains? Could there be 1 million different genes for variable regions that might be turned off and on by exposure to different antigens? Would this mean 1 million repeated constant region genes would go along with the variable region genes?

It became obvious that, by combining different light and heavy variable domains, specificities could be varied: 1000 different light chain variable region genes and 1000 different heavy chain variable genes would suffice ($1000 \times 1000 = 1$ million).

Research has indicated that the constant region genes follow Mendelian patterns; thus, there must be one gene pair for each of the constant regions (one pair each for the light and heavy chains). However, because the variable regions are coded for by multiple genes, there are apparently only hundreds, not the 1000 for each chain as previously speculated. The implication of this must be that the variable region genes are highly mutable, possibly changing in response to environmental conditions. There is also evidence that a **joining** type of genetic region (J gene) exists in the DNA which controls the joining of the variable and constant regions together. In fact, there are five different J genes and these recombine with the variable region gene in a random way. This mechanism multiplies the diversity of the variable genes by a factor of five. If 300 variable region genes were available, recombinations with the five J genes one at a time would produce 1500 different variable domains.

The most widely accepted model for antibody production states that a cell capable of producing the required antibody already exists in the body prior to exposure to an antigen. When the specific antibody is required, a cell that is capable of producing the proper antibody may be selected by some undetermined mechanism for rapid proliferation. Two genes are functioning (one for the light chain and one for the heavy chain), each with a distinct variable region that will react to the given antigen. The clone of cells rapidly produces the necessary antibody in large amounts and the foreign substance is attacked and removed.

Much is to be learned about the unique way that antibody genes operate, but the process serves as an amazing model for the production of a huge complex of molecules by an orderly and easily managed array of genetic units.

ADDITIONAL READING

ALLISON, A. C., "Sickle Cells and Evolution," *Scientific American* (1956), 195 No. 2, 87–94.

BARRY, J. M., *Molecular Biology: Genes and the Chemical Control of Living Cells.* Englewood Cliffs, N. J.: Prentice-Hall, 1964.

CRICK, F. H. C., "The Genetic Code: III," *Scientific American* (1966), 215, No. 4, 55–62.

CUNNINGHAM, B. A., "The Structure and Function of Histocompatibility Antigens," *Scientific American* (1977), 237, No. 4, 96–107.

DEBUSK, A. G., *Molecular Genetics.* New York: Macmillan, 1968.

LEHNINGER, A. L., *Biochemistry* (2nd ed.). New York: Worth Pub., Inc., 1975.

RICH, A., and S. H. KIM, "The Three-Dimensional Structure of Transfer RNA," *Scientific American* (1978), 238, No. 1, 52–62.

STAHL, F. W., *The Mechanics of Inheritance.* Englewood Cliffs, N. J.: Prentice-Hall, 1964.

STRYER, L., *Biochemistry* (2nd ed.). San Francisco: W. H. Freeman & Company Publishers., 1981.

SUTTON, H. E., *An Introduction to Human Genetics* (3rd ed.,). Philadelphia: Saunders College/Holt, Rinehart & Winston, 1980.

WATSON, J. D., "The Involvement of RNA in the Synthesis of Proteins," *Science* (1963), 140 17–26.

______, "The Double Helix" in *A Norton Critical Edition with Commentary, Reviews, and Original Papers*, ed. G. S. Stent. New York: W. W. Norton & Co., Inc., 1980.

______, *Molecular Biology of the Gene* (3rd ed.). Menlo Park, Ca: W. A. Benjamin, Inc., 1976.

WEISS, L. *The Cells and Tissues of the Immune System: Structure, Functions, Interactions*. Englewood Cliffs, N. J.: Prentice-Hall, 1972.

ZMIJEWSKI, C. M., and J. L. FLETCHER, *Immunohematology* (2nd ed.). New York: Meredith Corp. (Appleton-Century-Crofts), 1972.

ZUCKERKANDL, E., "The Evolution of Hemoglobin," *Scientific American* (1965), 212, No. 5, 110–18.

REVIEW QUESTIONS

1. The scheme symbolized by the superficial equation, DNA → RNA→ protein, has been called "the central dogma" by some molecular biologists. How could such a strong epithet be justified for this simple scheme?
2. Explain why the term *transcription* is appropriate for the process of RNA synthesis using DNA as a template (DNA→RNA).
3. Explain why the term *translation* is appropriate for the process of protein synthesis using mRNA as a template (RNA→protein).
4. Because many amino acids are designated by more than one triplet in the genetic code, it is said to be *degenerate* . Why does this seem to be a good word to describe the condition? Could the word *redundant* also be applied?
5. How many amino acids show only one codon? What are they? What is unusual about the codon for methionine in this respect?
6. Suppose that a portion of the nitrogen base sequence in a gene (chromosomal DNA) is known to be C T A T A C G A A A G A T.
   (a) What would be the sequence of bases in the complementary DNA half-chain on the other side of the molecule?
   (b) What would be the sequence of nucleotides in a strand of messenger RNA made under the direction of this gene segment?
   (c) What sequence of amino acids would be specified in a protein by the message in part (b)?
7. Mutations are changes in the nucleotide sequence of chromosomal DNA. They are heritable, since the new nitrogen base sequence is copied during replication. If a mutation were induced through the action of some external agent such as X-rays, it is conceivable that the DNA sequence in Question 6 might be changed to C T ↓T↑ T A C G A A A G A T (a base replacement or "switch" at the point indicated by the arrows) or changed to C T A T A ↓↑ G A A A G A T (a nucleotide deletion at the point indicated by the arrows). What effect would these two changes have on the protein? How might this affect the final activity of the gene product if it were an enzyme?
8. U A G is an mRNA codon for termination of a protein (synonymous for the punctuation mark called a period). It causes the translation to end on a ribosome and the new protein then drops free. What would happen if the normal messenger RNA sequence,

   C U C A U U G A U C U A U U U U U G C A U A A A

were mutated to

$$\text{C U C A U U G A U C U A U U U} \underset{\uparrow}{\overset{\downarrow}{\text{U}}} \text{A G C A U A A A}$$

or to

$$\text{C U C A U U G A U C U} \underset{\uparrow}{\overset{\downarrow}{\text{G}}} \text{U U U U U G C A U A A A?}$$

Why do you think this first type became known as a "nonsense mutation" when it was first discovered? Why do you think that the second type has been called a "silent mutation?"

9. If you identified three transfer RNA molecules with the following exposed and unpaired triplet nitrogen base sequences (anticodons) at the position which binds to mRNA, what amino acids would you expect them to carry to ribosomes during protein manufacture?

	tRNA 1	*tRNA 2*	*tRNA 3*
Anticodon =	A A A	U A C	A C C

10. Some antibiotics are known to inhibit growth by preventing protein manufacture. Puromycin accomplishes this because it resembles and substitutes for transfer RNA's during translation, but it is unable to attach an amino acid to the forming protein. What sort of gene products would you expect to find in a system to which puromycin had been added?

11. **(a)** If there are 1200 different light chain variable genes available in the immunoglobulin antibody-producing cells of a person's body and 1000 different heavy chain variable genes available, how many different antibodies could be assembled under the control of these 2200 genes?

(b) Suppose there are five separate genes (J or joining genes) for both the light and heavy variable regions that encode the last 13 amino acids in both variable regions. These five genes in each chain can be selected randomly to form the full variable region gene by recombining with the other variable region genes. To provide 1 million different antibodies, how many different variable region genes would be required to combine with the 10 J genes (five for the light, five for the heavy chain)?

12. Consider the histocompatibility antigens produced by genes at the HLA-A and HLA-B loci to be different proteins that are symbolized by Arabic numerals. Assume that in one parent antigen HLA-A 3 and HLA-B 12 are linked together (haplotype $\underline{3 \quad 12}$) and HLA-A 9 and HLA-B 8 are linked (haplotype $\underline{9 \quad 8}$). The person's genotype can be written $\dfrac{3 \quad 12}{9 \quad 8}$. The other parent's genotype could be given in haplotypes using the same style of notation, if enough information were provided by the children's phenotypes. The following HLA antigens are found in their three children. From this information, give the other parent's HLA genotype, showing correct haplotypes.

	Child 1	*Child 2*	*Child 3*
Series A antigens	3, 11	2, 3	9, 11
Series B antigens	12	5, 12	8, 12

12. Tay-Sachs disease is a very serious human nerve disorder that is caused by the lack of an enzyme (hexosaminidase A). Its absence is due to the fact that two mutant (recessive) genes are possessed by an individual. Such individuals invariably die, usually by 3 or 4 years of age. Persons with one or two normal genes produce adequate amounts of the enzyme. The mutant gene is found with a frequency of 1/60 among Ashkenazi Jews of northern European descent. Because there is selection against the mutant (persons acquiring two genes die), why do you think it occurs with such a high frequency in this population? Remember that the mutation rate is probably not higher than 1/10,000.

CHAPTER SUMMARY

1. Eukaryotic chromosomes are composed of DNA and histone proteins.
2. Transformation and transduction are two phenomena in which DNA is transferred from one bacterial cell to another and as a result hereditary traits are changed. Along with other evidence, these processes indicate that the hereditary material in chromosomes (the gene) is DNA.
3. Watson and Crick proposed a model for the structure of DNA in 1953 which has proven to be correct. This model is significant to genetics because it immediately suggests the underlying means of precise replication, as well as the means of controlling protein synthesis and hereditary traits.
4. The submolecules of DNA are inorganic phosphate, deoxyribose, adenine, guanine, cytosine, and thymine. The molecule is very large and consists of a double-stranded chain twisted into a helix. Across the two chains are consistently paired (by hydrogen bonds) nitrogen bases. Adenine is paired with thymine and guanine is paired with cytosine. The nitrogen bases can be arranged in any random linear sequence, thus the variability is virtually infinite.
5. It is possible for a DNA single strand to serve as a template for the production of another DNA strand (replication) with a precise sequence, due to the complementarity requirement for bases (A to T and G to C). The same general process also allows DNA to serve as a template for the production of RNA (transcription). In RNA, however, ribose is substituted for deoxyribose and uracil is substituted for thymine. RNA remains single stranded, except for some pairing within regions on the same strand.
6. Three types of RNA are produced using DNA as a template: messenger RNA, transfer RNA, and ribosomal RNA. All three types are required for the synthesis of proteins in the cytoplasm in the process of translation.
7. The linear sequence of amino acids in a protein is specified by DNA nitrogen base sequence in a gene. Messenger RNA carries the complementary sequence of bases where triplets of bases (codons) signify specific amino acids. The message of codons is read by ribosomes and transfer RNA molecules bring the specific amino acids to the site of the codon-ribosome interaction. Amino acids are thus connected by peptide bonds into a long chain (polypeptide) which is folded and bent into a characteristically shaped protein molecule.
8. All 64 possible codons have been identified with either specific amino acids or "punctuation marks" (start and stop signals); it is possible to construct synthetic sequences in messenger RNA for a given linear array of amino acids.
9. Hereditary material is able to change to provide for the abundance of diversity found among living things. The Watson-Crick model also suggests the possibility for change in the nitrogen base sequence through mutation. Not only may such mutations result in differences in the resulting protein structure, they are heritable, since the changed base sequence will be copied at the next replication.
10. Mutations are extremely rare, occurring on the order of about one in every 10,000 to 1,000,000 gametes for a given gene. Mutations are often due to uncontrollable environmental factors such as cosmic irradiation. Unnatural environmental factors, such as chemicals or X-rays, are also responsible for mutation.
11. The many different forms of genes that exist are due to mutations or changes in DNA structure at some point in the ancestry of an organism.
12. Hemoglobin is a good example of a human protein under the control of genes. Mutations have changed its form, as in hemo-

globin S (of sickle cell anemia), or caused its reduced production, as in thalassemia. Both mutants occur in certain populations at a rate well above mutation frequency, because they may confer a partial immunity to malaria.

13. It is apparent that genes may exist in many different forms, besides the two recognized by Mendel. This phenomenon is known as polymorphism. Some gene forms may not affect the functional properties of the gene product, while others may produce easily recognized phenotype differences.

14. A well-known series of human gene loci that are highly polymorphic is the HLA or histocompatibility series. The antigens produced by the genes result in tissue rejection in grafts and transplants. Because there are many forms of each gene and the loci are closely linked, the HLA system is an effective means of analyzing cases of disputed parentage.

15. Eukaryotic genes contain intervening sequences (introns) that are spliced from the mRNA before it is translated into protein.

16. The molecular structure of antibodies is well known and the underlying genetic mechanism for their production is complex. Antibodies are immunoglobulins composed of 12 domains or polypeptides about 110 amino acids long. Four of the domains are variable, while the other eight are constant in structure and amino acid sequence. The constant domains follow simple Mendelian inheritance patterns, but the variable domains are controlled by a large number of highly mutable multiple genes that can recombine with other component parts to provide an immensely variable selection of specific antibodies. Each antibody molecule consists of two long chains (three constant domains and one variable domain) and two short chains (one constant domain and one variable domain).

CHAPTER

5

HUMAN CHROMOSOMES

Thus have the gods spun the thread for wretched mortals. . .
Homer, The Iliad *(700 B.C.)*

Two lines of investigation began to focus on cell nuclei in the period between Mendel's original publication in 1865 and its rediscovery in 1900. These investigations led to the establishment of two new sciences important to genetics: **cytology,** the study of cells, and **biochemistry,** the study of the chemicals of life.

Germany in the late 1800s was the site of many studies important to the future of biology. In 1871 Friedrich Miescher characterized and named an important constituent of the nucleus, *nuclein.* This material, composed of nucleic acids (mainly DNA) which readily combined with proteins, was identified first in human white blood cells and then in the sperm cells of Rhine River salmon. The reason for its concentration in the nuclei of sperm cells seems obvious now, but it was a number of years before its significance became apparent.

In 1879 Walther Flemming in his studies of cell division noted the presence of easily stained nuclear material which he called **chromatin.** By 1882 he had worked out the sequence of steps in nuclear division and named the process **mitosis.** Another German, Wilhelm Waldeyer, refined the studies somewhat and labeled the discrete threads that became apparent during mitosis **chromosomes.**

E. B. Wilson, an American, noted in 1895 that chromatin was probably identical to nuclein and that its transmission was likely to account for inheritance. Thus, by about 1900 the groundwork had been laid for intensive and careful analysis of the constancy of the chromosomes within a species and their association with the hereditary mechanism. Microscopic studies continued as chromosome counts were made for a number of different organisms and investigations of the chromosomal differences between males and females were conducted.

By about 1912 Hans von Winiwarter had made a reasonably good attempt at counting human chromosomes. However, he reported that male cells contained 47 chromosomes and female cells, 48. In 1923 T. S. Painter, an American, cautiously reported that human cells, both male and female, carried a complement of 48 chromosomes. The report was based on Painter's earlier observations of 46 and he was not totally convinced of the accuracy of his analysis.

METHODS OF HUMAN CHROMOSOME ANALYSIS

The difficulty of counting human chromosomes was due to both technique and materials. Microscopic skills and equipment were adequate in the early 1900s, but human tissue for detailed analysis was difficult to obtain. Furthermore, when tissue was acquired, often from pathological sources, it had to be cut into very thin slices and stained for microscopic examination. Such thin slices presented problems with small chromosomes, as the nuclei were much thicker than the sections and it was necessary to trace a chromosome from one section to the next. More disconcerting was that the chromosomes were often clustered together, making the identification of discrete individuals nearly impossible.

In 1921 John Belling devised a method of squashing various types of cells on a microscope slide so that the cells could be spread out freely with the chromosomes well separated. This facilitated counts greatly for easily separated dividing cells, but no human tissue provided such cells until human tissue culture cells were grown in a liquid suspension.

Jo Hin Tijo and Albert Levan reported the correct diploid human chromosome number of 46 in 1956. Their report was also cautious, not so much because of doubts about technique, but because the number 48 had become so well established by that late date. Probably the most important feature of their technique was the source of cells used. By the 1950s **tissue cultures** of human cells were readily available and they provided the free suspensions of cells necessary for the squash technique. Tijo and Levan tilted the scales in their favor by using the plant alkaloid **colchicine.** This natural product arrests mitosis at metaphase when the chromosomes are at their shortest, most easily separated stage. By treating the cells with a weak salt solution, they were made to swell and spread well on the slides.

Many more tricks have since been learned to aid in the microscopic analysis of chromosomes. The technique of producing preparations has become routine for some technicians. Human blood is now used as a ready source of suspended cells that can be quickly induced to divide in culture. Red blood cells are of no value for this purpose since they have no nuclei, but **white blood cells** have proved to be ideal.

To remove the unwanted red blood cells, it was found that another plant product, **phytohemagglutinin** (PHA) was useful. This compound causes the red blood cells to coagulate in large, easily removable clumps, leaving suspended white blood cells in the serum. Fortuitously, PHA was found to have another important property: It stimulates cell division in the white blood cells. In other words, it acts as a **mitogen** (generator of mitosis) and is part of the protocol used to prepare human white blood cells for chromosome analysis.

The combination of these techniques provided much information about the detailed structure of chromosomes. With the recognition that DNA was the hereditary material came an intense interest in the minute molecular arrangement of components in the chromosomes.

While eukaryotic chromosomes proved to be excellent subjects for light microscopy, they have been unusually difficult to prepare satisfactorily for ultrastructural studies with the electron microscope. However, through the use of protein- and DNA-digesting enzymes, centrifugation studies, careful chemical analyses, and sophisticated electron microscope techniques their basic molecular structure has gradually been revealed.

HUMAN CHROMOSOME STRUCTURE

Some techniques show partially disassembled chromosomes to have a beaded appearance. The swollen areas, called **nucleosomes,** have been found to consist of a compact mass of exactly eight globular histone protein molecules (an *octamer*) with the strand of DNA wrapped twice around the complex (see Fig. 5.1) These beads are connected together by the continuous strand of DNA associated with another histone protein molecule. The beaded appearance can only be seen under certain experimental conditions and there is no intervening space at these connecting points. This natural configuration of chromatin in a compact coil of DNA around proteins forms a neat fiber of genetic material that is about 1/6 the length of the same free DNA strand. The strand of chromatin is about five times as thick as DNA strands alone, but it is still only 1/2,500,000 of an inch in diameter (Fig. 5.2).

When the genes are not actively transcribing their message, the chromatin is thought to be coiled further into a fiber that is two or three times the thickness of the nucleosome-containing thread. Thus, relaxed and outstretched interphase chromosomes exist in thicker strands than can be accounted for by the nucleosomal structure. During mitosis and meiosis the shortened condition of the chromosomes is due to a tight looping and folding of the strand into an even more compact structure. The chromosomes are much shorter and broader in metaphase than in interphase. It has been estimated that the amount of compaction represents a final metaphase packing ratio of about 5000 to 10,000 times that of free DNA.

It should be obvious that chromosomes are elegantly structured so as to store a huge amount of molecular material, while maintaining a well-preserved linear order. In spite of coiling and folding, the actual arrangement of the molecules constituting a chromosome results in a very long threadlike structure with the genes arranged in linear sequence along the length. During interphase the chromosomes are stretched out to long lengths; it is not convenient to study them at this stage if one wishes to make counts or identify specific chromosomes. All the chromosomes of a nucleus are entangled within the confines of the nuclear envelope when in the relaxed, elongated state. They are impossible to trace unless they are in the shortened condition; therefore, most descriptions of chromosomes are based on light microscopic studies of their compactly shortened appearance during cell division.

(a)

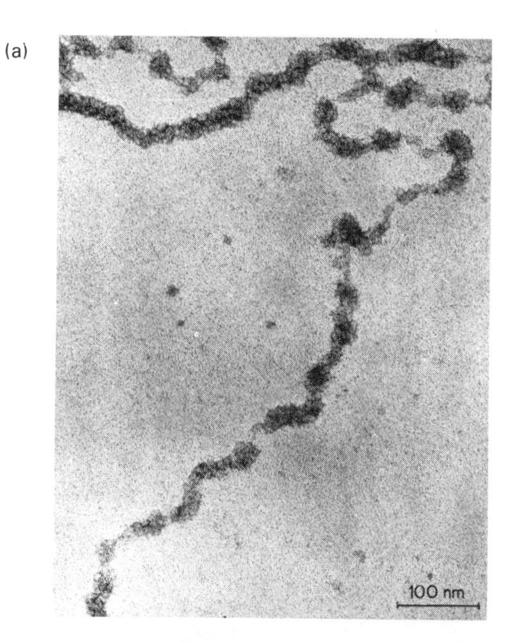

(b)

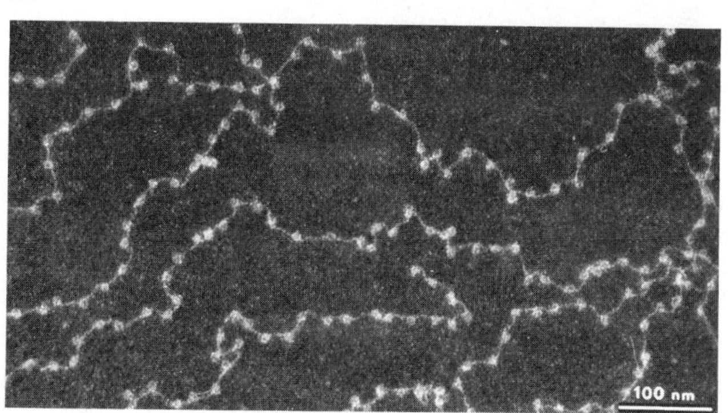

Figure 5.1. High resolution electronmicrograph of chicken blood cell chromosomes. (a) Chromosomes treated to maintain 20- to 30-nm (nanometer) fiber of natural conformation. (b) Chromosomes treated to release compaction between nucleosomes at one histone connecting point. The compacted strands would be of about 10-nm diameter, the diameter of individual nucleosomes. The 20- to 30-nm fibers of (a) are thought to represent the nucleosomal strand coiled into a shorter and thicker fiber. (One nm is equal to about 1/25,000,000 of 1 inch.) (Courtesy Donald E. and Ada L. Olins and *American Scientist*)

Early cytologists found that chromosomes had an attraction for certain stains. By using these stains, the chromosomes were readily studied during cell division. Even though the cells had been killed in preparation for observation, much could be learned about the dynamics of chromosome action within a cell. By treating the cells with certain chemicals, the chromosomes were preserved in the shape and arrangement which they possessed at the time of killing. By studying many different cells, it was possible to determine the sequence of stages in the arrested processes.

As important as the original chromosome-staining observation, T. Caspersson and coworkers found in 1969 that certain stains concentrated selectively in specific regions of chromosomes. The intensely stained **bands** that can be revealed with **Giemsa stain** or fluorescent **quinacrine mustard** are consistently the same from cell to cell and organism to organism within a species. The molecular basis for banding patterns resides in the G—C, A—T concentrations and it seems obvious that they represent fundamental structural subdivisions of chromosomes. These patterns allow un-

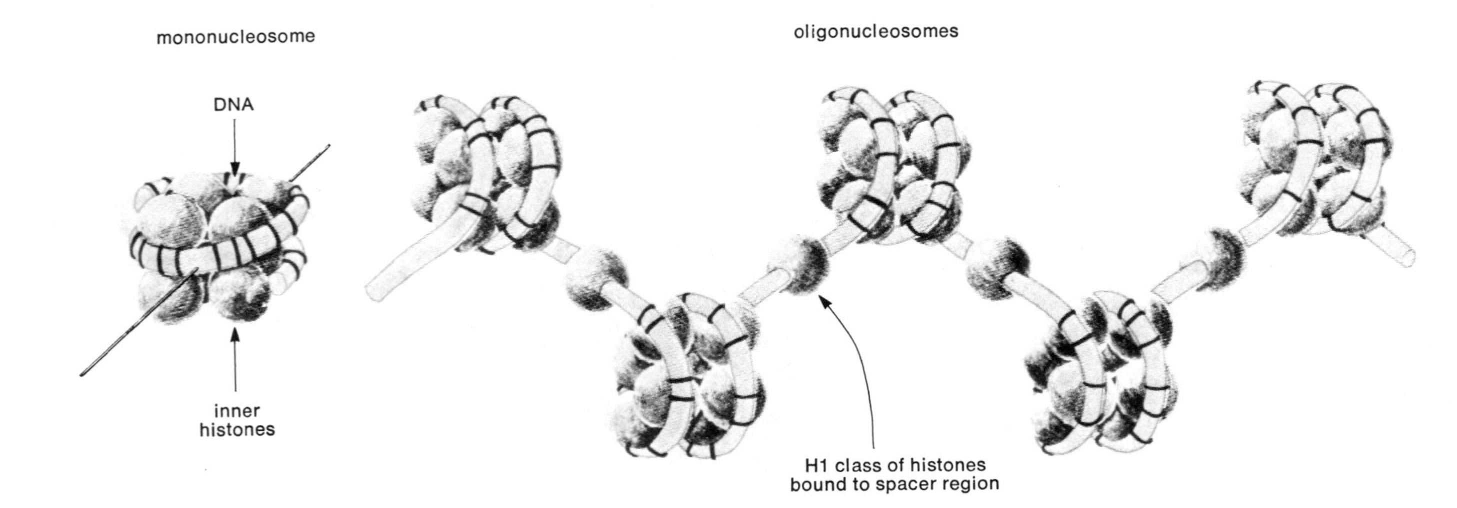

Figure 5.2. The nucleosomal structure of eukaryotic chromosomes. Diagram shows the eight globular histone protein molecules in each nucleosome. The chromosomal DNA is wrapped twice around each octamer of histones. There is a short strand of DNA and another histone molecule between adjacent nucleosomes. (Courtesy Donald E. and Ada L. Olins and *American Scientist*)

ambiguous identification of specific chromosomes, as well as a ready means of categorizing alterations or aberrancies in normal chromosomal structure.

These modern staining techniques have permitted the unique identification of each human chromosome. Beyond simple identification of the individual chromosomes, precise identification of regions within the chromosomes has become possible. Thus, a good technician can now detect various types of structural changes that have occurred within a given chromosome. Furthermore, such structural changes, along with sophisticated DNA-RNA pairing techniques, have even permitted the identification of certain regions as locations of given genes or blocks of genes.

Progress in such studies since the advent of Giemsa and fluorescent staining techniques has been astounding. Human chromosomes are becoming the most well described of all higher organisms. It has become a highly specialized and important laboratory technique to identify the human chromosomes from well-maintained tissue cultures with carefully prepared slides. In modern human genetic studies it is even possible to remove cells from an unborn fetus and examine the chromosomes for abnormalities. The technique is clearly valuable but the results often require difficult decisions.

During cell division it has been seen that the chromosomes shorten and become untangled by coiling up tightly. Areas that appear to be thicker and more darkly stained in relaxed interphase chromosomes are called **heterochromatic** (hetero = different), while the thinner, fainter regions are said to be **euchromatic** (eu = true). The heterochromatic regions are found not to coil as tightly as the euchromatin during cell division, sometimes causing distinct constrictions along a chromosome's length. Each chromosome is found to have at least one such constriction, the largest of which is called the **primary constriction.** The primary constriction has been discussed by its more familiar name—the **centromere.** The major function of smaller **secondary constrictions** found in some chromosomes is generally the formation of nucleoli.

The centromeres serve in cell division as the centers of chromosome movement, since they lead the way to the spindle poles during anaphase. At that stage, the arms of the chromosomes are seen to be trailing back toward the center of the cell as the centromeres move toward the ends. Various types of evidence indicate that the centromeres serve as formation centers and attachment points of spindle microtubules.

THE NORMAL HUMAN CHROMOSOME COMPLEMENT

To study the chromosomes of humans it is necessary to utilize cells that are dividing in order to find shortened, well-separated chromosomes. This was rather difficult until in 1960 it was discovered that white blood cells could be isolated and stimulated to divide and grow in test tubes. Such cells are ideal for chromosome studies, since they can be flattened and stained on microscope slides. Photographs may be taken of such preparations, and the individual images of chromosomes can be cut out and arranged in a sequence for study. With some effort the **homologous** pairs of chromosomes can be identified and then arranged according to length and centromere location. Such photographic arrangement, called a **karyotype,** provides a systematic means of understanding the chromosomal basis of human heredity (Fig. 5.3).

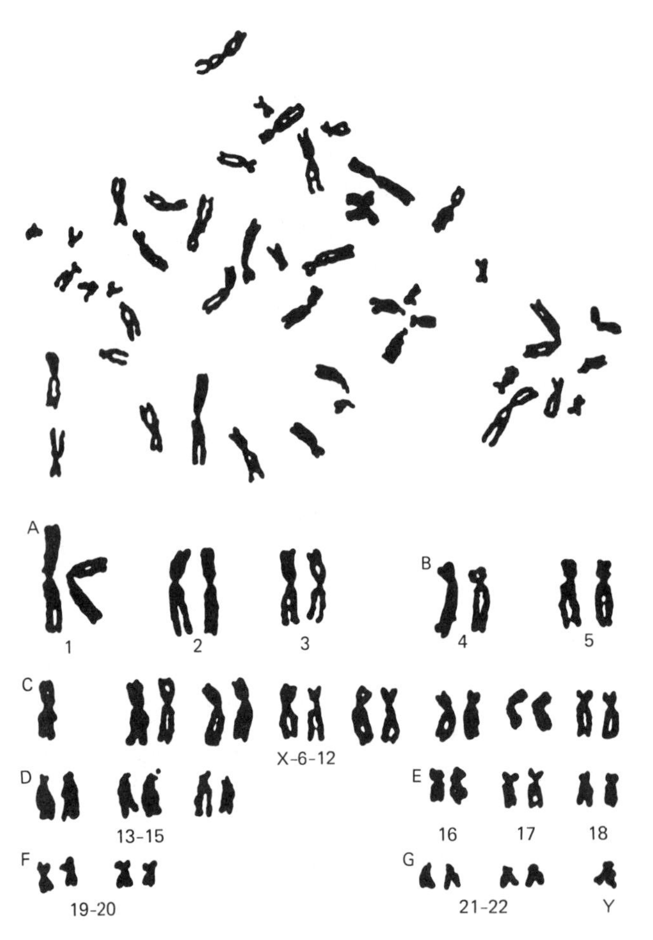

Figure 5.3. Intact metaphase spread of human male chromosomes (top) and a karyotype arranged from these chromosomes (bottom). (Courtesy J. D. Rowley, J. J. Yunis, and Academic Press, Inc.)

Because the chromosomes are in pairs in a normal cell (diploid), it is necessary to match each chromosome to its homologous member in constructing the karyotype. The process of karyotype construction was aided immensely by the 1969 Caspersson discovery of differential staining procedures that revealed consistent banding patterns. Because all the genes necessary to the functioning of a human are included twice within the 46 chromosomes displayed by the karyotype, it may be said to represent two complete sets of genes **(genomes).**

Although the chromosomal complement of a typical human can be understood by studying a karyotype, a diagrammatic interpretation of the chromosomes **(idiogram)** is probably even more effective for general information. The system of classifying human chromosomes according to length, centromere location, and banding patterns is based on a scheme adopted in 1971 at a Paris conference on standarization in human cytogenetics. The general framework for the system was based on techniques agreed to at earlier conventions in Denver (1960), London (1963), and Chicago (1966). With the new banding techniques it became necessary to refine the system to include this information.

In this system it is standard practice to group similar-appearing chromosomes into seven categories symbolized by the letters A through G. The individual chromosomes or pairs of chromosomes are referred to by number, starting with the longest

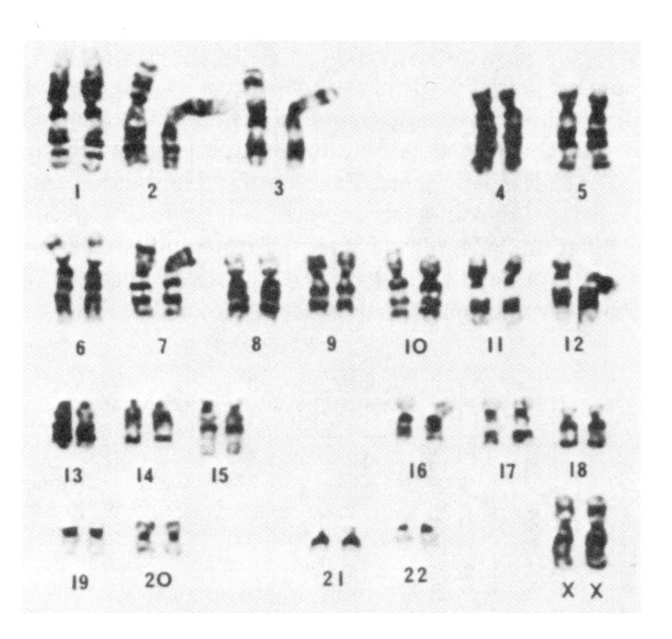

Figure 5.4. A Giemsa-stained preparation of human female chromosomes. The chromosomes exhibit what is referred to as the G-banded pattern. (Courtesy T. C. Hsu and Springer-Verlag).

as the number 1 pair and progressing to the shortest as number 22. This accounts for only 44 chromosomes, which are referred to as **autosomes.** The other two are involved in determination of sex and are called the **sex chromosomes.** The longer sex chromosome is labeled the **X chromosome** and the shorter is the **Y chromosome.** Females have two X chromosomes in each cell, while males have an X and a Y chromosome. In a karotype the medium length X chromosome fits into group C, based on length and centromere position, while the short Y chromosome would be aligned with group G. Both are unambiguously distinguishable, however, on the basis of the Giemsa (G) or quinacrine (Q) mustard staining patterns (Fig. 5.4).

To refer to specific locations in individual chromosomes, the Paris convention agreed to label the chromosomal short arms with a *p* and the long arms with a *q*. In most human chromosomes the long and short arms are readily identifiable, since the centromere is not located exactly at the midpoint of length (Fig. 5.5). In arrang-

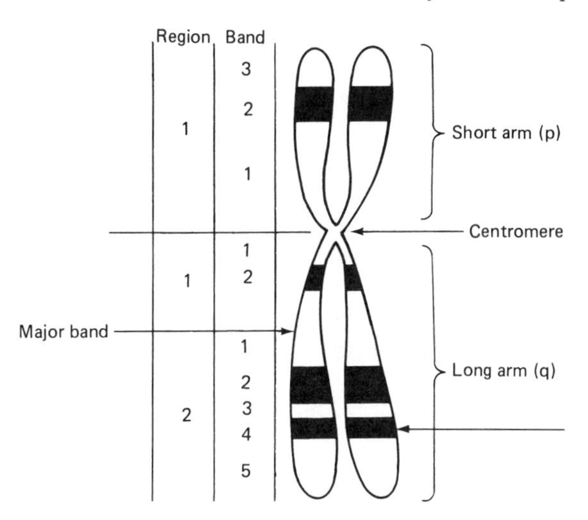

Figure 5.5. Diagram of human chromosome 17 showing banding patterns. Where major bands can be identified (as in long arm), regions are numbered (first column) sequentially outward; otherwise, there is only one region (as in short arm). Stained and unstained bands are numbered within regions (second column), again starting nearest the centromere and progressing toward the ends. Band indicated with lower arrow is labeled 17q24.

ing a karyotype the chromosome is placed vertically so that the short arm is above the long arm. Clearly identifiable large sectors or *regions* containing several bands between major landmarks are then labeled with Arabic numerals, beginning near the centromere and progressing to the ends in both arms. The stained and unstained bands within the large sectors are labeled in turn with Arabic numerals. To refer to a specific point on a chromosome, the notation 2p24 might be used. This would translate as follows: chromosome number 2, short arm sector (region) 2, subsector (band) 4. A haploid set of human metaphase chromosomes exhibits about 320 bands (Fig. 5.6).

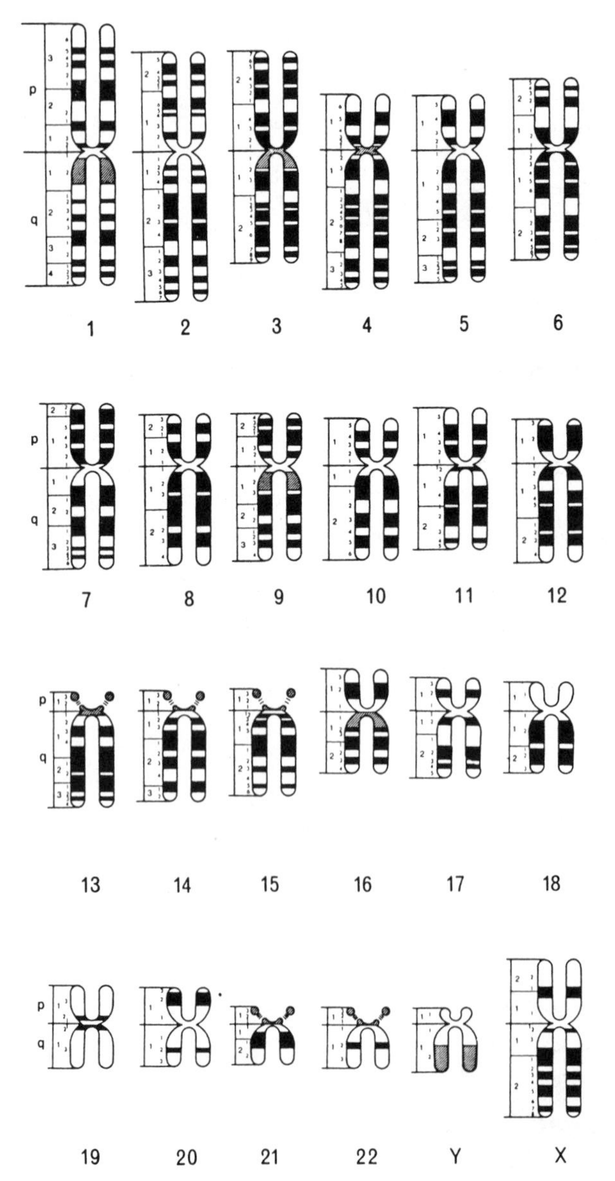

Figure 5.6. Diagrammatic representation of human chromosomes showing the banding patterns of Q-, G-, and R-staining methods. Centromeres are representative of Q-staining method only. Black areas represent positive Q and G bands and negative R bands. White areas represent negative or pale Q and G bands and positive R bands. Shaded areas show variable staining reactions. (Courtesy The National Foundation, New York: *Paris Conference (1971): Standardization in Human Cytogenetics)*

Using incompletely contracted chromosomes from a dividing cell (late prophase or *prometaphase*), over 1200 specific regions or bands can be labeled on a human karyotype.

Because there are many more genes than Q and G bands in a set of human chromosomes, it is logical to assume that more than one gene may be present in each band. An ultimate goal might be an increase in the resolution of the procedure to the point where the locus of each gene would be identifiable. Such an ideal will likely never be realized.

CHROMOSOMES AS GENE CARRIERS

It is obvious that the genes of an individual human are spread out within the nucleus of each cell among 23 pairs of chromosomes. Only a relatively small percentage of the genes have been discovered; the approximate location of an even smaller number of these is known. In fact, it is possible only to guess at the number of genes present in the human genome. To begin, one can estimate the number of genes that could be found on an average human chromosome.

There is a large discrepancy in the length of chromosomes in a human karyotype. The longest chromosomes might account for five or six times as many genes as the shortest chromosomes, based on their length. A figure for an average number of genes would provide a rough estimate of the gene capacity of medium length chromosomes of the C group (about half the size of the longest).

Rough estimates of the amount of DNA in a haploid set of human chromosomes indicate that there is enough to code for between 3 and 5 million moderate sized protein molecules of about 300 amino acids each. The largest chromosome contains enough DNA to direct the manufacture of over 200,000 such proteins alone! Using these figures, an average length chromosome in the C group might then be expected to code for over 100,000 proteins.

Such calculations imply that all the DNA is functioning as genetic material in the chromosomes, as in the prokaryotes. In fact, this implication is not true. As seen, some parts of the chromosomes are labeled heterochromatin. Heterochromatin is generally considered not to contain functional genes and, although the amount varies among individuals, a significant portion of the human chromosome complement is constantly heterochromatic. In addition a significant amount of nonfunctional DNA exists as introns.

With all this information, geneticists estimate that the human genome consists of about 100,000 genes. One can thus calculate that an average-sized chromosome may contain the directions for about 4000 proteins. Such figures are hard to comprehend, but the compact efficiency of the hereditary mechanisms is exceedingly impressive.

It should be obvious that not all the genes are functioning within a given cell at one time. In most instances only a relatively small number of the genes could be expected to be coding for gene products in a single cell. However, it is probably safe to assume that all the genes operate throughout the body in different cells and at different times. The human body is a complex, integrated organism with untold factors influencing the functioning of the hereditary material.

LOCATION OF SOME HUMAN GENES

Determining the physical location of specific genes within the genome has presented a challenge to cytogeneticists for many years. It has been an especially difficult task in humans, because of the problems associated with preparing chromosomes for study, as well as the small number of offspring produced within families. The key factor in locating genes within chromosomes is an unambiguous association of some trait with a particular chromosome. Once a gene has been definitely identified with a particular chromosome, other genes may be shown to be linked to the first gene using genetic data. To prove this often requires a sizeable number of offspring or families, but once a linked group of genes has been established and one of them has been associated with a particular chromosome, the entire group can be correctly assumed to reside in the same chromosome.

Recombination Frequency and Map Distance

Detailed **chromosome maps** have been constructed for some well-studied organisms such as fruit flies, mice, and corn plants. A large number of genes are known for these organisms and they are all capable of producing large numbers of offspring in controlled crosses. The technique most frequently utilized for mapping the genes has been purely genetic: It is based on (1) determination of **linkage groups** and (2) establishment of the frequency of **crossing-over** between genes. The principle underlying the system is that two distantly linked genes in a chromosome provide a better chance for a random cross-over to occur between them than do two genes that are close together. The logic of this basic principle is simple, and it has been exploited very effectively to determine the sequence of genes and their relative distances apart within a chromosome.

Numerical map data are based on comparisons to the figures expected for independent assortment between two genes. Mendel could predict the results of crosses involving separate gene pairs based on the statistics of independent assortment. If the classical Mendelian ratios are statistically distorted beyond expectations, linkage can often be assumed. When genes are physically linked together, they cannot assort independently.

Pairs of homologous chromosomes from an individual's two parents pair up in a close two-by-two association during **synapsis** in meiosis. At that point, the chromosomes have duplicated so that four strands exist. The two closely associated strands **(chromatids)** may then exchange parts by breaking and rejoining in a process called crossing-over. Crossing-over is a common occurrence in meiosis and probably occurs at least once in every synaptic pair of homologous chromosomes. Crossing-over between two linked genes can be diagrammed as follows:

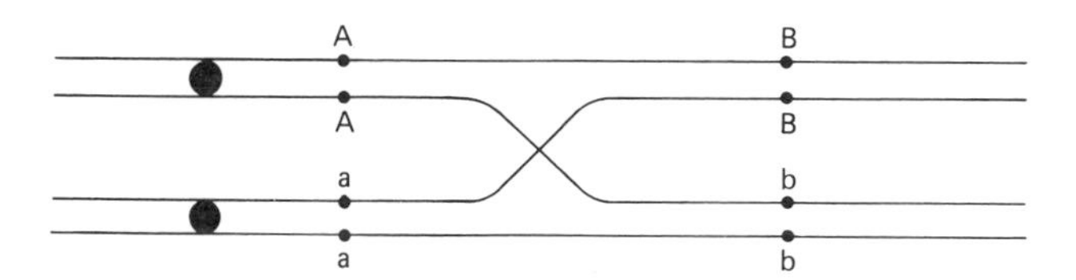

The result of such a cross-over is a recombination of the original parental sequences of genes in half the products of the division. If the two gene loci shown were far enough apart so that crossing-over always occurred between them, the results would be indistinguishable from independent assortment (1/4 of each type of gamete: AB, Ab, aB, ab). However, if they were linked closely enough, this would not be true. Although crossing-over occurred, the original combinations would be the same. This can be understood by considering a diagram of a situation in which the cross-over point is not between the two genes:

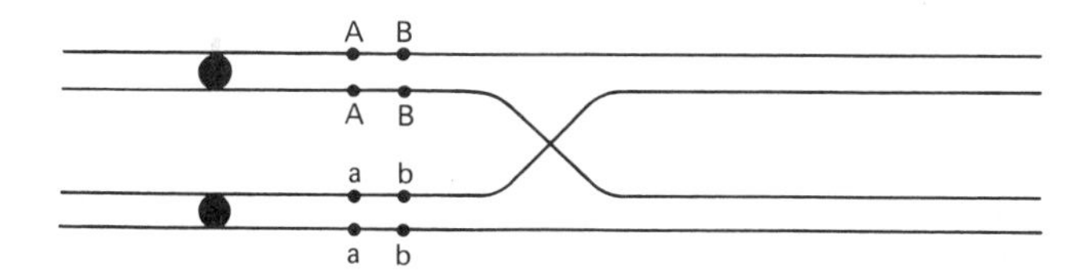

The combinations aB and Ab would occur considerably less frequently than AB and ab, since cross-overs between the two genes would not occur in every synaptic pair.

It should be apparent that the more closely linked two genes are within a chromosome, the less chance there is for crossing-over to occur. The basic assumption is that crossing-over is a randomly occurring process which usually takes place at least once in each homologous pair.

The physical distance between two genes can therefore be estimated from the percentage of **recombination.** In fact, such percentages have been traditionally read directly as **map distances.** The map units (percentages of recombinants) are called **centimorgans,** after Thomas Hunt Morgan who in 1911 demonstrated genetic evidence for crossing-over. One centimorgan (cM) represents 1% (centi = one-hundredth) recombination between two linked genes. If a single cross-over occurred between gene loci in every meiotic cell, one-half the products would be recombinants (see preceding diagrams). The gene loci would thus be 50 cM apart. This is the same as **random assortment.** Therefore, map distances for two loci must be considerably less than 50 cM to have any meaning, maps must be constructed by adding consecutively many small distances between adjacent gene loci. Based on the amount of crossing-over that occurs in human meiotic cells, the human genome has a total length of about 3000 cM.

Family Studies and X-Chromosome Mapping

Although humans were not as good experimental material for recombinant mapping studies as some other organisms, a start was made using recognized techniques for mapping some chromosomes. It was a slow, laborious process to find enough families to detect recombination between linked genes. The first chromosome to be mapped was the X chromosome. The first gene associated with a human chromosome was one for **colorblindness**, located on the X chromosome by E. B. Wilson in 1911. More than 100 X-linked genes are now known.

Note that females have two X chromosomes and males have only one. In Chapter 7 it will be shown that this results in the phenomenon of **sex linkage**, since males have only one of each X-linked gene and females have two. Males receive their single X from their mothers and a pattern of inheritance from mother to son to daughter can be detected.

It is also possible to determine readily the combinations in which genetic characteristics are linked in a male, since only a single X chromosome is present. From such information, it is possible to determine that recombination has occurred in his daughter. Remember that a female has two X chromosomes—one from her father, one from her mother—that are paired up in synapsis. The combination of X-linked genes present in the father will be preserved where no crossing-over occurs in the woman and she may pass this combination on to her son.

A grandson's genotype will reveal the crossing-over which occurred during meiosis in the woman. If a different combination of X-linked genes is observed in grandfather and grandson, it can often be determined that crossing-over between the genes in question occurred in the woman's reproductive cell formation. By accumulating data from a large number of such families where father, daughter, and grandson can be observed, an accurate map can be constructed for the X chromosome.

As more information is provided, the distances and even the presumed order of the X-linked genes may be changed. Figure 5.7, a sketch of such a map, will give you an idea of the manner in which chromosome maps are drawn, using recombination percentages as map distances (centimorgans). As more genes are studied that are close

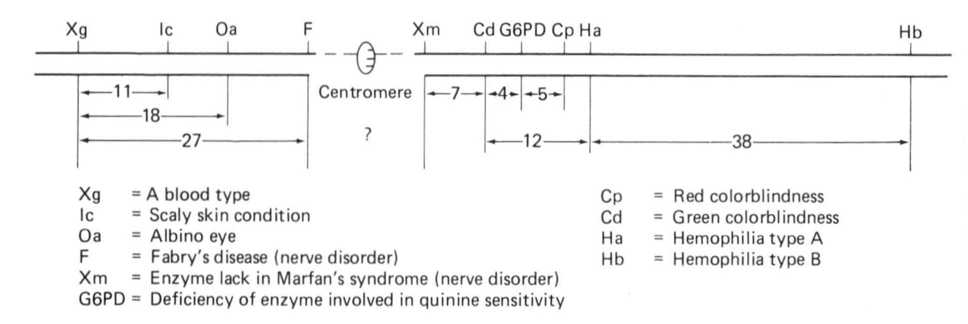

Figure 5.7. Partial human X-chromosome gene map. The figures given for map distances are expressed in centimorgans (cM). Each map unit (cM) represents 1% genetic recombination due to crossing-over which takes place during synapsis in meiosis.

together, the number of locations will increase and the known locations will be made more accurate. Chromosome maps are most accurately constructed by gradually adding information from many closely linked genes exhibiting short distances. For clarity, only ten loci are included in Figure 5.7, but many more could be added.

It is apparent that the construction of a genetic map of the X chromosome could only be accomplished with data from recombination in females. Because males have only one X chromosome, it is impossible to measure recombination in that chromosome during sperm formation. In addition it has been shown that crossing-over in the other chromosomes (autosomes) during meiosis in males is measurably less frequent than in females. Thus, genetic map distances are less in males than in females, but this does not mean that the physical distances are different. Only the absolute amount of crossing-over and recombination is lower in males, while relative distances remain essentially the same. The reasons for this phenomenon have not been determined, but it is known that crossing-over does not occur at all in *Drosophila* males.

Deletion Mapping

Because of the development of modern staining techniques that revealed banded areas in chromosomes, it has been possible to demonstrate deletions of short regions in some cases. It has been possible to show genetically that a phenotypic effect is correlated with the deleted segment. Such a genetic effect may be similar to that shown by a recessive gene, and easily demonstrated.

This sort of correlation is strong evidence that the gene in question is located in the region of the deletion. In addition the correlation can serve as a technique for constructing a cytological map, based as it is on cytological study, rather than a genetic map, which is based on recombination studies. The sequence of genes in the two maps is the same, but distances are not necessarily well correlated, due to variations in the amount of crossing-over in different regions.

Cell Hybridization Studies

For many years it appeared that chromosome gene mapping in humans would be an extremely difficult and slow task because of the problem of finding a convenient technique to associate genes with particular chromosomes. In 1960, however, a novel addition to the repertoire of tricks used by the tissue culture scientists was developed. Certain viruses were discovered to enhance the ability of different types of cells to fuse together in culture. Thus, it became a routine procedure to create **hybrids of somatic cells** from mice and humans. An interesting feature of the hybrid cells was that they gradually lost chromosomes from the human genome during subsequent divisions, so that only a few remained after a short time in culture. Some cultured cells were even found with only one human chromosome.

Astute workers were able to capitalize on this situation by associating the presence of certain human gene products (usually enzymes) that were produced in the cell cultures with a particular human chromosome. It was thus possible to associate unambiguously some gene products with definite chromosomes. As a result of this technique, there has been a virtual explosion of information in the last few years linking various genes together in groups that are located within specific human chromosomes. With time and a large amount of genetic recombination data from a number of families, it will be possible to produce accurate information about the linear sequence of such genes and their relative distances apart. Genes associated with every human chromosome have been definitely identified. The diversity of genes so identified covers a large number of enzymes, as well as such traits as blood types, antivirus agents, sensitivity to various diseases, ribosomal RNA production, and cataract development. Such information will be extremely valuable and may even serve a practical role in some medical context, such as the replacement of defective genes through genetic engineering.

The tissue culture cell hybridization technique has brightened the picture of human chromosome mapping in another way. The technique not only allows the ready identification of specific chromosomes as carriers of genes for distinct products, it often allows the localization of genes within given regions or sites on the chromosomes. By using radioactive messenger RNA molecules that will attach to their specific gene site or by finding aberrant broken chromosomes, it has been possible to associate genes with definite bands or other identifiable regions. It has been noted that mRNA hybridization with complementary DNA can be utilized to locate a gene and also to reveal the presence of introns and exons.

Maps made on the basis of physical location of genes in chromosomes are really cytological, not genetic, maps (Fig. 5.8). The actual distances between loci are not necessarily proportional to distances derived from recombination percentages. However, the sequences are the same and much information can be obtained by combining the two types of maps.

Continuing recombination studies of families and cytological identification of physical gene locations can be expected to make the human chromosome map more precise and meaningful in the years ahead. Although more than 400 mapped genes represent a major accomplishment, with some 3000 human genes already recognized and with potentially 100,000 genes in all, more information can be expected in the future.

Restriction Endonucleases

Starting in the late 1960s molecular biologists discovered almost as curiosities a marvelous series of tools for the analysis of genetic material. Today use of such tools is becoming routine. They are called **restriction endonucleases** since they are enzymes that attack DNA (nucleases) by cutting it within the chain (endo-) and they restrict the activities of invading viruses. Bacteria of many different species are capable of producing these protective enzymes. The endonucleases cut the foreign DNA of viruses at specific points recognized by the enzymes. They are designated by letters based on the name of the bacterial species that produces them. Thus, Eco RI comes from *Escherichia coli* (*R*estriction *I*) and Hind III comes from *Hemophilus influenzae* strain *D* (Restriction *III*).

Each enzyme recognizes a sequence of four to six nucleotides in the DNA chain that reads the same on both sides of the chain in different directions. Thus, Eco RI recognizes the following nucleotide sequence in DNA

```
 ↓
-G-A-A-T-T-C-
 | | | | | |
-C-T-T-A-A-G-
           ↑
```

and cleaves at the same point in the sequence on each side (indicated by the arrows). The unpaired sequence can then be thought of as a "sticky end," which will recombine with a complementary one on the same or another chain that has been cut by the same endonuclease:

```
-G              and  -A-A-T-T-C-.
 | | | |              | | | | |
-C-T-T-A-A-                    G-.
```

Another example is Hae III (from *Hemophilus aegyptius*) which recognizes and cleaves at

```
   ↓
-G-G-C-C-.
 | | | |
-C-C-G-G-
   ↑
```

Note that this symmetrical cut does not leave sticky ends and could not be used to splice in other pieces of unrelated DNA.

There are over 70 different restriction enzymes with more being discovered as the search continues. Many interesting tricks can be done with endonucleases, since they have the ability to attack different DNA molecules from diverse sources, including humans. Besides several other important factors, they serve as the basis for genetic engineering. Because those that leave sticky ends will cut and allow for the insertion of foreign DNA strands at given points, they are extremely important for the transplantation of genes from one organism into the genome of another. They have provided a tool for implanting human genes in bacteria, where human gene products, such as insulin, can be readily produced on a large scale.

Of no minor import is the fact that restriction endonucleases will cleave the 3.3 billion base pair human genome into smaller, more manageable pieces. A restriction endonuclease which cleaves DNA into segments of approximately 4000 base pairs will produce about 1 million different DNA fragments from the human genome. Although the number of such fragments is overwhelming, by placing them in a gelatin-like matrix in an electric field (**gel electrophoresis**), they can be spread out into unique positions, based on their molecular weight and size. This preparation can then be treated to allow the DNA to separate into single-stranded form and radioactive single chain DNA or messenger RNA representing specific genes is applied to the gel. The mRNA, which is a given gene product, or the exact DNA gene copy will bind to complementary DNA strands and their exact location (and that of the gene) can be determined because of the radioactivity of these probes. An example of this sort of technique was mentioned in the discovery of introns in mammalian genes, where mRNA was hybridized with genic DNA and then photographed with an electron microscope.

Mapping techniques for eukaryotes have traditionally been dependent on the recognition of phenotypic manifestations of a gene's presence. Whether recombination frequencies were determined from family studies, the physical location was ascertained from chromosomal features, or a gene was associated with a chromosome's presence in a hybrid cell, it was essential to identify the gene based on the synthesis

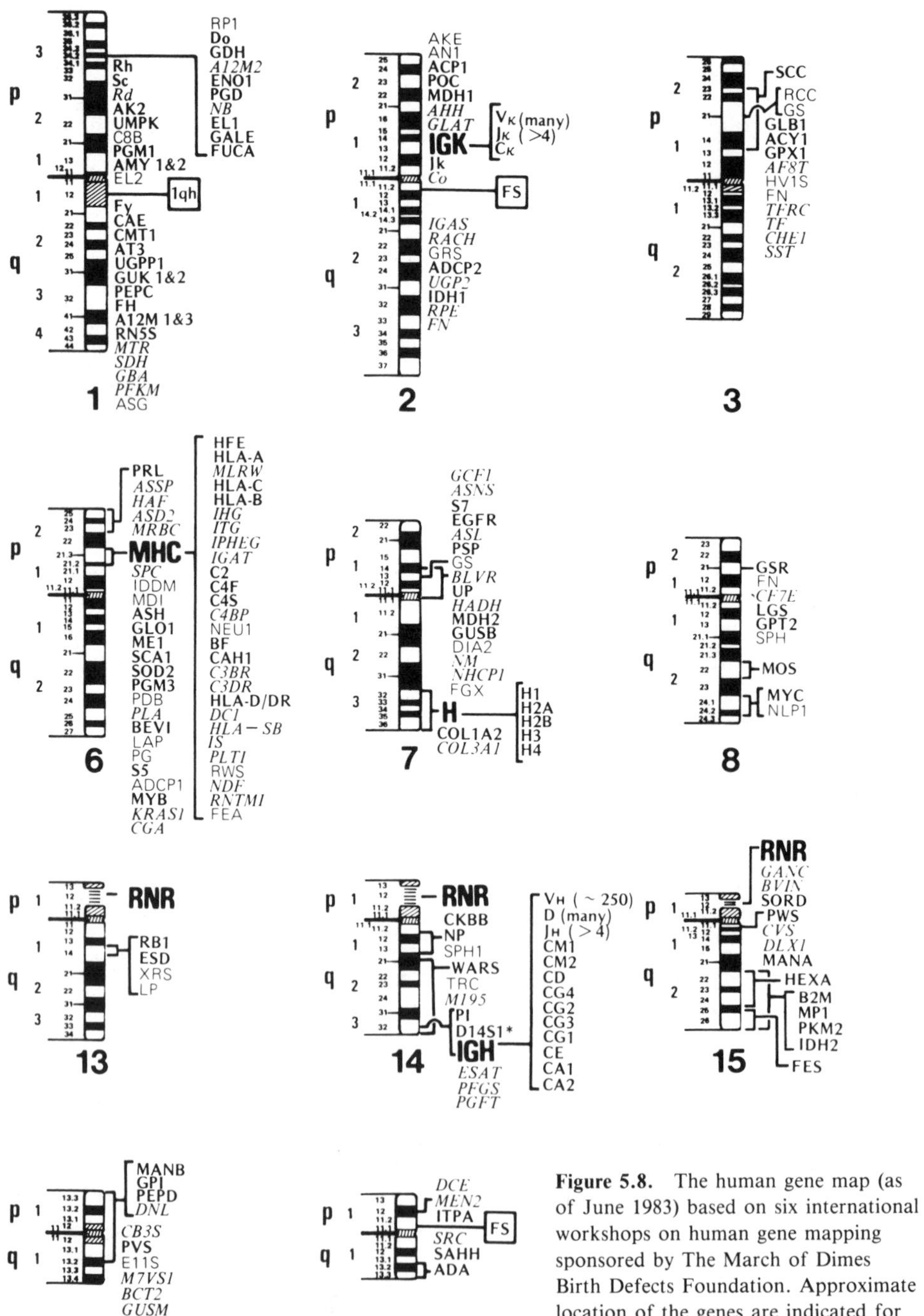

Figure 5.8. The human gene map (as of June 1983) based on six international workshops on human gene mapping sponsored by The March of Dimes Birth Defects Foundation. Approximate location of the genes are indicated for identifiable bands in the chromosomes. A key to the symbols used is maintained and updated by Victor A. McKusick (see his references at the end of chapter). (Courtesy Victor A. McKusick, M.D., Johns Hopkins University.)

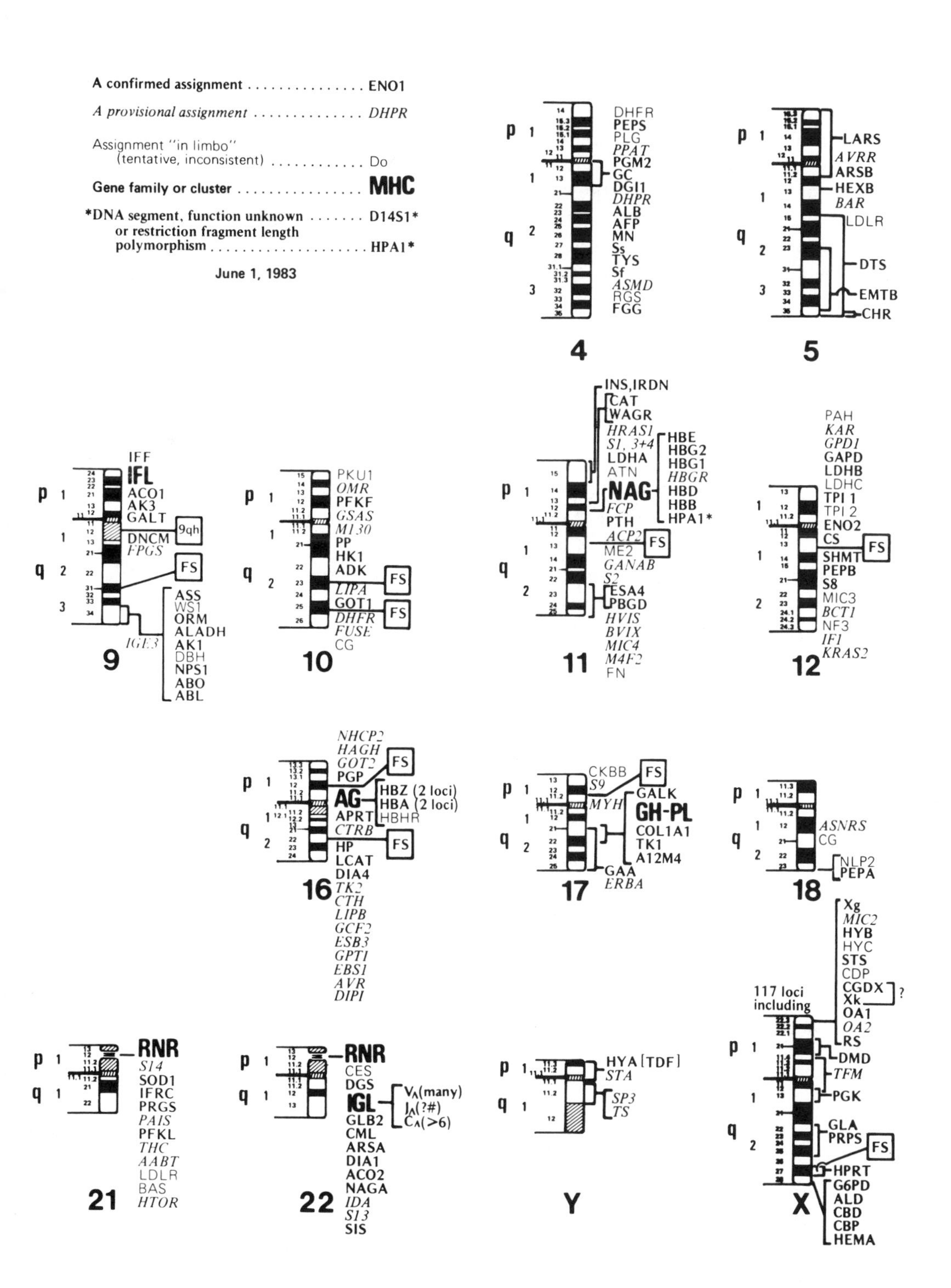

Figure 5.8 *(Continued)*

of its product. Modern techniques have provided the ability to fragment the genome into smaller segments and then visualize a gene's location by adding a complementary pairing substance (DNA or mRNA). Furthermore, the exact sequence of nucleotides can be routinely determined in the fragments.

By extremely accurate cleavage of DNA and identification of nucleotide sequences, the exact location of genes can be found in the human genome. Without the ability to cut the genome into manageable and recognizable segments, this process would be impossible. Thus, a new vista has opened up for chromosome mapping with incredible precision. The process is so precise that chromosome segments that represent recombinations due to crossing-over may even be recognized; careful comparisons to classical gene techniques are therefore possible.

Probably the ideal method for mapping genes would be to determine the sequence of nucleotides in the DNA chain and find each gene in a given position along the length. Knowing how many bases long each gene was with its introns and other nonexonic material, and precisely how many bases there were between genes, would be like knowing the exact numerical street address of a house. Such maps have been made with bacteria and viruses, utilizing other techniques where single chromosomes and smaller amounts of DNA are involved.

Through the use of restriction endonucleases, precise mapping is possible with human chromosomes; because of the extreme accuracy of the methods, the technique is becoming an important tool in the study of human genetics in spite of the huge amount of DNA involved in the human chromosomes. As these new maps are compared to those based on recombination frequency from family studies and other structural mapping information, the human chromosome map is quickly becoming the most well-documented gene catalogue.

ADDITIONAL READING

Chicago Conference: Standardization in Human Cytogenetics. New York: The National Foundation, Birth Defects: Original Article Series, II: 2, 1966.

FORD, E. H. R., *Human Chromosomes.* London: Academic Press, 1973.

GERMAN, J., "Studying Human Chromosomes Today," *American Scientist* (1970) 58, No. 2, 182–201.

HSU, T. C., *Human and Mammalian Cytogenetics.* New York: Springer-Verlag, 1979.

MCKUSICK, V.A., "The Mapping of Human Chromosomes," *Scientific American* (1971) 224, No. 4, 104–13.

______, "The Status of the Gene Map of the Human Chromosomes," *Science* (1977) 196, 390–405.

______, "The Anatomy of the Human Genome," *Journal of Heredity* (1980) 71, 370–91.

OLINS, D. E., and A. L. OLINS, "Nucleosomes: The Structural Quantum in Chromosomes," *American Scientist* (1978) 66, No. 6, 704–11.

Paris Conference (1971): Standardization in Human Cytogenetics. New York: The National Foundation, Birth Defects: Original Article Series, VIII: 7, 1972.

RUDDLE, F. H, and R. S. KUCHERLAPATI, "Hybrid Cells and Human Genes," *Scientific American* (1974) 231, No. 1, 36–44.

SWANSON, C. P., T. MERZ, and W. J. YOUNG *Cytogenetics: The Chromosome in Division, Inheritance, and Evolution* (2nd ed.). Englewood Cliffs, N. J.: Prentice-Hall, 1981.
YUNIS, J. J., *Human Chromosome Methodology*. New York: Academic Press, 1974.

REVIEW QUESTIONS

1. What physical or chemical features of chromosomes are responsible for the characteristic banding patterns which result from the use of specialized staining techniques involving quinacrine mustard or Giemsa stain? What was the chemical basis of the DNA stain reaction that Caspersson originally thought would reveal differently stained areas in chromosomes with fluorescent stains?
2. How can you account for the fact that only about 300 bands can be identified in the haploid set of human metaphase chromosomes, whereas more than 1200 can be found when human late prophase or prometaphase chromosomes are utilized?
3. Heterochromatin was defined in 1928 as chromosomal material which remains condensed during interphase, while euchromatin is stretched out in an uncondensed state. Why do some investigators consider heterochromatin to be transcriptionally inactive and unable to produce RNA?
4. Assume that genetic studies of a number of families indicate that gene loci for CBD (deutan or green colorblindness—D), HEMA (hemophilia—A), and CBP (protan or red colorblindness—P) are linked in the X chromosome. By observing the sons of females who are heterozygous for the three linked genes, it is found that recombinations for the D-A loci total 11.7%, A-P recombinants total 3.0%, and D-P total 9.0%. What is the apparent order of genes and what are the map distances in centimorgans? Draw a map showing order and distances. Could more than one interpretation apply? Can you think of a reason distance D-A might be lower than anticipated?
5. Observe the nature of the short arms of chromosomes 14 and 21 in Figures 5.6 and 5.8 on pages 113 and 120–21. The short arm is labelled *p* above the centromere. In both cases you will see a secondary constriction responsible for ribosomal RNA production (often called nucleolar organizing regions). Compare these two chromosomes to the others in the idiogram. Which other chromosomes share the same kind of structure, apparently duplicating the rRNA manufacturing function?
6. Chromosome number 1 in humans has the following series of genes linked in the order listed. The map distances shown are derived from recombination studies in females and are given in centimorgans (1 cM = 1% recombination). Considering that cross-over

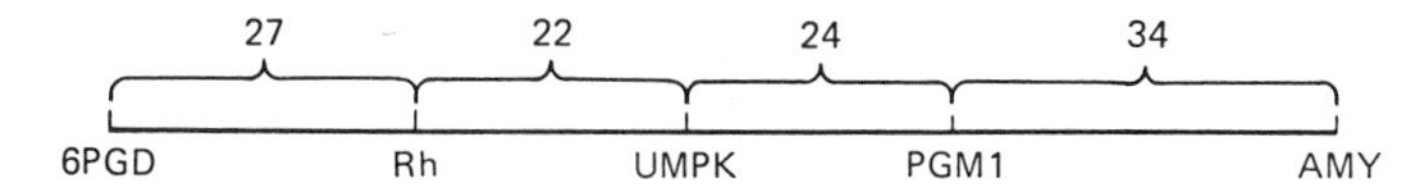

frequency is twice the recombination frequency (that is, if a cross-over occurs 100% of the time between two loci, 50% of the products will be recombinants), what percent of the synaptic figures viewed microscopically will show a chiasma between these points? Give figures for each of the adjacent gene loci shown. How do you account for the fact that the figure for all four add up to over 100%
7. Cross-over frequency is lower in males than in females, thus map distances are less. Assume that males show only 60% of the recombination frequency shown in females (e.g., cross-

ing-over occurs only 60% as often as it does in females). What would the corresponding map distance be if the data for the map in Question 6 had been based on males? The following map is based on male data. Are your calculations close to these figures. How do you account for discrepancies?

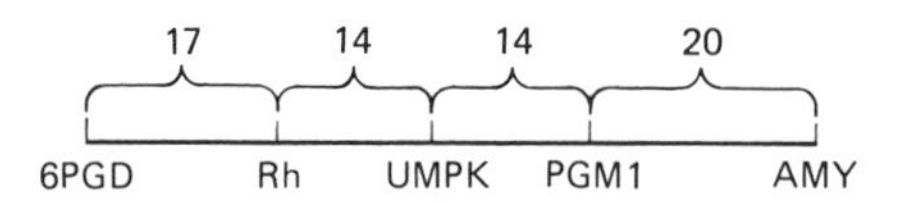

8. A gene for ichthyosis (Ic) is carried by the X chromosome as shown in Figure 5.7 on page 116. Assume that a man shows this scaly skin condition because he has the recessive gene for the trait. He has only one gene for it in each cell because he carries only one X chromosome. If he marries a woman who has two genes for normal skin genes (as a female, she has two X chromosomes), what will be the gene complement for the Ic locus for their daughter? If the man has a grandson (his daughter has a son), can you make any predictions about his chances to have the disease? Does it make any difference whether the grandson's father (the daughter's husband) has a dominant or recessive Ic gene?

9. Using Figures 5.6 and 5.8 on pages 113 and 120–21, list the gene symbols for the genes you might expect to be located at 7q32 and 1p34. How do these two loci and the genes cited differ from each other? Where are the genes G6PD and HEMA located in the map? Why are they listed at the same location?

10. How do you account for the fact that mRNA and gene DNA hybridization experiments take place? An example was the work done to show that the β-globin gene has introns. What mechanism operates when mRNA is placed with chromosomal DNA that has been heated?

11. Explain what is meant by describing two types of human chromosome maps: (a) Based on *family studies*; (b) Based on *cell hybridization techniques*. What sort of information would be utilized in preparing each type of map? In presenting the two types of maps, how would they appear different? What kind of discrepancy would you expect between the two types? How would you explain such discrepancies?

12. In preparing human genetic chromosome maps based on recombination frequencies from family studies, why are the map distances longer if data are collected following genes from mother to child than when following genes from father to child? Do you think such map distances should be recorded as separate male and female maps, as an average of the two, or only for females or males (with conversion factors available where necessary)? Which system would provide the most accuracy?

13. Why are genetic chromosome maps for the human X chromosome from family studies invariably based on data from females? Does this mean that the X-linked genes are not present in males at the same relative distances?

14. Suppose that a gene which is about 2000 base pairs in length has a small portion consisting of nine nucleotides that specify three amino acids in sequence: —PROLINE—GLUTAMIC ACID—GLUTAMIC ACID—.

The DNA base sequence for this part of the gene is

```
-C-C-T-G-A-G-G-A-G-
 | | | | | | | | |
-G-G-A-C-T-C-C-T-C-
```

If a point mutation changes the sequence to

```
-C-C-T-G-T-G-G-A-G-
 | | | | | | | | |
-G-G-A-C-A-C-C-T-C-
         ↑
```

the sequence of amino acids is changed to—PROLINE—VALINE—GLUTAMIC ACID—.

Assume that a restriction endonuclease recognizes and cleaves at the se-
-C-C-T-N-A-G-G-
quence -G-G-A-N-T-C-C- (N = any of the four nucleotides, only six-base specificity is involved). It would normally cleave this particular gene into two segments about 1000 base pairs long. What would be the effect of the point mutation on this endonuclease action? Can you think of a practical use for this phenomenon?

CHAPTER SUMMARY

1. The diploid number of human chromosomes is 46, as reported by Tijo and Levan in 1956. Males have 22 pairs of autosomes plus XY and females have 22 pairs of autosomes plus XX.
2. Eukaryotic chromosomes have been shown to have a "bead on a string" structure in artificially treated material. The beads are called nucleosomes and each consists of eight histone protein molecules inside a two-gyre coil of DNA. The beads are normally held close together by the continuous DNA strand and one more intervening histone molecule.
3. The unambigous identification of the 22 different autosomes and two sex chromosomes became a reality in 1969 with the discovery that certain stains caused characteristic banding patterns to appear in each chromosome.
4. The human chromosome complement is arranged and labeled according to a scheme adopted by cytologists at a 1971 Paris conference. Each chromosome has a number based on their arrangement in a karyotype that is arranged according to decreasing length. Each band also receives a unique number based on its position relative to the centromere and the arm in which located.
5. It is estimated that the human haploid genome contains 100,000 genes. In addition to intronic DNA there are other regions of DNA that do not serve as genetic determiners. Altogether there is probably 30 to 50 times more DNA than required for all human genes in a single nucleus.
6. Chromosome mapping is an important activity of human geneticists and cytologists. The determination of the physical location of genes in given linear array in specific chromosomes is the goal of many studies. Genetic maps are constructed on the basis of linkage and recombination frequency among genes as observed in family studies. Cytological maps are based on cell hybridization studies and restriction endonuclease cleavage of chromosomal DNA and molecular identification of specific genes in the various fragments. In some instances deletions have revealed the location of genes and have added information to the mapping process.

CHAPTER

6

CHROMOSOMAL ABNORMALITIES

Great blunders are often made, like large ropes, of a multitude of fibers.

Victor Hugo, Les Miserables (1862)

The work of the early geneticists proved beyond doubt to the scientific community that heredity could be studied as a precise mathematical science. Although exact and unfailing predictions could not be made because of the element of chance at the root of the mechanism, with large samples, the data followed statistical patterns consistently. Within the given parameters of chance, it was possible to analyze and predict with confidence what type of offspring could be expected and to analyze with great accuracy the genetic make-up of parents.

At the basis of all studies of heredity is the inherent presence of a large reservoir of variation. Genetic mechanisms exist for the shuffling and mixing of the huge number of genes in the human genome, adding to the diversity of the human population. As seen, the underlying cause of the variation within the genome is found in the mutability of the molecular structure of the genes themselves.

At another level in the structural foundation of the genetic mechanism there is a susceptibility to change and variation. This is at the level of the chromosomes constituting the genome. The human karyotype has been presented as following a predictable pattern as to numbers of chromosomes with genes arranged in distinct linear patterns. Although there is a definite form and pattern to the chromosomal arrangement of the genome, there are alterations and changes that can be recognized.

It might well be predicted that, with millions of years of adaptation in the past, the human genome can be considered to be in a highly balanced functioning state. Over the relatively few years since human chromosome structure and arrangement became known, a number of defects have been revealed which in fact support such a contention. It is generally true that chromosomal aberrancies or changes usually result in some sort of physical or physiological change that causes an individual to be somewhat less fit in terms of adjustment to the environment.

Considering the intricate processes of mitosis and meiosis, and the rigorous requirements placed on reproductive cells to transmit genes to the next generation, it is easy to understand that an error or inconsistency could develop from time to time. Although such exceptions to a consistent pattern are rare, when one considers large populations, these exceptions will be discovered.

Some types of modifications may occur that alter the standard arrangement of a normal genome. In a few cases there are distinct alterations in the pattern of development and functioning of individuals who possess such aberrancies. In many cases where aberrant genomes are present there is either no deviation from normal development or it is so subtle as to be unrecognizable. The only way of detecting such discrepancies may be through a microscopic examination of the chromosome complement of cultured cells, a most important technique in predicting or analyzing the transmission of such defects within families.

As discussed, radiation can cause a disruption and altered reformation at the molecular level of the gene; radiation can also cause chromosomal breakage that results in aberrant forms after subsequent natural repair. As with mutations, some changes can be categorized as *spontaneous,* often without knowing how much alteration has been caused by environmental factors.

The classification of some observations as "normal human karyotypes" implies that there are abnormal or aberrant karyotypes as well. The study of these unusual types is important in human heredity, no matter how rare they may be. Obviously, a study of the unusual leads to a better understanding of the normal. Furthermore, aberrant types often result in abnormal development and disabling physical conditions. From the medical standpoint, some of the most significant genetic conditions are associated with aberrancies in the chromosomal complement.

To understand the possibility that abnormal chromosomal patterns may develop in living organisms, one needs to view the chromosomal mechanism as an amazingly resilient, yet fragile, system. The nature of replication is a precisely ordered process and the DNA molecule supported by proteins in the chromosomes is basically a sturdy component of the cell, capable of self-repair. However, chemical reactions and molecules are subject to alterations due to changes in the environment in which they are located. Such factors as temperature or potentially interacting chemicals in the vicinity interfere with replication and the stability of the molecules of heredity. Ionizing radiations, such as X-rays, are notably active in causing breakages and alterations in large molecules like DNA.

Even the mitotic process must be considered fragile and dependent on a reasonably standard set of environmental conditions. Interference with the replication of chromosomes or their movement and apportionment to the two new cells can cause bizarre changes in the chromosomal complements of the daughter cells. It is not difficult to create artificial changes in the environment of dividing cells which will stimulate such changes. There is no doubt that natural conditions cause similar effects.

ABERRANT NUMBERS OF CHROMOSOME SETS

Because replication of chromosomes during the interphase preceding mitosis is so far removed from the actual segregation of the two sets of chromosomes into two separate nuclei, one might anticipate that the two processes could be interfered with separately. A classic example of this, recognized since the early 1930s, is the use of the chemical **colchicine,** an extract from the meadow saffron plant (*Colchicum*). Colchicine is a spindle poison that prevents the normal assembly of **spindle microtubules.** Only the movement of the chromosomes to the poles, not chromosomal replication, is affected. Colchicine was encountered as a tool in the study of chromosomes in Chapter 5. It was used as a means of arresting mitosis at metaphase to study chromosomes in their most shortened condition. The cells were arrested at metaphase, because without a spindle, they did not move but remained in the vicinity of the original nucleus.

When colchicine is applied to dividing cells, it is possible to induce the chromosomes to replicate one or more times without movement to the poles. The arrested mitosis progresses after the prolonged metaphase to modified anaphase and telophase stages, where the homologous chromosomes separate and lengthen. A single restitution nucleus reforms each time, with the nuclear envelope enclosing all the chromosomes. Thus, it is possible to form cells in one cycle that are **tetraploid** ($4n$) instead of diploid ($2n$). Various levels of **ploidy** are possible, not only from the use of colchicine, but also through the action of such processes as fertilization by unreduced ($2n$) gametes, experimentally or naturally produced, or even by penetration of the egg by two sperm. Thus, chromosome numbers may be haploid ($1n$), diploid ($2n$), triploid ($3n$), tetraploid ($4n$), pentaploid ($5n$), hexaploid ($6n$), and so on. Levels of ploidy above diploid are often referred to collectively as **polyploids.**

The typical sexually reproducing organism is diploid. **Haploids** are fairly common among primitive organisms but are generally so rare and weak among higher organisms as to be essentially nonexistent. However, most plant life cycles have haploid nuclei in various essential stages. With the exception of briefly independent free-living reproductive cells, they would not be anticipated in a genetically balanced and intricately structured organism as a human. Likewise, polyploids are not often encountered among animal species; this is not true among plants, however, where many of the largest flowers and best food-producing plants are tetraploids. Levels of ploidy above tetraploids are generally not particularly successful among plants.

Reports of triploids or tetraploids among humans are overwhelmingly from spontaneously aborted fetuses. In fact, among spontaneous abortions caused by aberrant chromosome numbers, almost one-fifth carry a **triploid** number. The extremely rare circumstance of a triploid live birth has most often resulted in death within 1 day. Some workers consider such live births to be essentially belated miscarriages.

Tetraploid zygotes and fetuses are even rarer than triploids, constituting about 6% of the spontaneous abortions due to chromosomal defects. Most live births have resulted in infants with birth defects and none has survived more than 6 months to 1 year. Most often such individuals are **mosaics** whose cells consist of two populations: one normal diploid line and an aberrant triploid or tetraploid line. The aberrant line of cells probably arises from an unusual duplication or nonseparation of chromosomes in a dividing cell of the embryo.

INDIVIDUAL CHROMOSOME GAIN OR LOSS

It is more likely that aberrant numbers of chromosomes will involve individual chromosomes rather than whole sets. Thus, it is rare but not unusual to identify organisms having one or two extra or missing chromosomes in their genomes. Such conditions probably result from aberrant cell division processes such as **nondisjunction** of a pair of chromosomes in meiosis.

There may be several reasons that homologous chromosomes might not separate normally. Notable among these are spindle defects or failure of centromeres to divide. With nondisjunction, or the failure of the homologous chromosomes to move independently into the two forming daughter cells, the result is that two chromosomes which normally segregate to opposite poles are carried to the same pole. Thus, two **homologous chromosomes** move into the same daughter cell, leaving a cell at the other end which is lacking this same chromosome.

If either of these products functions as a **gamete** in fertilization, the resultant offspring would have either three of the same chromosome or only one (assuming that the other gamete carried the normal haploid complement). Because only one chromosome and its linked group of genes would exist in an aberrant number (three or one), it is easy to see that the balance of genes is not so severely upset as in polyploidy or haploidy, where whole sets of chromosomes are involved. This is especially true if a short chromosome with relatively few genes is involved.

An individual who lacks a single chromosome, thus having only one of a type, is referred to as a **monosomic** (the condition is monosomy). If there is an extra chromosome (three of one type), the condition is called **trisomy.** If two extra chromosomes of a single type are present (four of one type), the condition is **tetrasomy.** In the case where two different chromosomes might be present singly or as three of each the individual would be classed as a **double monosomic** or **double trisomic.** If a genome is completely lacking a whole chromosome, the term **nullisomic** is used.

When nondisjunction occurs during mitosis in nonreproductive tissues, the result is a mosaic with two cell lines again. In this case one cell line in the tissue is trisomic for one chromosome, while the other cell line is monosomic for the same chromosome. If the nondisjunction occurs late in the development and affects a proportionately small number of cells, there is generally no recognizable effect. If it occurs early in development and influences a large number of cells, some obvious serious defects may arise.

Down Syndrome

The most common human disability associated with an aberrant genome is **Down syndrome,** named after Langdon Down from Great Britain who described it in 1866. The disability has been called Mongolism or Mongolian idiocy because of the almond-shaped eyes reminiscent of Oriental peoples. Such names are still encountered in the older literature, but they are no longer considered acceptable. Down syndrome has been recognized as a congenital defect since the late 1800s, but its chromosomal basis was not discovered until 1959 in France by Lejeune and Turpin. The identification of the cause came when it was recognized that Down syndrome individuals had 47

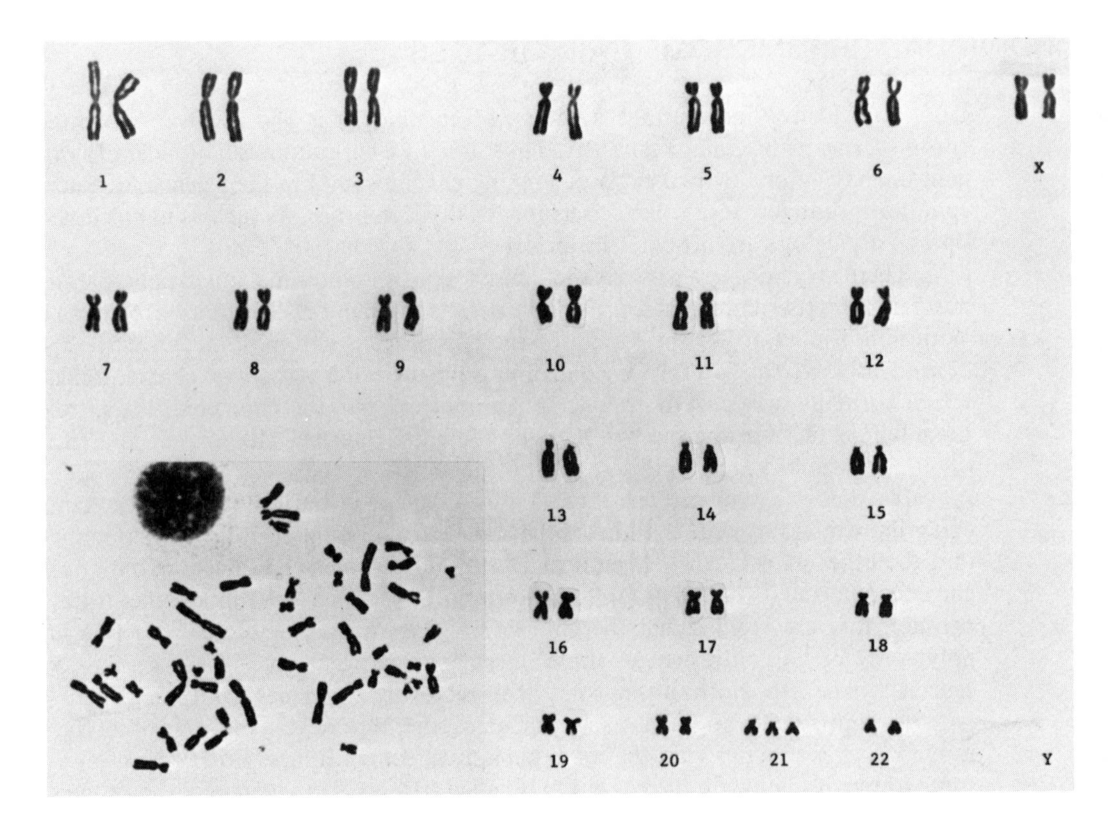

Figure 6.1. Karyotype of a child with Down syndrome. Note the presence of five chromosomes in the G group. There are three of chromosome 21. (Courtesy March of Dimes Birth Defects Foundation)

chromosomes. Because there is an extra number 21 chromosome, these individuals have a trisomy for number 21 (Fig. 6.1).

A syndrome is a group of associated symptoms characterizing a disease or disability. In Down syndrome the most pronounced symptoms observed are the following: the head is small and round with a flattened face and thick epicanthic eye folds; the mouth characteristically is held open revealing a thick, furrowed tongue; ears are set low and misshaped; the feet and hands are thick with a tendency to show webbing between the digits. Physiologically, there is also a spectrum of problems involving, among others, congenital heart defects and intestinal disorders.

One of the most important problems is the accompanying mental retardation, which is typically rather serious. Bodily coordination is generally weak, but patients are frequently educable in terms of performing routine tasks. It is generally conceded that such persons possess reasonably happy, pleasant personalities. Generally, an infant who suffers from the syndrome is easily recognizable at birth. Furthermore, the technique for identification of the syndrome prenatally has become routine for well-qualified medical teams. This technique will be explored in greater detail in Chapter 12.

The syndrome has proved to produce its most serious effects in terms of asso-

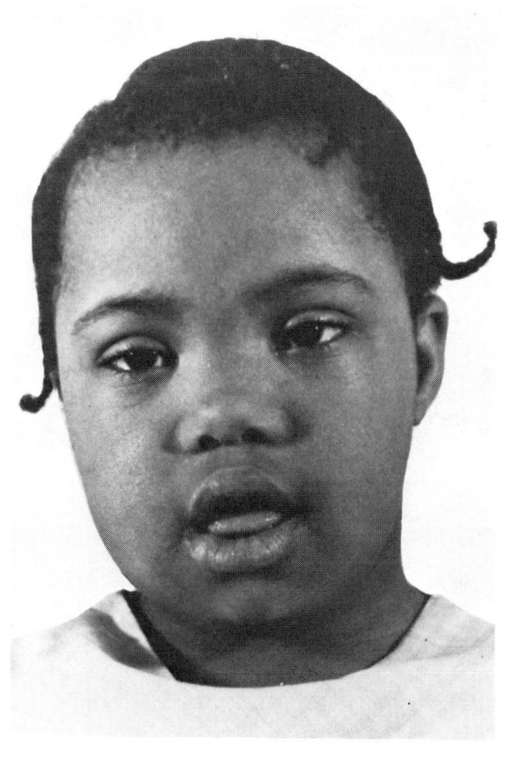

Figure 6.2. Two children with Down syndrome. Trisomy for chromosome 21 and the associated disability can occur in any race. (Courtesy March of Dimes Birth Defects Foundation)

ciated mental retardation, susceptibility to infectious diseases, heart diseases, and leukemia. There is clearly a spectrum of severity of some of the symptoms, but life span is usually significantly shortened. There are many recorded cases of successful training of Down individuals to lead useful lives (Fig. 6.2).

Because within the general population the incidence of occurrence is about one in 700 births, Down syndrome is one of the most commonly encountered genetically determined birth defects. One very important aspect of the syndrome is its relation to the age of the mother. The frequency of Down syndrome children born to mothers in their early twenties is about one in 2000 births; however, this rate rises quickly after maternal age 30 to about one in 50 births after age 45 (Fig. 6.3).

No one can state specifically why this maternal age correlation exists, but, based on a knowledge of the meiotic process in females, it is possible to speculate about the mechanism. It is probably significant that the future eggs enter into prophase of the first meiosis prior to birth of the female. They remain in this arrested condition until they complete the division, possibly 12 to 50 years later. It is not hard to imagine that a meiosis requiring 40 years or so to complete might be in more jeopardy of becoming defective in its operation than one requiring less than half that time,

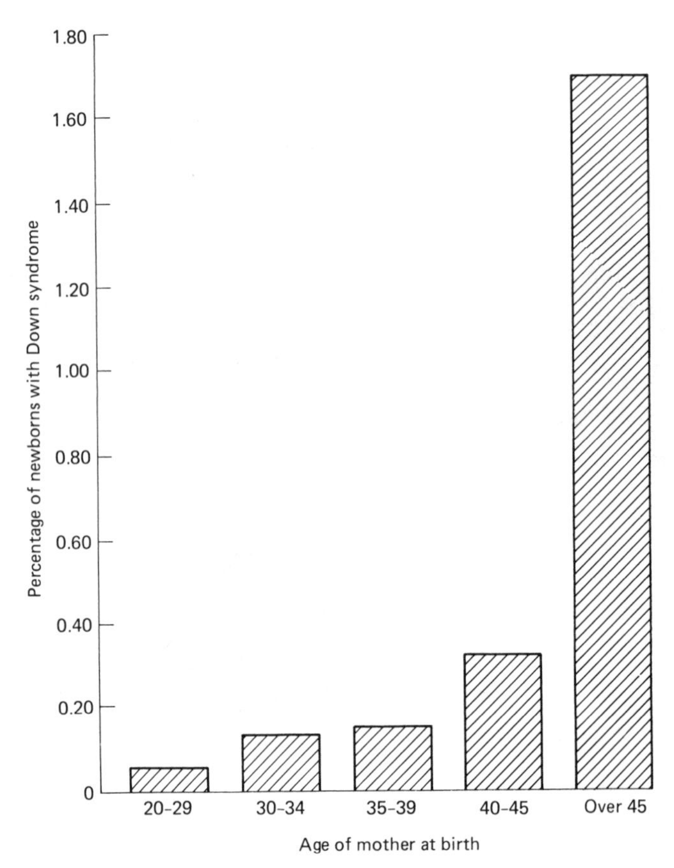

Figure 6.3. Histogram showing the incidence of Down syndrome at various maternal ages. Note the abrupt increase in incidence over age 40. (From Redding, A. and Hirschhorn, K., "Guide to Human Chromosome Defects," Bergsma, D., ed. White Plains: The National Foundation-March of Dimes, BD:OAS IV(4), 1968.)

especially as the hormonal and chemical balance within the woman's body may be changing during that time. It has been estimated that the incidence of abnormal chromosome numbers among newborn children would be reduced by one-third to one-half if women over 35 stopped having children.

Although it was assumed that paternal age was not a contributing factor in human trisomies, more recent work with chromosome banding indicates that nondisjunction of chromosome 21 does occur in males and that as many as one-third of Down syndrome births may be due to this source. Studies indicate that paternal nondisjunction of chromosome 21 increases after age 55.

The cause of Down syndrome must be a defect in the meiotic mechanisms, since what occurs is a nondisjunction of two number 21 chromosomes. If the centromere fails to divide properly during the second anaphase, two homologous chromosomes might go to the same gamete (see Figs. 6.7 and 6.8 on pp. 146 and 147). When a normal gamete bearing a single number 21 fertilizes this aberrant gamete with two 21's, there will be three of the same chromosome in the zygote—a trisomic.

The same sort of nondisjunction occurs with some other chromosomes where a correlation with the increasing age of the mother has been shown. **Trisomy 21** individuals are better able to survive as fetuses and infants than the other less common types of trisomy and they also show less severe symptoms. Chromosome 21 may represent a small block of genetic material or one loaded with genes that are not overwhelmingly important to the survival of the fetus and young individual. There is, however, a profound effect on the normal developmental process.

Based on the discussion of chromosomal aberrations, it is safe to predict that Down syndrome could result from other types of abnormal chromosome structures. For instance, if a large segment of chromosome 21 were translocated to another chromosome, it would be possible to have an abnormal combination passed along to offspring by either sperm or eggs. It is known that in some genomes a large piece of number 21 (all the long arm) has been translocated to chromosome number 14 in what is called a **Robertsonian translocation.** Both chromosomes 14 and 21 have lost their nonessential short arms and fused together at their centromeres. If a gamete carries the 14/21 translocation with a normal 21 and fertilization with a normal gamete occurs, the individual will receive the following chromosomes:

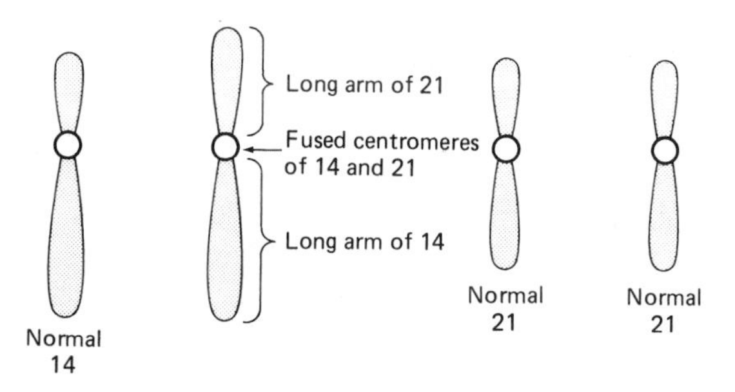

This individual will be trisomic for most of the genes on number 21 and thus show the symptoms of Down syndrome. It is of course much rarer, accounting for only about 3% of those having the syndrome. It can be predicted to be heritable if one of the parents is recognized as a carrier of the translocation; typical trisomy 21 cannot be predicted before conception. The carrier parent will have only 45 chromosomes, since the translocation incorporates 14 and 21 into one unit:

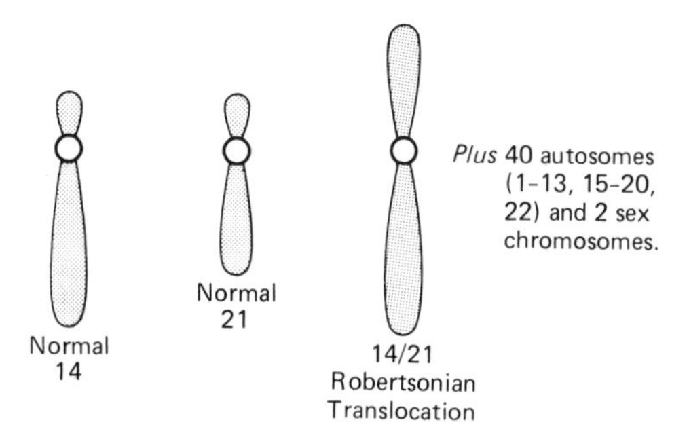

Unlike trisomy 21 Down syndrome, there is no correlation of translocation 14/21 Down syndrome with maternal or paternal age.

In about 2% of the cases of Down syndrome, it has been found that not all the cells in a person's body have three number 21 chromosomes. Some cells have the normal diploid number of 46 chromosomes. Thus, there are two cell lines in the patient's body: one normal and one trisomy 21. Such persons are **mosaics** for the two lines of cells and the severity of their symptoms depends on the proportion of trisomic cells actually present. Some persons have been shown to be mildly mosaic for the trisomic condition with virtually no symptoms of Down syndrome. If the cells constituting their reproductive tissues contain the trisomic cell line, however, it is obvious that there could be a tendency to pass on the condition to offspring. In fact, some families with several Down children and apparently normal parents have been found to have a parent who was a mosaic for trisomy 21.

There is another, but extremely rare, type of trisomy 21 that has been identified. A type of chromosome may form which is known as an **isochromosome** (iso = the same or similar). This unusual structure forms in acrocentric chromosomes, where loss of the short arm is genetically not significant, as seen in Robertsonian translocations. The mechanism involved in isochromosome formation is rather unusual and unexpected. When the centromeres divide in the anaphase separation during mitosis or the second division of meiosis, it appears to be possible for the centromere to divide aberrantly in a plane at right angles to the typical longitudinal direction. Thus, a transverse break in the centromere of the longitudinally duplicated chromosome number 21, followed by lateral fusion with the homologous long arm, might look like this:

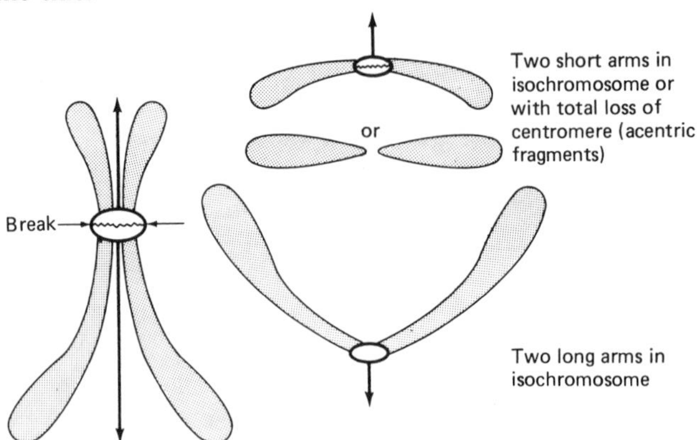

A gamete that has received the two long arms in a single isochromosome is essentially diploid for the bulk of number 21 and can survive without the nonessential short arm genes. When it joins in fertilization with a gamete carrying a normal 21, the resultant zygote will be trisomic for chromosome 21 and develop into a Down syndrome child.

Patau and Edwards Syndromes

Two more frequently encountered types of trisomy identified in humans are **trisomy 13 (Patau syndrome)**, which is found at the rate of about one in 6000 to 10,000 births, and **trisomy 18 (Edwards syndrome),** which occurs in about one in 3500 to 7000 births.

Both are severe, with infants rarely surviving more than a few months. Fewer than 5% of trisomy 13 live births survive to age 3 years and only 10% of trisomy 18 children who survive birth reach 1 year of age.

The two syndromes share some of the same development symptoms. Heart and kidney defects are common, with severe mental and developmental retardation. There is some similarity in physical features such as small defective eyes, narrow receding chin, and low set, malformed ears. Although not as pronounced as in Down syndrome, there is evidence indicating increased risk for both types with advancing maternal age. In some severely disabling cases double trisomies involving two different chromosomes have been identified.

In both syndromes it is apparent that many fetuses with the trisomies are aborted prior to birth. This prenatal loss affects males more than females and therefore more than twice as many females with Edwards syndrome are born than males. For Patau syndrome, there are only slightly more females than males.

Patau syndrome infants show a number of abnormal physical features (Fig. 6.4). The condition is often accompanied by cleft lip and cleft palate. Sometimes there is an excess number of fingers and toes (**polydactly**). Almost all appear to be deaf. There are a number of eye defects and some affected infants are blind. The defects are so severe that more than half those having trisomy 13 die by 1 month of age.

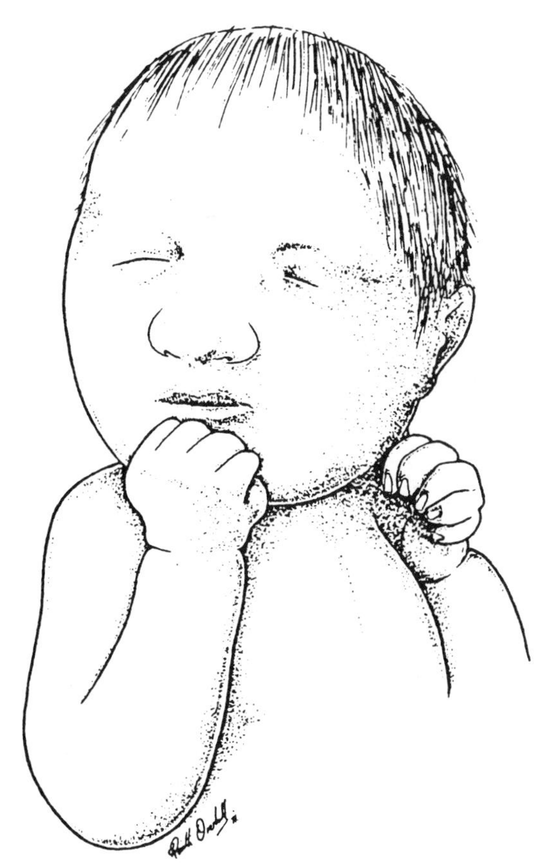

Figure 6.4. Drawing of some of the characteristics associated with Patau syndrome. (With permission from Ford, E. H. R., *Human Chromosomes*. 1973. Copyright: Academic Press, Inc. (London) Ltd.)

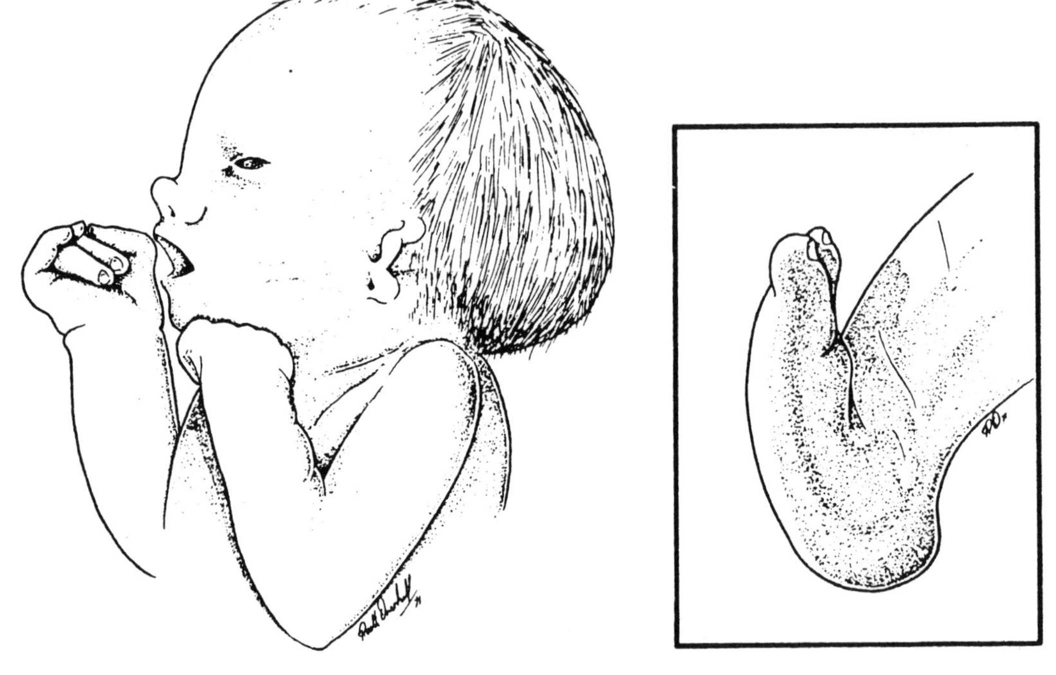

Figure 6.5. Drawing of some of the features associated with Edwards syndrome. Lower diagram shows the characteristic "rocker-bottom" feet abnormality. (With permission from Ford, E. H. R., *Human Chromosomes.* 1973. Copyright: Academic Press, Inc. (London) Ltd.)

Edwards syndrome infants are as severely disabled as those with Patau syndrome (Fig. 6.5). One characteristic feature of trisomy 18 seems to be an above average general muscle tone. The fingers are tightly clenched with the index finger folded across the third finger in a distinctive fashion. Other muscles are so rigid that the joints can hardly be manipulated. Many of those affected show upwardly curved feet ("rocker bottom").

GENE NUMBER CHANGES WITHIN CHROMOSOMES

Where missing or supernumerary genes are concerned, the least significant aberration in terms of the amount of genetic material involved would be the addition or loss of segments of chromosomes. Several factors in the natural or experimental environment of cells can cause breakage of chromosomes, and it is a normal feature of chromosomes to heal these breaks through natural processes. Thus, it is not unusual to find cells in which chromosome breakage and repair have occurred that have segments of chromosomes missing. This type of aberration is called a **deficiency** or **deletion.**

In similar fashion it is sometimes possible for a broken segment of a chromosome to be reattached to a part of the genome in another position, either in the same chromosome or an entirely different one. It may even be possible, through some unusual defect of the replication process, to find that a repeated section of a chromosome occurs in tandem with the normal arrangement. Additional segments of chromosomes in the genome are referred to as **duplications.**

Once again, effects on the organism carrying deletions or duplications become significant if such aberrancies are passed along to the next generation in the reproductive cells. When a deletion is present in a sperm or an egg, a monosomy for the missing genes is formed on fusion with a normal gamete. If the missing segment is not too large, the presence of a normal, complete chromosome is likely to cover for the missing genes with no significant untoward effects; a monosomic for the entire chromosome, however, would show quite detrimental effects on the characteristics of the individual. If a deletion affects the same region on both chromosomes in a homologous pair, the individual is nullisomic for genes in the affected region. In this case the complete lack of genes may produce an inviable situation. When a gamete carrying a duplication joins with a gamete carrying the normal chromosome, a trisomy for the repeated segment is the result. In general this type of aberrancy seems to cause no more serious effect than a deletion and is considered to be less deleterious in most cases. The aberrancies are often so difficult to identify that they go unnoticed. A traditional means of detecting them microscopically has been through the study of meiosis. During synapsis, as precise two-by-two pairing occurs, a **loop** or **foldback** may develop when the duplicated section is unable to pair with its homologous mate. The same sort of loop synaptic figure is also seen in a deletion as the normal chromosome is faced with a missing segment in the pairing process. Such loops represent the intact chromosome, consisting of histones and DNA, and they contain the material which cannot synapse with nonexistent homologous regions in the adjacent chromosome.

A loop formed during synapsis due to a deletion of gene loci 4 and 5 would be diagrammed as follows:

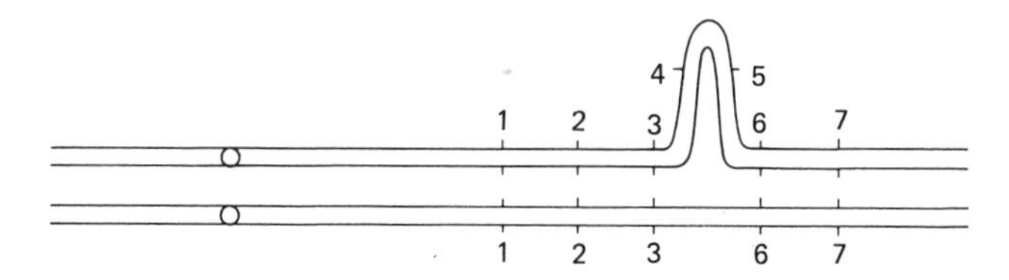

A duplication of gene loci 4 and 5 might produce a loop with a similar appearance:

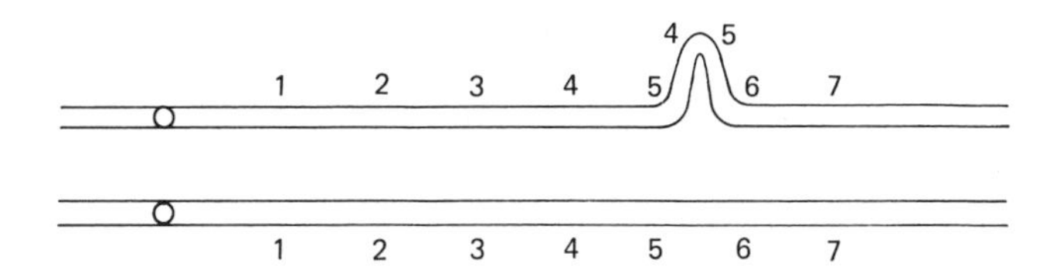

Cri du Chat Syndrome

Recognizable deletions are known to occur in humans as well as abnormal numbers of chromosomes. These may also have very serious consequences for the developing individual. Thus a deletion of about one-half the short arm of chromosome number 5 was discovered by LeJeune and coworkers in Paris in 1963. The **cri du chat** syndrome (French: cry of the cat) derived its name from the peculiar plaintive catlike cry of affected infants. (Fig. 6.6). It is rare—only 100 or so have been identified—and results in infants with small heads and severe mental retardation. The life span usually extends into adulthood, but mental ability is so low as to require institutionalization. Technically, such persons are monosomic for the bulk of the short arm genes on chromosome number 5.

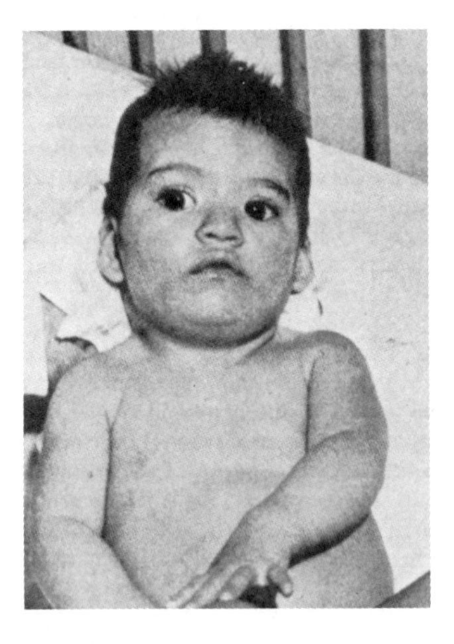

Figure 6.6. A child with *cri du chat* syndrome. (Courtesy James German and *American Scientist*)

Monosomy for a chromosome or for even a segment of one is extremely rare, not only among live births, but in spontaneous abortions as well. Obviously the condition is much more deleterious than trisomy. Monosomy for a chromosome has only been identified for chromosome 21.

CHANGES IN GENE LOCATION

The genes cannot be considered to be immutably locked into specific positions and sequences in rigid, unchangeable chromosomes. Although on a statistical basis there is very little chance of finding the exceptional out-of-position gene, millions of individuals are dealt with all over the world in scientific or medical contexts and the unusual combination, even though it affects only the chromosomes, often stands out from the normal situation. Thus, even though all genes may be present in the diploid condition in an organism, there have been many observations of atypical arrangements of genes within the genome.

It is apparent that the function of a gene may be changed somewhat if its position is altered. In some organisms a **position effect** has been described for genes that are situated adjacent to different genes in the same genome.

Most often the rearrangements that have been observed result in cytological effects that are encountered in meiosis during the two-by-two pairing of synapsis. A normally arranged sequence of genes in a chromosome whose homologous mate contains the same alleles differently aligned will produce some unusual synaptic figures in cells undergoing meiosis. Furthermore, there are likely to be some unexpected combinations of genes passed on to the offspring of such individuals, due to the abnormal anaphase separations that follow. There may be such large duplications or deletions of groups of genes that the combination is lethal or severely detrimental to normal development.

Reciprocal Translocations

As one might expect a deletion to result from two breaks in a chromosome, allowing a segment to fall out of the main axis of the structure, most gene rearrangements also result from two simultaneous breaks in chromosomes. A type of aberration that has been encountered in many organisms is known as a **reciprocal translocation.** It is called a translocation because a segment of a chromosome is moved completely from one chromosome to another. Generally a broken chromosome end is found to rejoin only at another broken end; thus, it is most common to find two different chromosomes with breaks exchanging segments. Because they have actually exchanged parts, the translocation is said to be reciprocal.

As an example of a reciprocal translocation, assume that a chromosome has a sequence of genes labeled with the letters A through I.

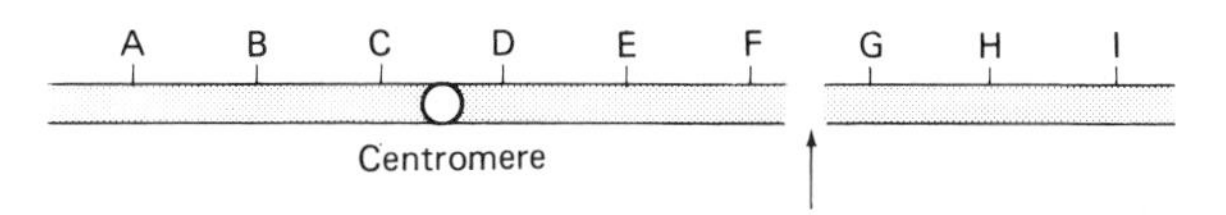

Assume also that another chromosome in the same genome has a sequence of genes labeled with the numbers 1 through 7.

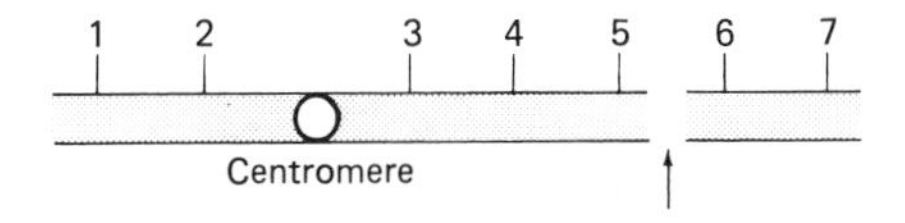

Now assume that breaks occur simultaneously in both chromosomes between genes F and G and between 5 and 6, with the broken ends moving out of position from one chromosome to the other reciprocally. The result of the translocation would be two chromosomes with the two new combinations:

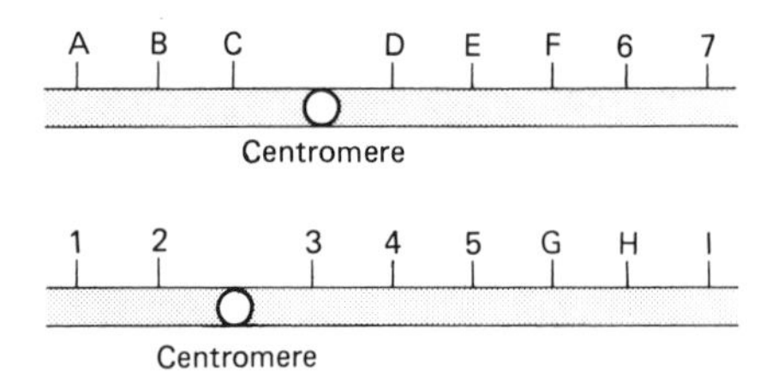

If a gamete having these two translocated chromosomes joins in fertilization with a gamete having the two normal chromosomes, the offspring will have two unmatched pairs of chromosomes. Observe that the individual is still diploid in terms of the full complement of genes for both chromosomes, even though they are arranged abnormally.

A difficulty arises during synapsis in prophase I of meiosis, when a **cross of four chromosomes** is required for two-by-two pairing:

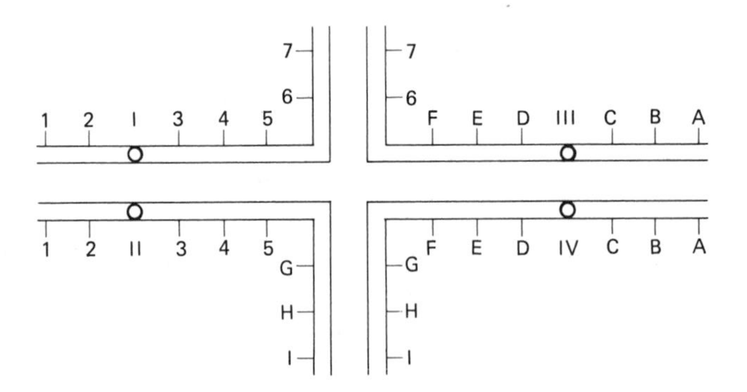

This synaptic pattern does not result in any defective characteristics, since pairing is still effective. It is during the anaphase separation that abnormal reproductive cells may be produced and cause aberrant forms in the next generation.

If centromeres I and IV move to one end of the dividing cell and II and III to the opposite end, the gametes will carry normally balanced complements of the two groups of genes. If, on the other hand, centromeres I and II move toward one end during anaphase, while III and IV move to the other, duplications and deficiencies will result in the reproductive cells. There will be two sets of genes 1 to 5 and no genes A to F in one new cell, and two sets of genes A to F with no genes 1 to 5 at the other end. Movement of I and III centromeres to one end, with II and IV going to the other would also result in duplications and deficiencies. One cell would have a double set of genes 7 and 6 while lacking genes GHI. The other product would have duplications for GHI and be deficient in genes 6 and 7. Large discrepancies involving indispensable genes might produce inviable gametes, unable to produce offspring. However, it is possible that a viable (but physically abnormal) offspring could result from the fusion of gametes carrying such duplications and deficiencies with normal gametes. The resultant offspring would be monosomic for some genes and trisomic for others.

Philadelphia Chromosome

What originally appeared to be a deletion of a portion of the long arm of chromosome 22 was shown with modern techniques by Janet Rowley in 1973 to be a translocation. The missing portion of chromosome 22 in most cases (but not all) is found attached to chromosome 9. The shortened number 22 was named the **Philadelphia chromosome**, because of its discovery in that city.

Persons with chronic myelogenous leukemia (a type of blood cancer) exhibit this aberrant genome in their bone marrow cells and among other cells that originate in the marrow. It has not been established whether this 9/22 chromosomal translocation is the cause or an effect of the cancer, but cancerous cells often show severe chromosomal abnormalities.

Cancer is so diverse in its possible causes and types of manifestation that it is difficult to group the many examples of the disease under a single category. The general feature of the disease that accounts for such an oversimplified approach is uncontrolled cell division and undifferentiated resultant tissues. Many studies of the chromosomes of cancerous cells have been made with no apparent pattern of aberrations.

A cancer of the eyes that affects children will be discussed in Chapter 8. The malady, **retinoblastoma**, is caused by a single dominant gene in most cases, but a **deletion** in the long arm of chromosome 13 is also known to result in the same cancer. Based on the fact that the loss of chromosomal material in the deletion has the same effect as a mutation that gives rise to a dominant gene, it has been hypothesized that the aberrancy involves the loss of function of some specific gene locus.

In most cancers there is no simple association with an aberration, as in the retinoblastoma deletion or the Philadelphia chromosome. Frequently there are discrepancies in numbers and configuration of chromosomes in cancer cells, but it is not clear whether such patterns result from the disease or are actually a part of the mechanism responsible for the unusual physiology and structure of the tissue. Some observations have indicated that the aberrancies change during the progress of the disease. There is no doubt that agents which interfere with the genetic material, such as radiation or disruptive chemicals, may be **carcinogenic**. It can be shown that such agents are often the cause of chromosomal breaks or rearrangements.

There are a number of known cancer-causing genes **(oncogenes)** that are normal components of the human genome. Fortunately, in the majority of persons these genes remain latent, producing no phenotypic effects. However, they may be "turned on" by a number of means. One way may be a breakage or reorganization of the chromosome carrying the oncogene. Possibly, the Philadelphia chromosome is an example of this effect. The explanation of the phenomenon might be that there is a **promoter** region on the chromosome that is necessary to turn on the oncogene. If a mutation occurs that activates the promoter, or if the oncogene is moved into the vicinity of an active promoter on another chromosome, it may produce its devastating effect. Viruses carrying oncogenes and their own promoters have been shown to cause cancerous effects in other animals or tissue cultures. As in most aspects of the study of cancer, more research will be required to provide a better understanding of the situation.

Robertsonian Translocations

In 1916 a cytologist in the United States, named William Robertson, described a type of translocation that bears his name. These aberrations were first noted in grasshoppers, but as seen in one type of Down syndrome (p. 133), they have proved in recent years to be significant in humans. Robertson originally called the rearranged sequences *compound chromosomes* because whole arms from two different chromosomes appeared to break and rejoin in the centromere regions. Although such rearrangements are sometimes called **centric fusions**, they are more commonly known as **Robertsonian translocations**.

It is conceivable that any two chromosomes could be broken in the region of their centromeres with reciprocal arms rejoining. In fact, centric fusions that are found naturally invariably seem to involve two chromosomes with centromeres placed close to one end (acrocentric). The imbalance created in whole arm translocations may be too great to permit viability of products in anything other than acrocentric chromosomes. The arm lengths in acrocentric chromosomes are extremely asymmetrical: There is a short arm with very little genetic material (probably none that is indispensable) and a long arm containing a number of significant genes.

When first observed, it was commonly assumed that the two centromeres each broke in the middle and refused reciprocally. However, more thorough banding studies have shown that breaks may occur in two other ways. The break can be in the short arm of one chromosome, close to the centromere, with the other break in the long arm of the other chromosome, also close to the centromere. Refusion in both types yields a chromosome with the two long arms from both chromosomes attached by the centromere and the other centromere joining the two short arms having no indispensable genes.

A third type of mechanism involves breaks in each short arm adjacent to the centromeres. Reciprocal refusion of such breaks results in the two short arms joined together with no centromere **(acentric)** and the two long arms joined together in a chromosome having two centromeres **(dicentric).** It is easy to see how the acentric chromosome would be lost in subsequent cell divisions, but it is interesting to note that only one centromere functions in the dicentric. The other one is no longer seen associated with a constriction in microscopic preparations.

Even when the two attached short arms have a centromere, the tiny, genetically unimportant chromosome is usually lost. Neither the presence nor the absence of the attached short arms seems important to the development and functioning of the individual.

The human genome has five acrocentric chromosome pairs that are associated with Robertsonian translocations: group D chromosomes (13, 14, 15) and group G chromosomes (21, 22). The most common types involve chromosome 13 with chromosome 14 and combinations of 14 with 21; other combinations have been reported with a much lower frequency.

Inversions

It is possible for two breaks to occur simultaneously at points some distance apart in the same chromosome. If the segment which is thus formed is not lost as a deletion, it may be turned around or inverted and the broken ends rejoined in reverse sequence. Again consider the chromosome with genes A to I:

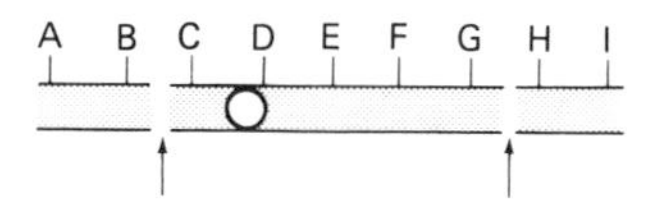

Assume breaks between genes B to C and G to H, followed by an **inversion.** Now visualize an inverted order for a block of genes therein:

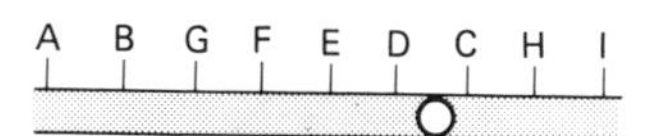

Because the centromere is involved in the inversion, it may appear to move to a different position along the chromosome. This type of inversion is called a **pericentric inversion**, since it rotates around the centromere.

A **paracentric inversion** does not include the centromere between the two breaks, because they both occur in one arm simultaneously. Assume in this case that the inversion has resulted from breaks between D to E and H to I:

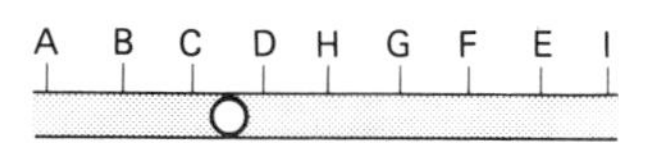

Once again, other than changes in the physical appearance of the inverted chromosome, there may be no apparent genetic effect on the individual carrying an inversion. It is during meiotic synapsis in a cell with a normal and an inverted chromosome that problems begin to develop. Pairing between a normal chromosome and an inverted one requires that a **loop** be formed:

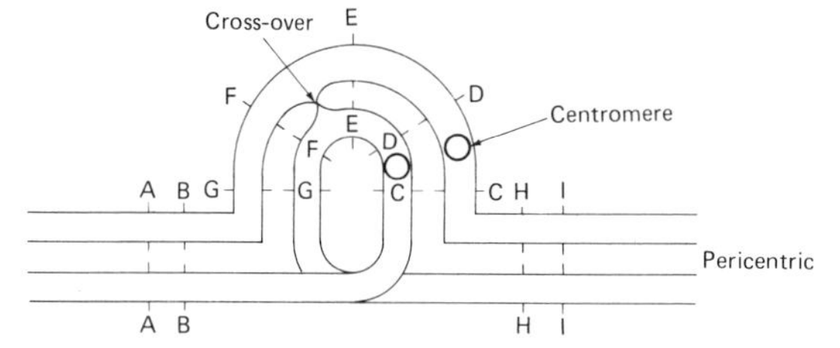

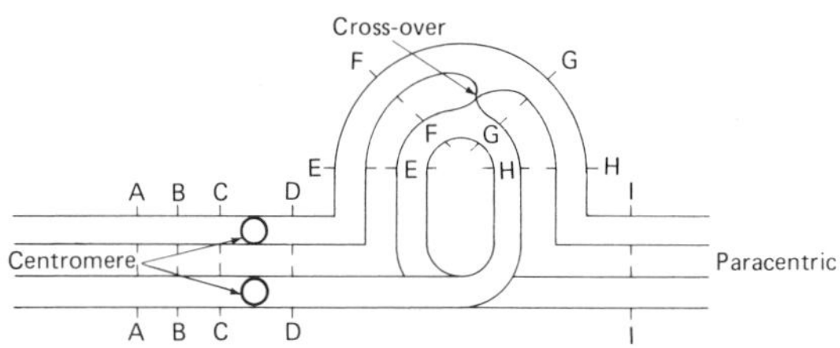

The chromosomes are double, due to replication prior to meiosis. Note that crossing-over is indicated in the diagrams, and it involves only the adjacent strands. Pairing is satisfactorily accomplished by the loop formation but the separation process is significant to the formation of gametes in both of these cases.

Crossing-over may occur in the synapsed chromosomes. If a cross-over occurs within the loop of one of the meiotic figures as shown, when the chromosomes separate toward the poles, large aberrancies are likely to result.

In the case of pericentric inversions there will be duplications and deficiencies, as in the case of crossing-over between genes E and F:

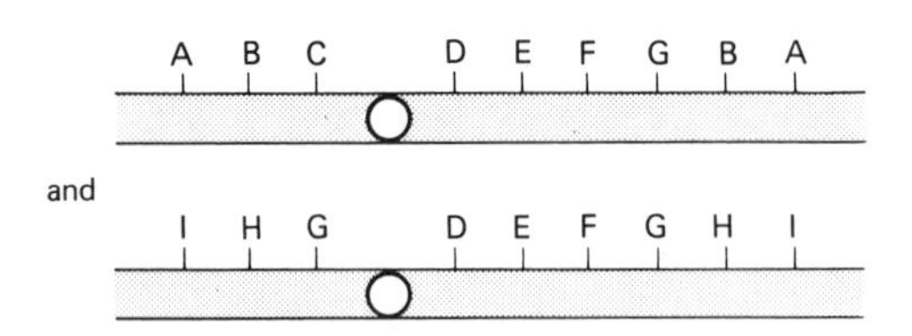

If crossing-over occurs in the loop of the paracentric inversion, one of the resulting products would have two centromeres (dicentric) and the other would have none (acentric), as in the case of an exchange between genes F and G:

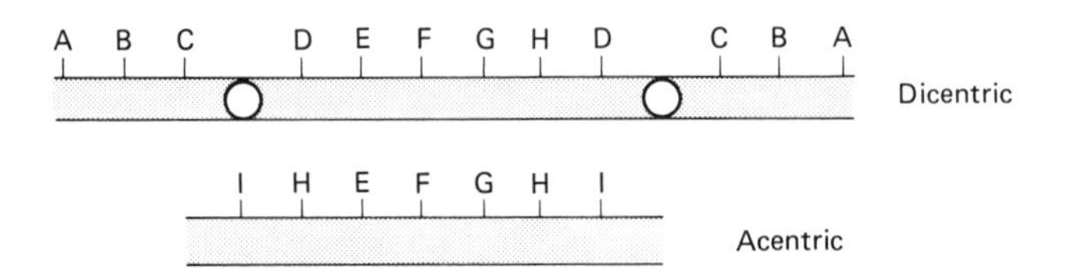

Not only are there duplications and deficiencies among the products, but as seen in one type of Robertsonian translocation, the acentric will be lost to the forming gametes, since its motive center is absent. The dicentric chromosome may also present a further complication because of the possibility that, if both centromeres remain functional, they may move in opposite directions while attached to each other and neither is then able effectively to enter one of the new reproductive cells.

Needless to say, even though inversions do not generally interfere with the balanced functioning of a somatic cell, they are a significant factor in the production of eggs and sperm required for the next generation. Because the products of a meiosis, which involves crossing-over within a synaptic inversion loop, carry major duplications and deficiencies, viable offspring may not be produced, or they will carry monosomies and trisomies for blocks of genes, as seen in some reciprocal translocations. In fact, because many products of such crossing-over are inviable, statistical studies indicate that recombination frequency was reduced in the region of the inversion.

Deficient Duplication Syndromes from Inversions

One form of chromosomal abnormality that was difficult to demonstrate with human genomes until the advent of banding techniques was an inversion. Some pericentric inversions were identified because the centromere had shifted far enough to be noticeable in the karyotype. With banding patterns, it is now possible to locate paracentric inversions as well.

An inversion carried by an individual along with a normal chromosome requires the formation of a loop for synapsis during meiosis, and crossing-over can result in gametes carrying duplications and deficiencies (p. 143). Penny Allderdice and coworkers at the Memorial University of Newfoundland identified a pericentric inversion in chromosome 3 in 1975 and a paracentric inversion in chromosome 9 in 1980. Both chromosomes resulted in duplications and deficiencies among offspring of inversion carriers. Profound physical and mental disabilities are associated with the syndromes. A distinct array of developmental problems was observed with obvious physical abnormalities like those seen in the trisomies, as well as problems with internal organs, limbs, musculature, and coordination. Shortened life spans generally are associated with the defects and there is much mental anguish for the families involved.

SEX CHROMOSOME (X AND Y) ABNORMALITIES

Modifications of the number of sex chromosomes are the most commonly encountered type of chromosome variants in humans. As will be discussed, the cause for this probably lies partly in the natural ability of the human genome to inactivate all X chromosomes except one in a cell. In addition because the Y chromosome is not essential for fetal development, its absence or supernumerary presence usually results in less severe physiological or physical symptoms than the autosomes.

Klinefelter Syndrome

During cell divisions that produce human gametes, it is known that in rare instances nondisjunction may occur with two synapsed chromosomes. The profound results of this process have been demonstrated in such genetic abnormalities as Down syndrome and other trisomies. During ovarian meiosis it is possible for two X chromosomes to remain together and move into a forming egg. This could occur during anaphase of either the first or second meiotic division (Fig. 6.7). The resultant zygote would then contain two X chromosomes from the mother. If the sperm cell involved in the fertilization carried a Y chromosome, the offspring would carry an **XXY** chromosomal complement and have 47 chromosomes.

Likewise, nondisjunction could occur during meiotic anaphase I in sperm formation. If X and Y chromosomes both moved into the same secondary spermatocyte during the first division and then went through a normal anaphase II, two products of the four would carry an XY complement (Fig. 6.8). After fusion with a normal X-carrying egg cell, the zygote would possess an **XXY combination**. It has been estimated that about 60% of XXY zygotes result from nondisjunction in females, whereas about 40% are derived from male nondisjunctions.

Mitotic nondisjunction occurs rarely in the developing embryo and can result in **mosaic** individuals for abnormal complements of sex chromosomes. If the first division of a zygote, or a very early embryo division, showed a mitotic nondisjunction for the X chromosome, one of the products would receive an XXY complement, while the other cell would have an inviable Y__ combination. The latter cell would die, allowing the XXY cell to continue dividing and it would then produce the entire embryo, resulting in the XXY combination in all cells.

The average age of mothers of XXY offspring is about 32, while it is 28 for the mothers of normal males. Again there appears to be a correlation between ma-

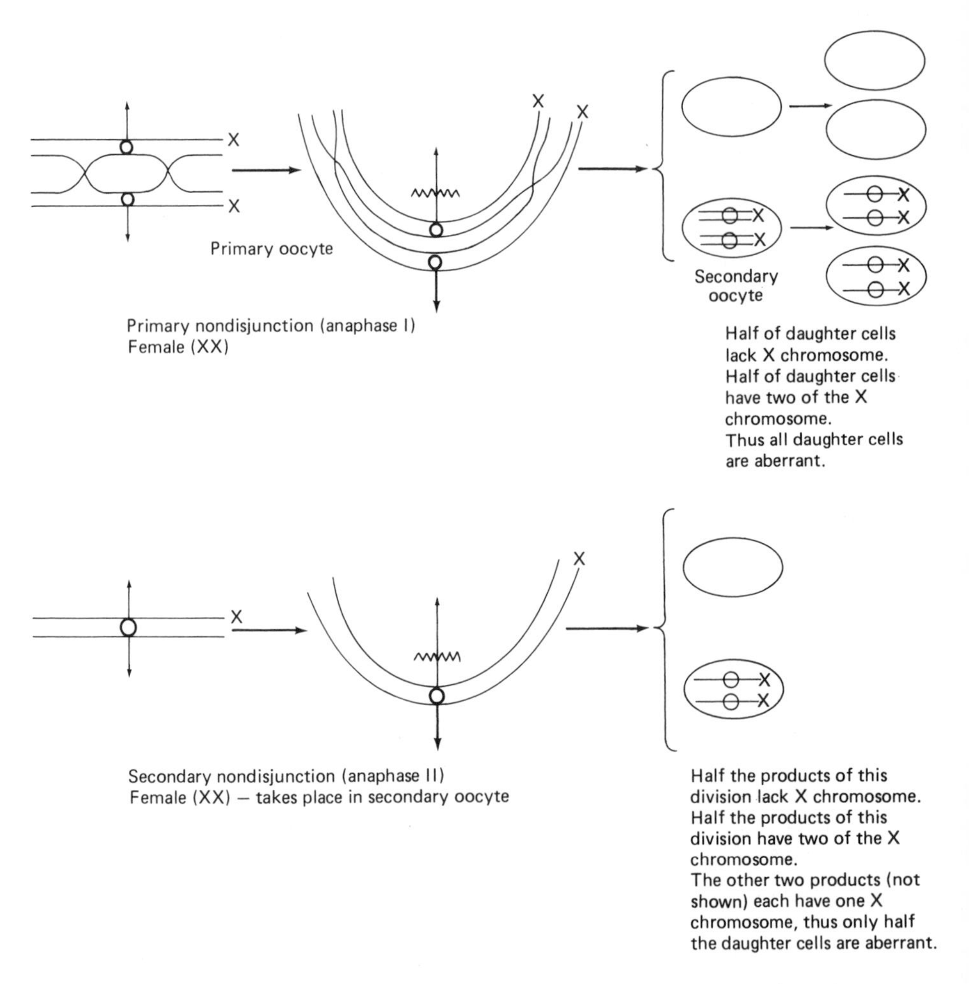

Figure 6.7. Diagram of nondisjunction of the X chromosomes in egg formation during anaphase I of meiosis (primary) and anaphase II of meiosis (secondary). Note that in both processes the result is that there are division products with either two X chromosomes or none.

ternal age and nondisjunction, but it is not as pronounced as in Down syndrome.

The XXY individual is a phenotypic male who is usually **sterile** because of very low sperm counts and exhibits such secondary female body characteristics as enlarged breasts. A larger than normal proportion are judged to be of lowered intelligence and are sometimes institutionalized.

The series of symptoms associated with the XXY condition was first identified medically and became known as **Klinefelter** syndrome after Harry Klinefelter of Boston, the medical investigator who first described its symptoms in 1942 (Fig. 6.9). The chromosomal basis for the condition was not identified until 1959 when Patricia Jacobs and J.A. Strong working in Scotland recognized the correlation. It was the first case of an abnormal chromosome number discovered in humans. The year 1959 does not seem so late when one considers that the human chromosome number was

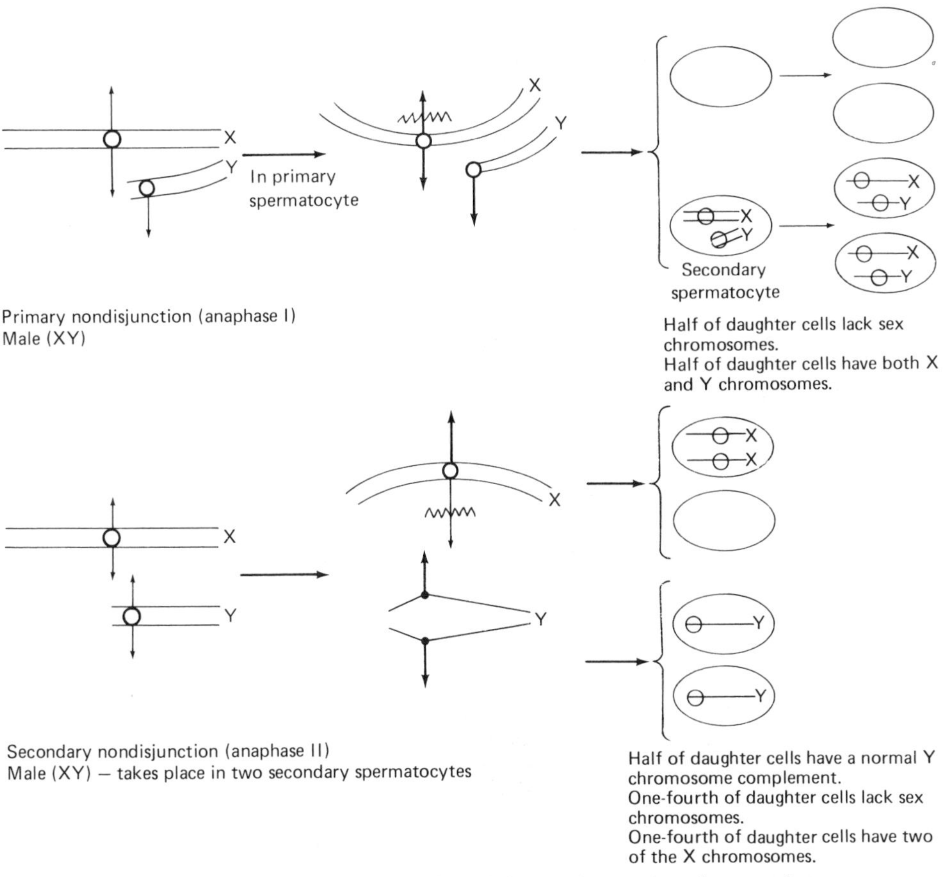

Figure 6.8. Diagram of nondisjunction of the sex chromosomes in sperm formation during anaphase I of meiosis (primary) and anaphase II of meiosis (secondary). Note that in primary nondisjunction some products have no sex chromosomes while some have both X and Y. Note that in secondary nondisjunction some normal products have a Y chromosome, while one-fourth have two X chromosomes and one-fourth receive no sex chromosomes.

considered to be 48 until 1956 when Tijo and Levan perfected the technique of counting human chromosomes.

Although about one in every 700 male live births shows the syndrome, Klinefelter syndrome accounts for about 1% of male patients institutionalized for mental defects. It is interesting that if a sample of men over 6 feet tall is examined, the frequency of the syndrome is about one in 260. This indicates that persons with the syndrome are above average height.

Obviously, if a means of preventing the syndrome could be discovered, it would be of considerable interest to medical science. Notice that even though the phenotype is based on an abnormal genome (47 chromosomes), it is totally unpredictable. The consequences of such a situation can be serious, but there is no way to anticipate the birth of an affected individual. It is possible through modern techniques to identify a fetus with an XXY chromosomal complement, but there are no clues to suggest that such a prenatal analysis should be made.

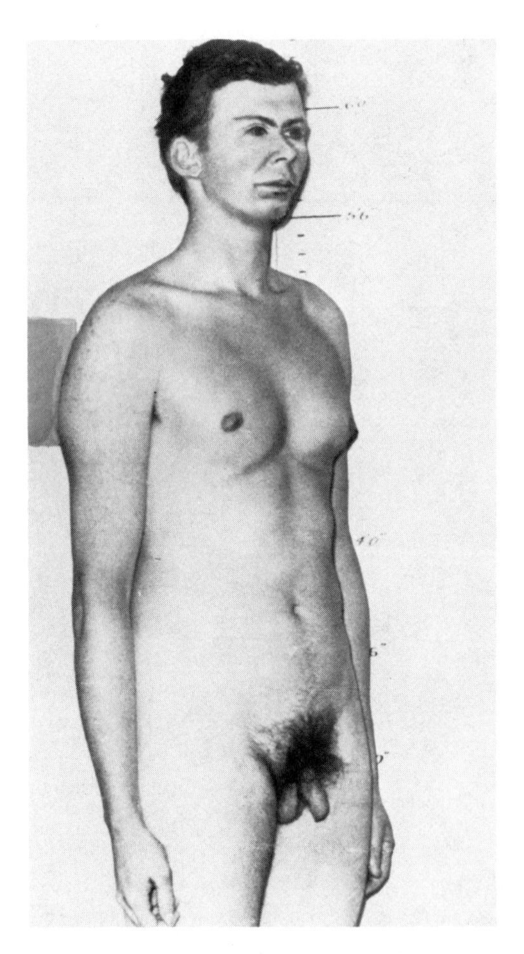

Figure 6.9. An individual exhibiting the symptoms of Klinefelter syndrome. (Courtesy Lester V. Bergman, New York)

Despite the serious detrimental nature of Klinefelter syndrome, it has contributed to knowledge of the chromosomal basis for human sex determination. Much knowledge of genetic mechanisms has come from research with the fruit fly *Drosophila* and the earliest discovery of sex chromosomes was in insects (Chapter 3). In *Drosophila* an XXY individual having two sets of autosomes would be female (X to autosome ratio = 1:1). But the fact that the sex of the human XXY is male, although generally abnormal, refutes the idea that sex determination is similar to that in the fruit fly. Based on the phenotype of Klinefelter syndrome, it is apparent that in humans, male sex is determined by the presence of a Y chromosome and not simply by a genic balance mechanism between X's and autosomes. This type of observation supports the theory expressed in Chapter 3 that the presence of the Y chromosome in prenatal development results in the formation of testes.

Naturally, the temptation is to oversimplify, reject the fruit fly data for humans, and propose that sex determination is based solely on the Y chromosome. If it is present, normal development leads to a male; if it is absent, normal development leads to a female. It is wise to simplify as much as possible to understand scientific phenomena, but in the case of such a complicated process, great caution should be exercised. Unfortunately, on closer inspection everything is more complicated than generally supposed. The Y chromosome may well carry the initiating factors for maleness, but many other genes on other chromosomes are likely to be essential to the completion of the process.

Turner Syndrome

If nondisjunction can lead to gametes carrying two X chromosomes, as in Klinefelter syndrome, it is also possible for a gamete to carry no X chromosome, as the other product of the aberrant division. The fertilization of such an egg by a sperm cell carrying a Y chromosome, would not produce a viable zygote. The Y__ combination would result in early death of the zygote and nondevelopment of the fetus. As seen, the X chromosome carries a significant number of important genes and when an embryo does not possess these genes, it will not proceed with a normal developmental sequence.

If a sperm cell with an X chromosome fertilizes the egg cell with no sex chromosomes, however, the resultant zygote and fetus may continue development and give rise to an individual with an abnormal sex chromosome complement (X__, 45 chromosomes). In similar fashion a conception involving a normal X-bearing egg cell and a sperm cell that carries neither an X nor a Y chromosome would also produce an X__ zygote. An individual developing from such an X__ zygote is phenotypically a female, based on the juvenile external genitalia, but the ovaries are absent or incompletely developed, so that the individual is generally **sterile.** Secondary sex characteristics are incompletely developed, there are skeletal abnormalities resulting in short stature, and the neck shows lateral folds of flesh (''webbed''). Such persons are said to exhibit **Turner syndrome** (Fig. 6.10). The syndrome was named for Henry Turner of the United States who described it in 1938. Mental development is apparently not as seriously affected as in Klinefelter syndrome, although this is difficult to ascertain. Studies have indicated that some individuals with Turner syndrome have difficulties with problems requiring spatial and mathematical skills. Approximately one in 2500 female births exhibits Turner syndrome, so that it is not encountered as frequently as many other types of human chromosomal abnormalities. This birth rate figure does not reveal the whole story, however, since possibly 95% of **XO conceptions** do not survive to birth. It is a common abnormality and about 20% of the fetuses which have been aborted naturally with obvious chromosomal abnormalities are of the XO type. In fact, approximately 1% of all recognized conceptions may have this aberrant chromosomal complement. Because normal males have only one X chromosome, it is somewhat surprising that the survival rate for XO individuals is not higher.

Other X-Y Chromosome Abnormalities

It is probably obvious that other types of aberrant sex chromosome combinations are possible. A double-X gamete joining with a normal, single X-bearing gamete would produce an offspring with an **XXX, 47-chromosome complement**. Such triplo-X individuals have been found and they are females, often of normal intelligence and fertility, although in some cases the situation seems to be correlated with decreased

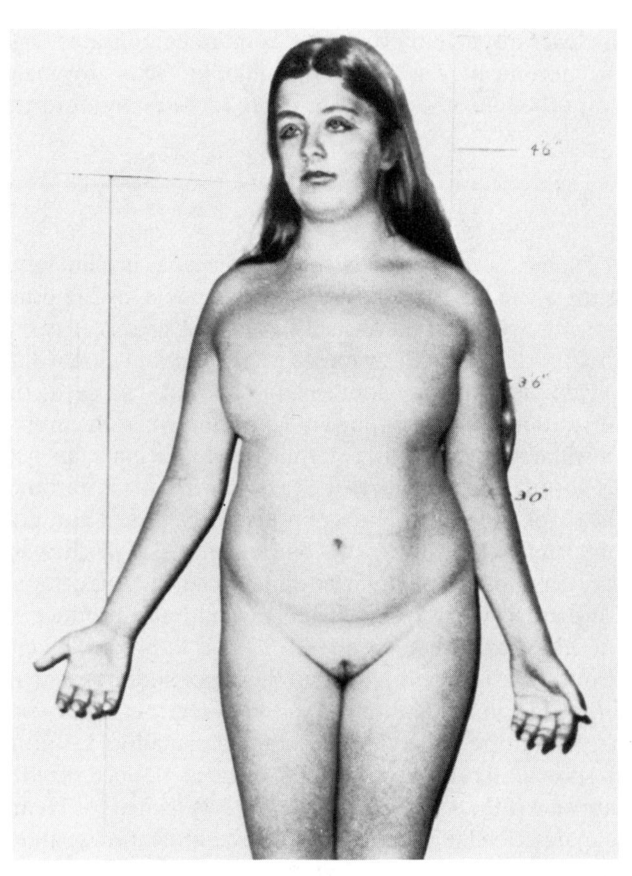

Figure 6.10. An individual exhibiting the symptoms of Turner syndrome. (Courtesy of Lester V. Bergman, New York)

intelligence and fertility. In addition XXX persons are found at a higher than expected frequency among persons classified as being mentally defective, but this does not actually prove that the XXX condition is the cause.

It is also possible for nondisjunction of Y chromosomes to occur during the second meiotic division of sperm formation (see Fig. 6.8 on page 147). This results in double Y-bearing male gametes. If one of these fertilizes a normal X-bearing egg cell, an **XYY** individual results. These 47-chromosome individuals are males who do not seem to exhibit any constant, easily identifiable phenotypic characteristics. However, they are almost always quite tall, develop heavy cases of acne, and seldom have a high intelligence level. Such individuals apparently are normally **fertile**.

Considerable debate has taken place concerning research which has shown a correlation of the XYY genome with above average height and a tendency toward antisocial behavior. Several reports have indicated a higher proportion of XYY males convicted of crimes of violence than XY males, but such reports have been disputed on several grounds. Some criminals have used the XYY genotype as a legal defense for their actions and have actually received reduced sentences in England and Australia.

Although the arguments over the validity and necessity for analysis of XYY individuals have been publicly conducted and heated at times, there will surely be more complete and comprehensive data in the future. Until more definite evidence is available to demonstrate such a biological determination of criminal tendencies, there should be reservations about identifying and defining preventive measures. Even if evidence is convincing, an ethical dilemma will be presented for everyone. Should XYY boys be identified and educated in a context different than XY boys? Will potential parents be justified in identifying XYY fetuses prenatally, with the possible option of abortion? Some feel that this issue does not warrant further investigation. Others feel that it is immoral to remain ignorant of the facts, because of the moral issues involved. There have been serious questions about the propriety of surveying all male infants to learn whether or not they carry the XYY combination. More potential harm might be done to such individuals as a reaction than if they were never identified. In fact, programs of this type have been halted in the past because of outside pressure.

A poorly understood aberration affecting the X chromosome has been identified through careful karyotype analysis of cultured cells. For some unknown reason, when cells are grown in particular types of media (deficient in some nutrients), an X chromosome breaks very close to the end of the long arm, releasing an **acentric fragment**. If the person with this "fragile X chromosome" is a male, the result is usually severe mental retardation. If the carrier of the fragile X is a female, it is likely that the heterozygous presence of a normal X will prevent a severe mental defect. Unfortunately, the protection does not appear to be complete, since mild mental retardation has been observed among females with the heterozygous condition.

An estimate of the significance of this unusual trait has indicated that it is second only to Down syndrome as a known cause of retardation. It appears that 2 to 4% of institutionalized males suffer from the **fragile X syndrome**. The syndrome is not accompanied by severe physical deformities, although there are such characteristic features as prominent jaw, large ears, and a smiling, cooperative attitude.

Unlike Klinefelter and Turner syndromes, predictions can be made about future children with the fragile X syndrome. If it occurs in one child, an analysis of the mother's karyotype will reveal that she is a carrier. With this knowledge it can be predicted that one-half the sons will be retarded and one-half the daughters will be carriers, since there is a one-half chance of transmitting either the fragile X or the normal X chromosome. No prenatal test is available for the fetus, except to determine its sex. Some parents might wish to abort a male fetus, rather than run the 50% risk of having another retarded son.

One of the most bizarre aberrations in sex chromosome complements is the XY combination found in some apparent phenotypic females. Technically, the XY combination is the normal complement for a male, and testes are produced in the fetus. However, the testis is incapable of organizing a male phenotype in the developing individual. The effect has been called *testicular feminization,* but a more appropriate name might be *androgen (male hormone) insensitivity.* It is considered to be gene determined, since it is found to run in families. The testes do not produce sperm and, while the external genitalia and secondary sex characteristics are distinctly feminine, no uterus, oviducts, or ovaries are present.

Such individuals have married and then sought medical assistance because no children were produced. The medical or genetic counselor for such an individual is probably correct to explain that there is an incomplete development problem resulting in sterility. Usually the deficient testes are removed surgically and female hormone application is utilized. To explain that the marriage was actually one involving two males on the basis of the XY genome seems to be gross and unfair to those involved. In fact, it is incorrect from a developmental standpoint.

THE BALANCING MECHANISM FOR X-CHROMOSOME NUMBER

What is the significance of different numbers of X chromosomes in human males and females? The examples used to demonstrate the basis for sex determination (Klinefelter and Turner syndromes) indicate that the number of chromosomes is important to the development of the individual. Even more pronounced effects have been noted with monosomic and trisomic situations involving autosomes (e.g., Down syndrome). While the Y chromosome seems to be the prime factor in sex determination, the X chromosome carries genes affecting many other characteristics. With this is mind, the obvious imbalance in number of X-carried genes between the sexes poses an important question. One might anticipate greater differences between sexes on the basis of sex chromosome abnormalities than actually found in situations like XXY (Klinefelter syndrome) or XO (Turner syndrome).

Human male and female cells only differ essentially in the possession of a Y chromosome and its very few functionally essential genes. A discovery that supports this contention was made by Murray Barr in 1949. He found that cells taken from females contain a dark-staining **chromatin body** within the nucleus that male cells do not have. Subsequently it was shown that this stained chromatin body was a condensed, **inactivated X chromosome** (called a **Barr body**).

In 1961 Mary Lyon explained the situation satisfactorily by proposing that one of the two X chromosomes is inactivated in each female cell. Thus, there appears to be only one active block of X chromosome genes, as in males. Inactivation of one of the X chromosomes occurs fairly early in the embryonic life of the individual. Once an X chromosome has been inactivated, that same one remains inactive in all cells produced by mitosis from the original. When the chromosome is first inactivated it is apparently selected randomly, so that all cells of the female do not contain the same inactive member of the pair. As a result, alleles present on one X function in one group of cells, while a nearby group may contain the other functional alleles. Thus, each female is technically a mosaic of cells with two different functional X chromosomes. In toto the products of both alleles are produced in the female and the result is essentially what would be expected in a heterozygous individual.

A significant feature of the X-inactivation process is that all X chromosomes are condensed and heterochromatic except one which remains functional (Fig. 6.11). This is true no matter how many X chromosomes are found in the genome. It is interesting to note that X-chromosome inactivation is apparently an irreversible process. Once one of the X chromosomes becomes inactive, it essentially never again reverts to an elongate functioning form. During replication prior to mitosis the inactivated X chromosome lags behind the other chromosome and acts much like other examples of heterochromatin.

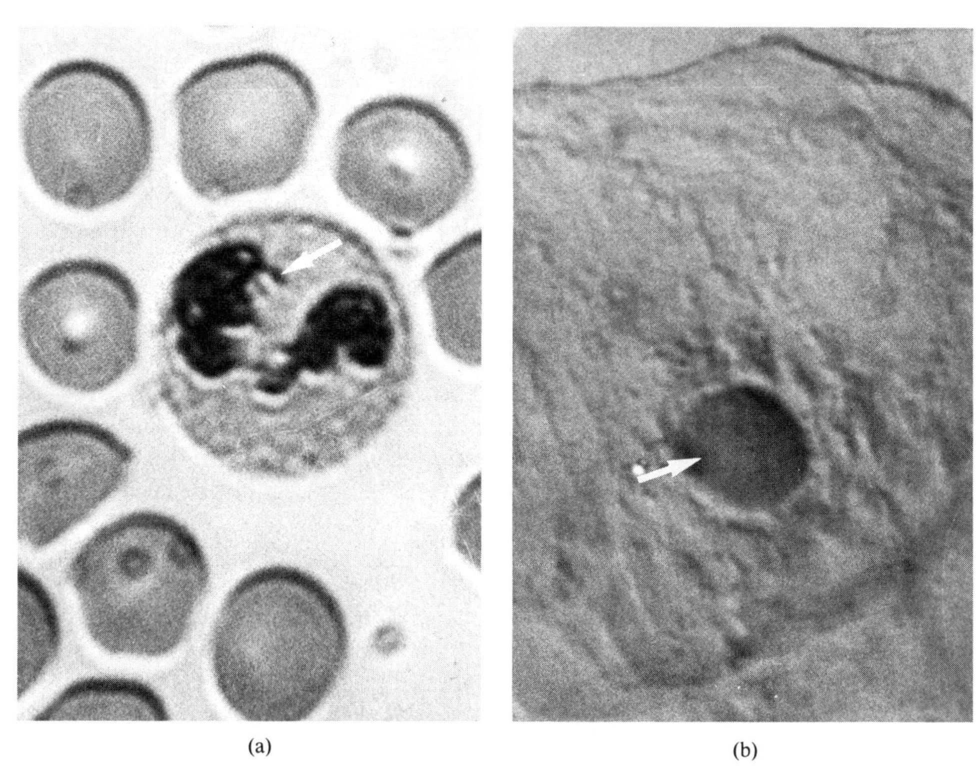

(a) (b)

Figure 6.11 Condensed or heterochromatic X chromosome in female (XX) nuclei. (a) White blood cells showing Lyonized X chromosome or "drumstick" (arrow). (Because Mary Lyon correctly described the condensing of all X chromosomes except one as heterochromatin, the process has come to be known as "Lyonization.") (b) A cell from a tissue culture of human female cells. Note the presence of the condensed heterochromatic Barr body at the periphery of the nucleus (arrow).

THE PHENOTYPIC EFFECT OF AN INACTIVATED X CHROMOSOME

There is good evidence to support the **Lyon theory** that the genes on one X chromosome have been inactivated. Such evidence also provides an understanding of the effects of X inactivation on the phenotype of individuals. To make observations of this sort, it is essential that a gene located physically on the X chromosome is used.

Although a number of genes have been identified with the X chromosome, the gene involved in the production of the enzyme glucose-6-phosphate dehydrogenase (G6PD) demonstrates best the genetic effects of the inactivation phenomenon. The enzyme is significant, since its absence due to a mutation results in a sensitivity to sulfa drugs, synthetic quinine (primaquine), and broad beans (fava beans), among other agents. Exposure to these materials results in anemia due to the breakdown of red blood cells. Incidentally, this is another good example of a gene-environment interaction, since persons without G6PD can live normally as long as they avoid exposure to agents causing the anemia.

To demonstrate the effects of gene inactivation, it is necessary to grow in a tissue culture human cells from the skin of a woman and test such cells for the G6PD enzyme. There are several different functional forms of the enzyme resulting from minor mutations; these polymorphic forms can be recognized chemically.

When tissue culture cells taken from the skin of a woman who is heterozygous for two forms of the enzyme are analyzed, it is found that one or the other form of G6PD (for example, type A or B) is produced by the culture, but not both types. A cell removed from another part of the skin to start a culture may give rise to a line of cells that produces only the other form of the enzyme. This effect is because the X chromosome carrying an allele for one form of G6PD is inactivated in one area but functioning in another. Thus, the body is made up of groups or patches of cells with one or the other inactivated X-chromosome.

An important question may arise after an understanding of the effects of the Lyon theory of X-chromosome inactivation. If one X chromosome is inactivated normally in females, why does the XO female (Turner syndrome) show such pronounced effects as infertility, webbed neck, and short stature? The answer probably is found in the fact that it is necessary for both X chromosomes to remain active and functional for the early period of embryonic development. Without the two functioning chromosomes in these early stages, normal female development does not occur. The fact that some genes remain active in the condensed chromatin of the **Barr body** has not been ruled out. In fact, a gene for one blood type (Xg) does not seem to be inactivated with the other X-linked genes of the Barr body. It could be that a female is functionally diploid for some of the X-chromosome genes, while functioning as a haploid for the others.

CYTOLOGICAL IDENTIFICATION OF BIOLOGICAL GENDER

It is possible to identify individuals with Turner and Klinefelter syndromes through the presence or absence of Barr bodies. An individual with only one X chromosome would not possess Barr bodies; thus, the Turner individual (XO) would be a phenotypic female without the inactivated X chromosome in her cells. On the other hand, the Klinefelter male would carry a Barr body in each nucleus. An XXX female has two condensed chromatin bodies in each cell, indicating that only one X is functional.

A test can be made to determine the number of Barr bodies in the cells of an individual. By simply scraping the inside of the mouth, living, intact, epidermal cells may be removed and placed on glass slides for observation with a microscope. If Barr bodies are present, most of the cell nuclei will show them after a simple staining procedure.

It is probably unfortunate that this technique has been utilized as a sort of "sex test." Why such a technique is deemed necessary when the physiological determination of males and females should be fairly simple is difficult to answer. The identification of a person as a functional male or female should be more meaningful than the presence of one, two, or three X chromosomes. As discussed, the Y chromosome probably has more to do with sex determination than the X. A fairly simple staining procedure using hair follicles is also available to demonstrate the presence of Y chromosomes.

In the early 1980s it was discovered that a large part of the terminal region of the Y chromosome flouresces strongly under ultraviolet light after staining with quinacrine mustard. The exceptional brilliance of the Y chromosome, even in interphase nuclei, will undoubtedly provide a ready means of identifying its presence.

If a functional female is found to carry a Y chromosome (which is possible), does that imply that she is a male? What difference could such a situation make and under what circumstances would there be concern?

In the International Olympic Games there has been considerable attention given to the idea that some males are "masquerading" as females, allowing them to win in competition with women at games originally designed for men. Even though there may not be an intentional attempt at deception, some officials are worried about the participation of individuals who are not biologically females (two X chromosomes). It is interesting that there seems to be less concern shown about biological females competing as males. Are females more graceful ice skaters than males? Should male skaters be checked for the XY combination?

To provide a definite answer about female participants, cells have been examined and contestants ruled eligible or ineligible on the basis of sex chromosomes. Presumably, if no Barr body is found, the person would be ruled ineligible to participate as a female. With no Y chromosome, a person would likewise not be a male, and the Turner syndrome individual would be ineligible to participate as either sex, no matter what degree of ability was possessed. In one case a contestant was ruled ineligible because she had three X chromosomes. Such a procedure seems to be a gross misuse, possibly even incorrect, of scientific knowledge. Modern techniques permit investigators to survey contestants for both X and Y chromosomes, but to define someone's sex strictly on the basis of the XX or XY combination must be seriously questioned.

ADDITIONAL READING

BEARN, A. G., and J. L. GERMAN, III, "Chromosomes and Disease," *Scientific American* (1961), 205, No. 5, 66–76.

BORGAONKAR, D. S., *Chromosomal Variation in Man: A Catalog of Chromosomal Variants and Anomalies* (2nd ed.). Baltimore: The Johns Hopkins University Press, 1977.

CROCE, C. M., and H. KOPROWSKI, "The Genetics of Human Cancer," *Scientific American* (1978), 238, No. 2, 117–25.

DE GROUCHY, J., and C. TURLEAU, *Clinical Atlas of Human Chromosomes*. New York: John Wiley & Sons, 1977.

GERMAN, J. (ed.), *Chromosomes and Cancer*. New York: John Wiley, 1974.

HELLER, J. H., "Human Chromosome Abnormalities as Related to Physical and Mental Dysfunction," *Journal of Heredity* (1969), 60, 239–48. (also reprinted in Mertens, T. R. (ed.), *Human Genetics: Readings on the Implications of Genetic Engineering*. New York: John Wiley, 1975.)

HOOK, E. B., "Behavioral Implications of the Human XYY Genotype," *Science* (1973), 179, 139–51.

LYON, M. F., "Possible Mechanisms of X Chromosome Inactivation," *Nature: New Biology*. London: Macmillan Journals Ltd. (1971), 232, 229–32. (available in Birth Defects Reprint Series, The National Foundation—March of Dimes.)

MITTWOCH, U., "Sex Chromatin," *Journal of Medical Genetics* (1964), September 1, 50. (available in Birth Defects Reprint Series, The National Foundation—March of Dimes.)

SUTTON, H. E., *An Introduction to Human Genetics* (3rd ed.). Philadelphia: Saunders College, 1980.

THERMAN, E., *Human Chromosomes: Structure, Behavior, Effects.* New York: Springer-Verlag, 1980.

VALENTINE, G. H., *The Chromosome Disorders* (2nd ed.). Philadelphia: Lippincott, 1975.

WISNIEWSKI, L. P., and K. HIRSCHHORN, *A Guide to Human Chromosome Defects* (2nd ed.). New York: The National Foundation, Birth Defects: Original Article Series, XVI: 6, 1980.

WITKIN, H. A., MEDNICK, S. A., SCHULSINGER, F., BAKKESTROM, E. CHRISTIANSEN, K. O., GOODENOUGH, D. R., HIRSCHHORN, K., LUNDSTEEN, C., OWEN, D. R., PHILIP, J., RUBIN, D. A., STOCKING, M., "Criminality in XYY and XXY Men," *Science* (1976), 193, 547–55.

YUNIS, J. J. (ed.), *New Chromosomal Syndromes.* New York: Academic Press, 1977.

REVIEW QUESTIONS

1. Assume that chromosome 1 has the following normal order of genes:

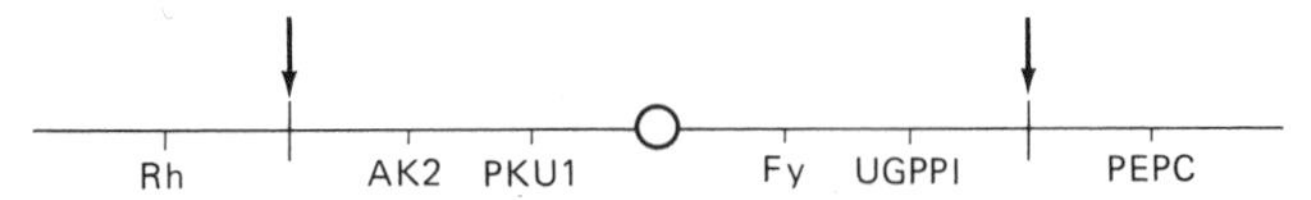

Assume further a paracentric inversion occurs causing the following order:

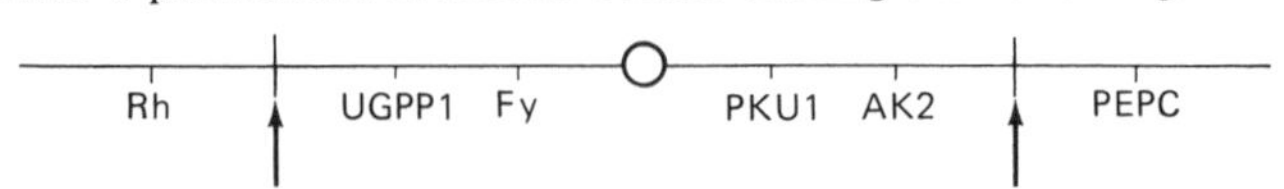

Show by diagram how these chromosomes might pair up in synapsis in a person whose genome carried both forms.

2. If crossing-over occurs between genes AK2 and PKU1, what would be the gene complements of the two cross-over products?
3. Does the information provided by the histogram in Figure 6.3 on page 132 suggest a means of lowering the number of Down syndrome births? Elaborate on how this might be put into effect.
4. Show by diagrams a mechanism which will account for the origin of the Philadelphia chromosome, (see pages 139–140 for information concerning reciprocal translocations).
5. Allderdice discovered in one family a pericentric inversion in chromosome 3 with a break in the long arm at q21 and a break in the short arm at p25. In another family she found a paracentric inversion in chromosome 9 in the long arm in the region q22 to q34. These were discovered because children in both families manifested distinct physical and mental defects due to duplications and deficiencies of some genes. Diagram the synaptic pairing configuration that would occur in meiosis of a person carrying both inversions along with a set of normal chromosomes. What would cause gametes to form with duplications and deficiencies in these chromosomes? Show by diagram how this would occur.

6. Consider Review Question 5 at the end of Chapter 5 (p. 123). Does an answer to that question provide a better insight into the manner in which a Robertsonian translocation 14/21 is possible? Why? Explain your answer. On this basis, what other human chromosomes appear to be good candidates for Robertsonian translocations?
7. Diagram the chromosomal arrangement for the chromosomes 14 and 21 in a person who has a balanced diploid array of essential genes for these two, but who carries a Robertsonian translocation 14/21 with normal 14 and 21 chromosomes. Diagram the synaptic pairing pattern for this combination of chromosomes during meiosis. Would crossing-over in such a synaptic configuration cause any further aberrancies? Diagram the chromosomal complements of gametes that might be produced by such a person for chromosomes 14 and 21. Which gametes might produce a Down syndrome child when fertilized by a normal 14-21 gamete? Which might produce a lethal situation?
8. Speculate why trisomies or monosomies for either whole chromosomes or segments containing blocks of genes might cause mental and physical developmental problems in individuals. Can you think of any medical approaches to such problems that should be explained to prevent birth defects of this nature?
9. Although other trisomies have been observed, usually for parts of chromosomes, the only ones found with any significant frequencies in humans have involved chromosomes 13, 18, and 21. Why? Explain the various aspects that may be involved.
10. Besides trisomy 13, 18, and 21, the only other one found in a nonmosaic situation was trisomy 22. Does this support your answer to Question 9?
11. Explain why mosaic trisomies may be viable for chromosomes 3, 7, 8, 9, and 10, whereas full trisomic conditions are not. Why would such conditions (mosaic trisomies) usually result in mental defects?
12. In a mosaic individual if gonadal tissue is trisomic for chromosome 21, fertile but aberrant gametes may be produced. Show by diagram how synaptic pairing might occur with the three homologous chromosomes of the trisomy (remember that synapsis is a precise two-by-two process). Diagram the types of gametes that might be anticipated in meiotic divisions of this type. Which of the possible gamete types could be expected to result in a Down syndrome child?
13. What would be the phenotypic sex of persons with each of the following sex chromosome complements (all have 44 autosomes)?

    (a) XXXY **(b)** XXYY **(c)** XXXXY **(d)** XXXX

    Would you anticipate that any of these would manifest the symptoms of already described syndromes?
14. By observing stained cells from persons with the following sex chromosome complements, how many Barr bodies would be found?

    (a) XXY **(b)** XO **(c)** XYY
    (d) XXX **(3)** XXXY **(f)** XXXX
15. The fragile X syndrome is responsible for severe mental retardation only in males, but varying degrees of mental disability have been observed in women heterozygous for normal X/fragile X. Some geneticists claim that the effect in women is due to Lyonization. Explain what is meant by this hypothesis.
16. Assume that a woman had a mentally retarded brother who was diagnosed with the fragile X syndrome. She appears not to be afflicted with any serious mental defects. Would it be possible for this woman to have children with the same defect? Is it possible to determine whether or not she is a carrier of the defect prior to having any children? If she is found to be a carrier, what advice would you provide regarding her having children?

Would the threat be greater for sons or daughters? What are the mathematical chances that she will have children with the fragile X syndrome?

17. Show by diagrams how nondisjunction of X chromosomes in sperm cell formation could give rise to Turner syndrome or triplo-X syndrome individuals. Do the same for nondisjunction of X chromosomes in egg formation. Based on the idea that the Y chromosome is smaller than the X chromosome, thus allowing Y-bearing sperm to swim more effectively and fertilize eggs more often, some workers have provided an explanation for the higher number of XY zygotes than XX zygotes. Could this same hypothesis be used in explaining more X__ individuals than XXX? Do you think it could also be applied as an explanation if proportionately more X__ individuals resulted from male nondisjunction than did XXX individuals (as compared to those resulting from female nondisjunction)?

18. Diagram as many ways as possible that an XXXX or XXXY individual could arise. Based on your diagrams, can you give a good reason why these types of conceptions are much rarer than types XXX, XXY, XYY, or X__?

CHAPTER SUMMARY

1. Typical sexually reproducing organisms are diploid, but exceptions are found in some organisms, for example, aberrancies in numbers of sets of chromosomes. Thus, three sets constitute triploidy, four sets, tetraploidy, five sets, pentaploidy, and so on. Collectively, such aberrancies are called polyploids.
2. It is more likely that an aberrancy in chromosome number will involve individual chromosomes, rather than full sets. Thus, nullisomics, monosomics, or trisomics may be identified.
3. Breakage and reorganization of chromosomes is a more common type of aberrancy than changes in total number of chromosomes. Types of chromosomal aberrancies that have been recognized are duplications, deficiencies, translocations, and inversions. Robertsonian translocations, the most commonly observed chromosomal aberration in humans, sometimes result in trisomies for most of the genes in chromosome 21 (Down syndrome).
4. Cytologically, aberrancies are most easily recognizable during meiosis, where unusual synaptic patterns may be seen. Crossing-over in inversion heterozygotes may result in duplications and deficiencies in meiotic products, while some patterns of disjunction in translocation heterozygotes may result in similar effects.
5. Down syndrome is the most common human disability which is based on an aberrant genome. The condition is caused by trisomy for chromosome 21. The trisomy may result from nondisjunction in gamete formation, a Robertsonian translocation (i.e., 14/21), or an isochromosome. A correlation of trisomy 21 with increasing maternal age has been shown and the risk for over 40-year-old women is much higher than for younger women.
6. Two other examples of trisomy in humans are Patau syndrome (trisomy 13) and Edwards syndrome (trisomy 18). Both are severely disabling conditions that result in early death.
7. The cri du chat syndrome results from a deletion of a large segment of chromosome 5. Mental retardation is the major disability of the syndrome.
8. Aberrant chromosomes have been discovered in association with chronic myelogenous leukemia (chromosome 22 translocation or Philadelphia chromosome) and retinoblastoma (chromosome 13 deletion). In rare instances duplications and deficiencies resulting from cross-overs in inversions have been found to cause serious human developmental problems.
9. Aberrancies in sex chromosome complement have been identified and they result in a spectrum of phenotypes from virtually no effect to severe mental retardation. XXY and XXXY males (Klinefelter syndrome) show mental retardation, sterility, and distinct physical abnormalities. XO females (Turner syndrome) are of essentially normal mentality, but they have distinct features and are sterile. XYY individuals show some common physical characteristics, and there is research indicating that they show criminal tendencies. However, there is considerabe debate about the validity of XYY observations. A number of other, much rarer sex chromosome aberrancies are known.
10. In human cells with more than one X chromosome all X's except one are condensed into inactivated heterochromatic Barr bodies. The Lyon hypothesis explains that females are mosaics for active X chromosomes, since one is randomly inactivated in each cell at an early fetal stage and all cells derived from a given cell maintain the same X chromosome as a Barr body.

CHAPTER

7

MENDELISM: LAWS OF CHANCE IN BIOLOGY

Although men flatter themselves with their great actions, they are not so often the result of a great design as of chance.

LaRochefoucauld, Maxims (1665)

In 1865 Mendel had no concept of the chemical reasons for variation among living things. He neither knew that chromosomes nor molecules were involved in the transmission of characteristics from one generation to the next. Because Mendel lacked a basic grounding in the elementary knowledge of cells, his knowledge of the actual mechanics of reproduction was restricted. Nevertheless, he was able to recognize the rules governing the transmission of genetic traits from parent to offspring. In addition he found that he could predict accurately the types of offspring that would result from various crosses and, with that information, he could determine the genetic make-up of the parents in a cross.

Mendel succeeded in fathoming the genetic mechanism where others had failed because his experiments were simple: He considered only one characteristic at a time until he understood its operation thoroughly. When the operation of a single gene was clear, he went on to additional traits. Once he understood the operation of several genes, he was able to combine two or three of them together and study their inheritance in the same organism.

Whether or not Mendel realized that the same mechanism determines hereditary patterns in peas (his experimental material) as in other living things, one shall never know for certain. It is probably safe to assume that he did anticipate a similar basis for all plants and animals. His methods of study and his system of notation remain effective today for the analysis of genetic problems.

REMNANTS OF MENDEL'S GENETICS IN MODERN SCIENCE

Mendel realized that the genes are in pairs and he recognized that there are two forms of each gene. When he determined that one form had an overshadowing or masking effect on the other, he said that this gene was **dominant**. The form which was masked was called the **recessive** gene. As seen in Chapter 1, the term **allele** came to apply to these two different forms of the same gene.

Dominance is frequently due to the manufacture of a functional protein in enough quantity by one allele to permit a characteristic to be exhibited. The other allele may be present but may not dictate the production of a functional protein; its presence therefore is masked by the dominant gene. In most cases the cells of an organism with two dominant genes probably produce a greater quantity of the protein in question, but quantity is not always significant to the exhibition of a characteristic. A tiny amount may suffice, with a surplus not adding visibly to the trait. This is especially true when the gene product is an enzyme. Although it is generally not possible to distinguish easily whether there are one or two dominant genes in an organism, using sophisticated chemical techniques, the distinction could probably be made. Obviously Mendel and other early workers had no means for such an analysis, so the mechanism of dominance and recessiveness remained a mystery. Today the mechanism of action of some human alleles still remains an enigma.

In genetics the physical appearance of an experimental organism is referred to as a **phenotype**. The term **genotype** describes the types of genes the organisms possess. Although these terms were not invented until the early 1900s, Mendel recognized two phenotypes of peas in regard to their height, tall and dwarf, and he knew that there were three different genotypes for these characteristics. Because the dwarf plants possess two recessive genes, they are said to be **homozygous** (homo—the same; zygous—zygote) for the **recessive** trait. Even though the tall plants show only one phenotype, there are two different possible genotypes. A plant can carry two genes for tallness or it can carry genes for tallness and dwarfness at the same time. The plant with two of the same allele is again called **homozygous** (this time for the **dominant** gene), while the plant having one dominant allele and one recessive allele is said to be **heterozygous** (hetero—different).

MENDEL'S NOTATION SYSTEM

In studying genetic transmission of traits, it is helpful if some shorthand method is utilized. Several methods have been proposed but geneticists are not unanimous in their choice of one over another. The system most frequently encountered involves the use of uppercase and lowercase letters to represent dominant and recessive genes. Mendel invented this system when he used *A*'s, *B*'s, or *C*'s for the dominant genes and *a*'s, *b*'s, or *c*'s for the recessive genes. Today geneticists attempt to choose a letter which symbolizes the trait in question. If possible, the first letter of the recessive phenotype is generally chosen. Thus, the gene for dwarfness would be *d* and that for tallness would be *D*. Often textbooks choose the first letter of the dominant trait as the symbol, rather than the recessive characteristic. Each method is merely an abstraction or notation and either technique will suffice, if one is careful to record what the symbols mean. Some geneticists prefer to use a plus symbol (+) for the nonmutant form, rather than a letter. The recessive mutant symbol still remains as a lowercase letter. Less often the mutant allele is dominant, in which case a capital letter would be used, with the + representing the recessive wild type.

MENDEL'S TECHNIQUE OF CONTROLLED CROSSES

The first step in Mendel's process was to secure **pure breeding lines**, or plants to be used in crosses that were homozygous for the traits being studied. Mendel did this by crossing pea plants to themselves **(selfing)** and selecting the offspring showing the desired trait for selfing in the next generation. When this is done for several generations, the desired homozygous plants can be obtained. Mendel's choice of the pea plant was particularly good for this technique, since it is naturally self-pollinating if the flowers are left undisturbed.

Once Mendel had established pure lines, two plants with different traits were crossed. For instance, he crossed homozygous tall (*DD*) and homozygous dwarf (*dd*) plants and observed that all offspring were tall. To understand the situation, it is necessary to remember that the paired chromosomes and genes are separated from each other during meiosis in the formation of reproductive cells. Therefore a gamete carrying a *D* gene from the tall parent joined with a gamete carrying a *d* gene from the dwarf parent. The resulting offspring was tall in phenotype, but it had a genotype that was heterozygous (*Dd*). It made no difference whether one or 1000 offspring were produced from the mating, the only possibility was for heterozygous tall individuals to be produced.

According to Mendel's notation system, the two parents in the cross made up the $\mathbf{P_1}$ (first parental) generation. The heterozygous offspring constituted the $\mathbf{F_1}$ (first filial) generation. By crossing the F_1 offspring to themselves or to others genetically alike the $\mathbf{F_2}$(second filial) generation was produced. The same notation system is still used conventionally in routine genetic analyses. Mendel referred to the cross of the F_1 individuals as the $\mathbf{P_2}$, a term which is not often used.

In crossing two heterozygous individuals the situation was more complex, because two types of reproductive cells were produced by each parent plant. In a plant of the genotype *Dd*, gametes carrying a *D* gene were produced, as were gametes carrying a *d* gene. It is important to note that the two types were produced with equal frequency. Thus one-half of those produced carried a *D* gene and one-half carried a *d* gene. Furthermore, the two types of gametes had *equal* opportunities to engage in a fertilization. One type had no advantage over the other.

MATHEMATICS TO MAKE GENETIC PREDICTIONS

At this point the element of **chance** is introduced and most genetic mechanisms to be discussed subsequently will be based on this phenomenon. An important feature of the chance process is the **principle of simultaneous independent events**: The chance for the simultaneous occurrence of two independent events is equal to the product of the chance for each separate event to occur.

If the separate, independent chance for parents to produce gametes carrying certain genes is known, the chance for them to join together simultaneously in fertilization can be calculated. For example, consider Mendel's P_2 mating, a cross of a *Dd* plant to another *Dd* plant. The chance for *D*-carrying gametes from each parent is known to be one-half (1/2). The chance for these two independently produced gametes to occur together in the same cell through fertilization (*DD*) is therefore $1/2 \times 1/2$ or one-fourth (1/4). Likewise, *d* gametes have a one-half chance in each parent's gametes and a *dd* offspring has a chance of occurrence of $1/2 \times 1/2$ or, again, one-fourth. A *Dd* offspring can occur in two ways: *D* from the female gamete and *d* from the male gamete ($1/2 \times 1/2 = 1/4$) or with a *D* from the male gamete and *d* from the female gamete ($1/2 \times 1/2 = 1/4$). These last two types are indistinguishable, since they are genetically alike (*Dd*). Therefore, they constitute one-half the offspring of the cross ($1/4 + 1/4 = 1/2$).

Through the use of this simple method involving the principle of independent events, one can predict that the offspring of the cross of a *Dd* plant to another *Dd* plant will result in a family of offspring consisting of 1/4 homozygous tall (*DD*), 1/2 heterozygous tall (*Dd*), and 1/4 homozygous dwarf (*dd*). With a small number of offspring produced in such a cross, the possibility is small that this ideal prediction will come true. For instance, if only four seeds were produced, it is not likely that one would be *DD*, two *Dd*, and one *dd*. However, with a large number of offspring, the ratios would come closer to the prediction. If 400 seeds were produced, the seeds would grow into approximately 100 *DD*, 200 *Dd*, and 100 *dd* plants. One would not be able to distinguish the *DD* plants from the *Dd* plants by looking at them. In reality the 1/4:1/2:1/4 proportion is a 3/4:1/4 proportion where phenotypes are concerned.

Mendel's actual results were 787 tall pea plants and 277 dwarf. These figures are very close to his 3:1 prediction (2.84:1). The data provided strong support for **paired unit factors, dominance and recessiveness,** and **segregation**. If any one of the principles had not been valid, his results would not have approximated the prediction so closely.

PUNNETT SQUARE

To show the simple Mendelian cross graphically, a **Punnett square**, named after R. C. Punnett, who repeated and expanded Mendelian studies in the early 1900s, is used. The method is simple but a bit cumbersome; Mendel did not use this technique. It is shown only as an aid to the understanding of the chance process in genetics, not as a suggested method of solving problems. Analysis with a Punnett square is neither as quickly done, nor as effective as the mathematical method, but the various combinations of gametes can be seen in their proper proportions in a simple graphic fashion.

In setting up the Punnett Square one lines up the male gametes of a cross horizontally across the top of the chart and the female gametes vertically down the left side. Four blocks will be formed when lines are drawn, each representing an equal chance for a fertilization. The gamete types for the male are transferred down and the female types are transferred across into the proper block. Each block then represents a separate fertilization and a type of offspring. Furthermore, the proportion of each type can be readily determined.

Using as an example the cross of two heterozygous tall plants (*Dd* × *Dd*), the Punnett square would look like this:

FEMALE ↓	MALE ⟶ *D*	*d*
D	*DD*	*Dd*
d	*Dd*	*dd*

The parents each carry the two different alleles and these segregate from each other during gamete formation. As a result, each parent produces two types of gametes, one type carrying the dominant form and the other carrying the recessive. It is important to remember that these two types of gametes are produced in equal numbers.

Four different types of offspring occur with equal frequency. Two are alike in genotype (*Dd*), but they arise in different ways. One gets its *D* gene from the female parent, the other gets its *D* gene from the male parent. One offspring of the four will be *DD*, two of four will be *Dd*, and one of four will be *dd*. In phenotype three of four will be tall and one of four will be dwarf. For this reason, it is said that the cross results in a 3:1 phenotypic ratio. If there were 4000, one could predict that approximately 3000 would be tall and 1000 would be dwarf.

Because the cross involves two parents that are **hybrid** for a single trait (*Dd*), it is called a **monohybrid cross**. It is a classic example of a 1:2:1 **genotypic ratio** and a 3:1 **phenotypic ratio**. It will serve as the basis for more complex crosses and must be understood thoroughly.

APPLICATION OF MENDELISM TO HUMAN STUDIES

As emphasized earlier, the same basic mechanism which controls heredity in pea plants underlies the genetic patterns found in humans. A common thread unifies all living things, plant or animal, and that is the system of determining characteristics that are passed on from one generation to the next. Because the same underlying mechanism operates in humans, one might expect the same "tricks" that Mendel used with peas to be applicable to an analysis of human heredity.

The genetic basis of human traits, however, cannot be studied by a process exactly similar to Mendel's work with pea plants. Human families always produce a small number of offspring by statistical standards and parents cannot be selected for controlled crosses. Instead of breeding the types of parents required for a cross and then making carefully controlled crosses, it is necessary to collect information on a large number of families exhibiting inheritance of the given trait in various patterns. By studying crosses that have already been made (i.e., families that already exist), it is even possible to collect statistical data that may be analyzed as effectively as if the experiment had been set up for this purpose.

In human genetics it is useful to think about only one trait at a time—Mendel's approach—in the analysis of patterns of inheritance. Once the mechanism of operation for a single gene is understood, it is then possible to consider several traits simultaneously, as long as the mechanism for each one is known. If the hereditary mode of action were not known for each gene, the situation would be so complex in studying only a small number of traits that the exercise would be futile. Still, there is always a chance for error if the trait being studied is affected by more than one gene, or if the gene product interacts with the product of some other gene. Nevertheless, there are a number of human traits that are straightforward in their mode of action and can be followed as conveniently as in Mendel's peas. In any event it is essential to develop an understanding of basic Mendelian patterns before going on to an analysis of the complications and exceptions that are often encountered.

There are numerous human hereditary traits that might be selected for examples, but one that is easily observed involves the **shape of the ear lobes** (Fig. 7.1). Most persons are mildly surprised to find that ear lobes of humans can be rather conveniently categorized as **free** or **attached**. In the free condition the ear lobes hang unattached quite visibly, whereas in the attached condition the ear lobes are joined directly to the head with no intervening space. For the individual, such a trait shows no apparent advantage of one form over the other. There also seems to be no preference

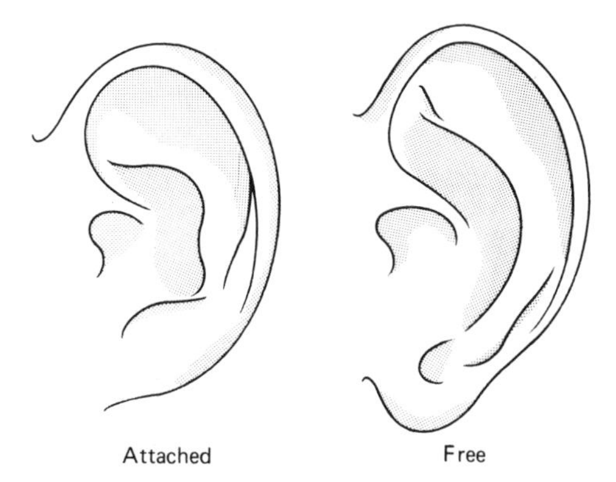

Figure 7.1. Human ear lobe shape. Basically two shapes of ear lobes are found among humans, free and attached. A single autosomal gene with two alleles controls the trait, with free being dominant over attached.

in mate selection of one type over the other. Artificially determined likes or dislikes appear not to be involved, since most persons do not even notice the difference until it is demonstrated. Such a trait, without selective advantage or disadvantage, presents exceptionally good material for routine genetic studies.

To establish a shorthand system of notation, the nature of inheritance patterns for the trait must be known. Dominance and recessiveness must be distinguished for the analysis. One type of mating is important to such an analysis, for example, the monohybrid cross described for pea plants. If one can find matings of persons in which both have free ear lobes, it is likely that there will be some families where there are children with attached ear lobes. If the simplest explanation possible is selected, it should be obvious that the way in which two parents may have a child showing a given trait, when neither of them has it, is for them both to be carrying a recessive gene for the trait. A child with attached ear lobes whose parents have free ear lobes demonstrates that attached is the recessive characteristic.

Thus, an analysis can be made by studying the family unit, sometimes working backward from children's to parents' genotypes. Once such an analysis has been made successfully and the mechanism is understood, it is possible to establish a method of notation that will allow further investigations of other types of matings.

Because it is most appropriate to select the first letter of the recessive characteristic for the shorthand system, one would probably select an *a* for attached; *A* would then be utilized as the symbol for the gene for free ear lobes. It is important to reiterate that these letters simply indicate that there is a sector of DNA somewhere on a chromosome in each cell that determines ear lobe shape. At this point it is not known how the *A* or *a* alleles exert this control. By analogy one can assume that one or the other alleles, or both, are involved in dictating the manufacture of a protein that exerts some form of control on the development of ear lobe shape in embryonic stages. Beyond the desire to understand the mechanism, it is not important to know exactly how the gene operates. The fact that it does operate allows analysis of persons and families and predictions about future children.

To analyze the monohybrid cross more completely, the child with attached ear lobes has a genotype written *aa*. Each parent has free ear lobes and therefore each carries an *A* gene. Because the child is *aa*, it is obvious that it had to receive one *a* gene from each parent, making each parent's genotype *Aa*. According to the conventional system, the cross would be diagrammed as follows:

$$\begin{array}{c} Aa \times Aa \\ \downarrow \\ aa \end{array}$$

Other types of children are possible from this cross, as shown by previous experience with a similar cross. There is one-half chance that each parent will produce a gamete which carries an *A* gene, thus there is a one-quarter chance ($1/2 \times 1/2$) that an offspring will be *AA* and have free ear lobes. Likewise, there is a one-half chance that an offspring will be *Aa* and have free ear lobes (1/2 chance for *A* from father $\times 1/2$ chance for *a* from mother, + 1/2 chance for *A* from mother $\times$ 1/2 chance for *a* from father). The prediction can be stated in several ways. For example, one could say that child has three-quarters chance to have free ear lobes from this mating and one-quarter chance to have attached ear lobes. One could predict that three-fourths of the children will have free ear lobes, one-fourth will have attached.

Such other types of crosses are possible as the following:

$$\begin{array}{c} AA \times aa \\ \downarrow \\ Aa \end{array}$$

This is like Mendel's original cross using pure line parents, and always gives rise to heterozygous free ear lobe offspring. In this case the chance for free ear lobes among offspring is said to be *one* or *unity,* since all the offspring will be alike.

It should be apparent that matings involving homozygous parents of either type could only result in children similar to the parents in regard to ear lobe shape:

$$\begin{array}{ccc} AA \times AA & & aa \times aa \\ \downarrow & \text{or} & \downarrow \\ AA & & aa \end{array}$$

Again, the chance for type of offspring in both cases is one. There is not even a fractional possibility of another type, barring the unexpected occurrence of a mutation. Remember that mutations can and do occur, but only once in about 100,000 gametes at best for any given gene. Such a chance would be too rare to justify a bet on the part of most gamblers!

Still another type of probable mating has been called a *test cross* in some genetics experiments using organisms other than humans, since it has been used to find out whether the organism is homozygous dominant (AA) or heterozygous (Aa). Remember that the phenotypes of such individuals are indistinguishable, with both showing the dominant trait. The cross is made by mating the unknown to the homozygous recessive individual and would be diagrammed like this:

$$\begin{array}{lll} Aa \times aa & \text{or} & AA \times aa \\ \downarrow & & \downarrow \\ 1/2\ Aa \quad \text{Free ear lobes} & & \text{All } Aa \quad \text{Free ear lobes} \\ 1/2\ aa \quad \text{Attached ear lobes} & & \end{array}$$

The resulting proportions of offspring are 1/2:1/2 when the free ear lobe parent is heterozygous, since that parent has one-half chance of contributing an A to the offspring and one-half chance of contributing an a. The other parent, however, can pass on only an a (the chance is 1). Thus, the chance for an offspring that is Aa is $1/2 \times 1$, and the chance for one who is aa is also $1/2 \times 1$.

Although the manipulation of problems involving a single pair of genes is basically simple, it provides the essential tools needed for understanding much more complex situations involving more than one pair of alleles. The method of handling single-factor problems must be clearly understood, as it is the basis of all other mathematical predictions or analyses in genetics.

SIMULTANEOUS CONSIDERATION OF MORE THAN ONE HUMAN GENE PAIR

As Mendel was able to go on to the study of more complex crosses once he had mastered the understanding of single-factor problems, a cross involving two factors simultaneously can now be considered. To set up such a scenario, it is necessary to select another human trait known to be due to a single gene that is independent of the first example, ear lobe shape. A good example, and one that is even less well known than ear lobe shape, is taste ability for the chemical **phenylthiocarbamide (PTC)**. As in ear lobe shape, one can assign no presently recognized advantage either to having or lacking the ability to taste PTC. Given the opportunity to taste it, once ability is discovered, one would probably agree that it would be better not to possess the trait. Most **tasters** describe PTC as very bitter. The ability is impossible to disguise, since only an extremely small amount applied to the tongue will cause a noticeable reaction on the part of a taster. A **nontaster** receives no taste stimulus at all.

The ability to taste PTC is based on the presence of a dominant gene, while the nonability is due to the recessive allele. The alleles can be expressed by using the uppercase letter P for the PTC taste ability gene and lowercase letter p for the PTC nontaste ability gene. Again, the mechanism of operation is obscure. It is not known whether a chemical component of the saliva is involved, whether the development of the taste buds is responsible, whether it is based on a response in the nervous system, or whether some other factor is controlled by the gene. Nevertheless, the pattern of inheritance is exactly like the other dominant-recessive alleles seen and can be analyzed in the same fashion.

Two heterozygous taster parents have the same possibilities in terms of mathematical chance as seen previously:

$$Pp \times Pp$$
$$\downarrow$$
$$\left.\begin{matrix} 1/4\ PP \\ 1/2\ Pp \end{matrix}\right\} \quad \text{3/4 Tasters}$$
$$1/4\ pp \qquad \text{Nontasters}$$

Now it is possible to think about two independent traits inherited simultaneously within the same family. The shape of ear lobes has nothing to do with the ability to taste PTC, but both are human hereditary traits and can be considered at the same time within a single person or an entire family. As a matter of fact, by following two traits at the same time one can see that a single pattern might be derived as the product of two independent but simultaneous gene patterns.

Suppose that two parents differ in regard to these two traits. One may be homozygous dominant for ear lobe shape and PTC taste ability and the other homozygous recessive for the same two traits. The first would have free ear lobes and be able to taste PTC and the second would have attached ear lobes and not be able to taste PTC. One can apply the same shorthand techniques and analyze for the type of offspring:

$$AA\ PP \times aa\ pp$$
$$\downarrow$$
$$Aa\ Pp$$

The children of this mating would all have free ear lobes and be heterozygous for the trait, as seen previously, and they would likewise all be heterozygous tasters of PTC. This is true because there are actually two separate crosses occurring at the same time in the same family:

$$\begin{matrix} AA \times aa \\ \downarrow \\ \text{Aa} \end{matrix} \qquad \text{and} \qquad \begin{matrix} PP \times pp \\ \downarrow \\ \text{Pp} \end{matrix}$$

The situation becomes more difficult to follow if a mating involving two persons who are both heterozygous for both traits is considered. In reality it is no more complex than a single-factor problem, except two things at once are considered:

$$Aa\ Pp \times Aa\ Pp$$
$$\downarrow$$
$$?$$

To answer the question concerning the types of offspring possible in this cross and the chance for their occurrence, remember the results for the two separate crosses that are taking place independently:

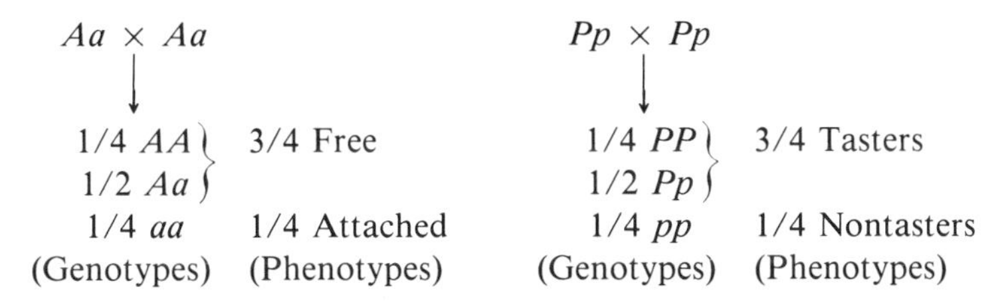

One can now predict that there can be four types of offspring when considering these two characteristics. A person with free ear lobes could either be a taster or nontaster, while someone with attached ear lobes would also be a taster or nontaster, since the traits are physically unrelated. As a matter of fact, there is a three-quarters chance for a child to have free ear lobes and three-quarters chance for the same child to be a taster. The product of the chances for these two independent events then is the chance for their simultaneous occurrence in the same individual. Mathematically the chance would be $3/4 \times 3/4$ or $9/16$.

In similar fashion a child has a chance of having free ear lobes and being a nontaster ($3/4 \times 1/4 = 3/16$), having attached ear lobes and being a taster ($1/4 \times 3/4 = 3/16$), or having attached ear lobes and being a nontaster ($1/4 \times 1/4 = 1/16$). The phenotypic proportions for this cross would be 9/16:3/16:3/16:1/16 for the four possible classes of offspring. With any two independent gene pairs showing the typical dominance-recessiveness pattern in a cross like this, one will always obtain this same ratio: 9/16 for both dominant traits, 3/16 for one dominant and one recessive trait, 3/16 for the reverse situation of one recessive and the other trait dominant, and 1/16 for both recessive traits. Of course this ratio is based on the chance or random distribution of independent genes to offspring with equal opportunities for both alleles. It exemplifies Mendel's fourth principle—independent assortment—discussed in Chapter 1.

Through similar mathematical analysis the genotypic ratio for the children can be predicted as easily. The chance for *AA* to be combined in an individual with the *PP* combination is $1/4 \times 1/4$ or $1/16$, because the chance for each of the independent genotypes to occur is 1/4. The *AA* could be combined with *Pp* (1/2) or *pp* (1/4), yielding chances of 1/8 ($1/4 \times 1/2$) for the *AA Pp* and 1/16 ($1/4 \times 1/4$) for the *AA pp* combinations.

The other genotypic combinations and their chances of occurrence are as follows:

Aa PP	$1/2 \times 1/4$	=	$1/8$
Aa Pp	$1/2 \times 1/2$	=	$1/4$
Aa pp	$1/2 \times 1/4$	=	$1/8$
aa PP	$1/4 \times 1/4$	=	$1/16$
aa Pp	$1/4 \times 1/2$	=	$1/8$
aa pp	$1/4 \times 1/4$	=	$1/16$

One might have predicted that there would be nine different genotypes among the offspring, since there are three different genotypes for each independent trait ($3 \times 3 = 9$).

A more traditional means of solving two-factor problems has been often used in the past, but it is more cumbersome, is fraught with numerous possibilities for error, and tends to induce the student to forget what is being done. Actually the method is an amplification of the **Punnett square** method demonstrated earlier for single-factor problems. At this point because of its historical importance and the preference shown for it by some students of genetics, the Punnett square method will be discussed. Keep in mind that it is not the most efficient method of solving two-factor problems.

In utilizing the Punnett square (or "checkerboard" as some call it) it is most important to determine the types of gametes which may be produced by a parent. Determination of possible gametes and their frequency is based on the fact that genes for separate traits like ear lobe shape and taste ability are independent and assort randomly. Thus, it is not difficult to understand that four gamete types are produced in equal frequency by an individual with the genotype *Aa Pp*. As a result of meiosis, gametes can carry either *A* or *a* and each of these would be joined randomly by either *P* or *p*. The four types of gametes then would be *AP, Ap, aP,* and *ap*. Because they occur with equal frequency, the chance for each is expressed as 1/4. Of course this implies that no type has any advantage over any of the other types in the fertilization process.

After the gametes have been determined, the Punnett square may be set up to determine the offspring as in a **monohybrid cross**. As before, the male gamete (sperm) types are lined up horizontally across the top of the chart and the female gamete (egg) types are lined up vertically down the left side. By transferring each sperm or egg genotype into the appropriate block in the chart, the possible types of children can be determined.

MALE ⟶ *Female* ↓	*AP*	*Ap*	*aP*	*ap*
AP	*AAPP*	*AAPp*	*AaPP*	*AaPp*
Ap	*AAPp*	*AApp*	*AaPp*	*Aapp*
aP	*AaPP*	*AaPp*	*aaPP*	*aaPp*
ap	*AaPp*	*Aapp*	*aaPp*	*aapp*

Once again, each block in the Punnett square contains the genotype of a type of offspring. To predict the character of and proportions in the phenotypes, one must translate each genotype into a phenotype and summarize the results. All the blocks in the diagram represent types that occur with equal frequency (1/16 each). There is duplication, but those having the same genotype arise in different ways. In one case a dominant gene is contributed by the female parent, in another, by the male parent. In any event a summary of the phenotypes represented will reveal the same result gained by using the **fraction method** or law of independent simultaneous events: 9/16 free ear lobes with PTC taste ability, 3/16 free ear lobes without PTC taste ability, 3/16 attached ear lobes with PTC taste ability, and 1/16 attached ear lobes without PTC taste ability.

A three-factor problem provides a more complicated analysis and illustrates the hereditary basis for the great diversity among humans. A problem involving three separate monohybrid crosses occurring simultaneously in the same individual would result in a Punnett square with 64 blocks (4 × 4 × 4). A cross of two parents heterozygous for four traits would result in an unwieldy 256 blocks! Errors in such analyses are almost assured if the Punnett square method is used.

If three factors are considered simultaneously, the fraction method may be used by considering the cross as three separate monohybrid crosses. One can add another rather innocuous human trait to those already considered. As in ear lobe shape and PTC taste ability, the presence of **hair** on the **back of the fingers** between the **first** and **second joints** seems to be due to a simple dominant allele, while the lack of hair is based on the recessive allele. (Fig. 7.2). Once again this is a trait that is likely to

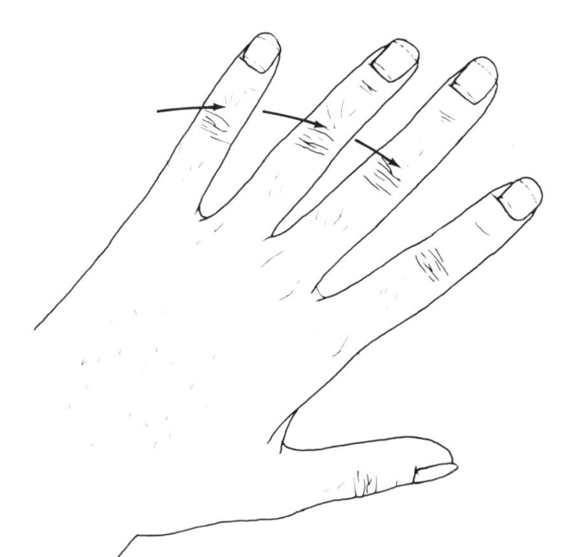

Figure 7.2. Middigital finger hair in humans. A single autosomal gene with two alleles controls the presence or absence of hair on the fingers (excluding thumb and index finger) in the middle segment (indicated by arrows). The presence of middigital hair is dominant to its absence.

be possessed by a person who has not noticed either its presence or that others may have it. Frequently, persons having the dominant gene will not show the trait because the hair is worn away due to such environmental factors as occupational use of hands. A magnifying glass, however, will usually reveal the presence of hair follicles in the middigital area. There appears to be no selective advantage for either the possession or lack of the trait.

For the shorthand symbol, one could select the letter *H*. Now a "trihybrid" cross can be written as follows:

$$Aa\ Pp\ Hh \times Aa\ Pp\ Hh$$
$$\downarrow$$
$$?$$

There are three separate monohybrid crosses occurring in this mating:

$Aa \times Aa$	$Pp \times Pp$	$Hh \times Hh$
↓	↓	↓
3/4 Free ear lobes	3/4 PTC Taste ability	3/4 Middigital hair
1/4 Attached ear lobes	1/4 PTC Nontaste ability	1/4 No middigital hair

For such a single cross involving three different genes, eight combinations are possible among the offspring:

Free lobes, taster, hair	$3/4 \times 3/4 \times 3/4 = 27/64$
Free lobes, taster, no hair	$3/4 \times 3/4 \times 1/4 = 9/64$
Free lobes, nontaster, hair	$3/4 \times 1/4 \times 3/4 = 9/64$
Free lobes, nontaster, no hair	$3/4 \times 1/4 \times 1/4 = 3/64$
Attached lobes, taster, hair	$1/4 \times 3/4 \times 3/4 = 9/64$
Attached lobes, taster, no hair	$1/4 \times 3/4 \times 1/4 = 3/64$
Attached lobes, nontaster, hair	$1/4 \times 1/4 \times 3/4 = 3/64$
Attached lobes, nontaster, no hair	$1/4 \times 1/4 \times 1/4 = 1/64$

Again, it would have been possible to predict that there are eight potential phenotypes, because there are two for ear lobe shape, two for PTC taste ability, and two for middigital hair ($2 \times 2 \times 2 = 8$). In similar fashion one should be able to predict the number of possible genotypes in this **trihybrid cross**. As in the other examples, each separate single-factor cross results in three genotypes (*AA, Aa, aa*). The three crosses considered simultaneously would be capable of producing 27 different genotypes ($3 \times 3 \times 3 = 27$).

If asked to predict the chance for an offspring in the preceeding cross to be homozygous recessive for all three traits (*aa pp hh*), the mathematical calculation is simple: $1/4 \times 1/4 \times 1/4 = 1/64$. The proportion of offspring who would have genotypes exactly like their parents for these three traits (*Aa Pp Hh*) would be $1/2 \times 1/2 \times 1/2 = 1/8$.

Careful systematic use of the fraction method in solving genetics problems requires an understanding of the basic reproduction mechanism and the transmission of genetic traits. Once this understanding is developed, it is seldom necessary to resort to the cumbersome Punnett square method. A large variety of different problems may be solved and accurate predictions may be made rapidly. The cornerstone of the method is the simple monohybrid cross. When the details of a single-factor problem are understood, one may go on to much more complex situations, always keeping in mind that there is a series of independent events occurring simultaneously in the same organism.

Figure 7.3. Facial freckles. This girl is a good example of facial freckles, a trait that appears to be caused by the presence of a single autosomal dominant gene. (Courtesy A. M. Winchester.)

To demonstrate the simplicity and usefulness of the fraction method, suppose that two other gene-determined traits are considered: **facial freckles** (Fig. 7.3) and **widow's peak** (Fig. 7.4) (both simple dominants). It is usually best to set up a sort of table for easy reference to gene symbols:

A = Free ear lobes	*P* = PTC taster	*H* = Middigital hair
a = Attached ear lobes	*p* = PTC nontaster	*h* = No middigital hair

C = Facial Freckles	*S* = Widow's peak
c = Clear skin	*s* = Straight hairline

Assume that a five-factor mating is observed:

$$Aa\ Pp\ Hh\ Cc\ Ss \times Aa\ Pp\ Hh\ Cc\ Ss$$
$$\downarrow$$
$$?$$

The chance that offspring will look like their parents (free ear lobes, PTC taster, middigital hair, freckled, and with widow's peak) is

$$3/4 \times 3/4 \times 3/4 \times 3/4 \times 3/4 = 243/1024 \text{ (a chance of about 24\%)}$$

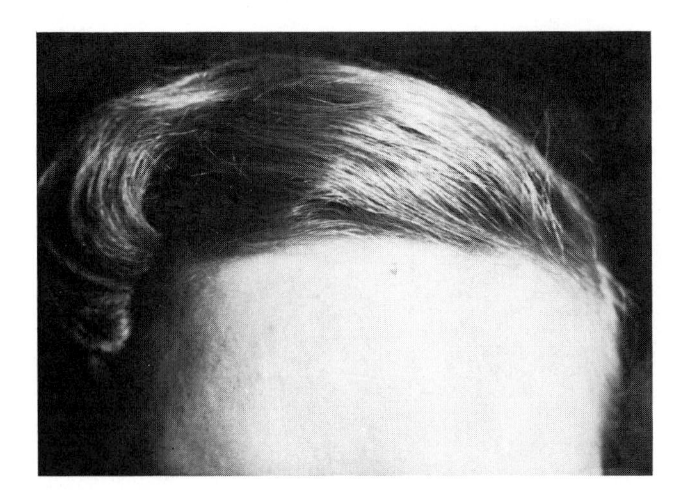

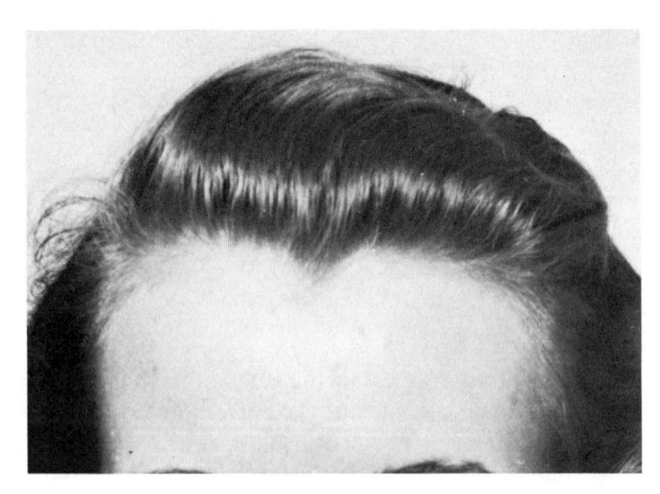

Figure 7.4. Human forehead hairline. Shown at lower left is widow's peak hairline, superstitiously thought to predict widowhood. A widow's peak seems to be caused by an autosomal gene which is dominant to that for a straight hairline, shown above. (Courtesy A. M. Winchester.)

One can even mix up the types of crosses somewhat and make predictions as effectively. Suppose that a five-factor cross such as the following is observed:

$$Aa\ Pp\ hh\ cc\ SS \times Aa\ Pp\ Hh\ Cc\ Ss$$

One parent is like the two in the previous cross, but the other has free ear lobes, PTC taste ability, no middigital hair, clear skin, and widow's peak. The chance for a child to have attached ear lobes, PTC taste ability, no middigital hair, freckles, and widow's peak can be determined as

$$1/4 \times 3/4 \times 1/2 \times 1/2 \times 1 = 3/64$$

Notice that one parent is homozygous for the gene for widow's peak and can only pass along an *S* gene to the next generation. Mathematically, it is said the chance is one. Logically, it is apparent that all children must have widow's peaks from this mating. The chance for children to have attached ear lobes, PTC taste ability, no middigital hair, freckles, *and* straight hairlines would have been zero.

Although a detailed analysis of several traits simultaneously in human genetics studies is seldom made, a consideration of more complex problems is helpful to understand the great potential for variation. Each person and each family is a distinctly individual genetic unit; considered on a global scale, definite predictable genetic patterns can be seen in all human populations. When several genes are considered simultaneously, the basis for diversity that is based on the action of thousands of genes acting together in a single person can be understood.

SOME GENERAL FACTORS INFLUENCING MENDELIAN INHERITANCE IN HUMANS

The small number of progeny in a human family makes it difficult to determine the mode of action of single genes. As seen, a recessive allele can be identified if one finds a child exhibiting the trait when his or her parents do not possess it. If this sort of pattern is identified in several families, there is no difficulty in assuming a case of dominance and recessiveness. However, it is not always as clear-cut as the patterns Mendel observed in the seven traits studied with pea plants.

Penetrance and Expressivity

Problems often arise in analyzing and understanding the mechanisms of inheritance involved in human traits. It has been found that some genes display varying levels of **penetrance**. Even though the alleles follow the principles of inheritance noted in Mendel's work, they do not always exhibit the expected phenotypic characteristics. The trait may only be seen in a fraction of those having the determining genotype. The term penetrance is thus descriptive of the incomplete effect of some genes in populations. Sometimes penetrance is expressed numerically as a percentage of the full potential. If a dominant gene were 50% penetrant, only half those having the gene would exhibit the phenotype for it. Naturally, partially penetrant genes present problems of analysis when one attempts to understand their mechanism of action.

Another phenomenon sometimes observed is classified as **expressivity**. If a gene-controlled trait is variable in its expression or phenotype from individual to individual, it is said to show variable expressivity. Some genes exhibit a reasonably constant form of expression, such as those for ear lobe shape. On the other hand, many human genes are variable in phenotype. The dominant gene for middigital hair on the fingers exhibits a variable expressivity. Hair may be found on the middle, ring, and little fingers in several combinations involving one to three fingers. The gene may even have a different expression in the right and left hands. It is usually difficult to quantitize variable expressivity, which may be due to a number of different reasons and even interactions with other genes. It may be particularly significant in genetic diseases, where the severity may vary greatly from one person to another.

Gene Interactions and Environment

Sometimes two different independent gene pairs affect a single phenotypic characteristic. Complications of analysis are almost insurmountable when this type of mechanism is involved. Some examples of such interaction are discussed in Chapter 10. There is no question that some human traits are based on the interaction of several gene pairs, presenting situations that are not easily resolved. The overall principle that all genes are operating in the presence of all other genes in the genome should always be kept in mind.

Along this same line, remember that the environment in which an organism exists is important to the function of all genes. It is particularly noticeable when several gene pairs affect the same trait. In truth the phenotype is a product of an interaction of genes and environmental factors. It is possible that a gene cannot even produce a phenotypic effect unless the proper environmental circumstances exist. In humans some genes may not produce a phenotypic response in one sex, while they are fully penetrant in the other. The age of an individual often determines whether or not a gene will function. Clearly hormones and diet, for example, are important to the function of certain genes.

Any experiments on gene action should make allowance for the effects of **environment** on gene function. Mendel's tall pea plants undoubtedly grew taller in a healthy environment than when environmental conditions were unfavorable. As students of human genetics, the importance of environment in human heredity must be recognized as well.

CODOMINANCE

There are some clear examples of genetic traits which are dependent on the principle of dominance and recessiveness; however, a common variation in the basic mechanism exists where two functional gene products are produced, as in sickle cell heterozygotes. In such cases both alleles exert an essentially equal effect on the phenotype of the organism.

Another example of this effect may be seen in the genes determining the human blood types that constitute the MN system. These characteristic blood types were discovered by Karl Landsteiner and Philip Levine in 1927. Landsteiner had previously discovered the ABO blood system in 1900 in association with medical difficulties arising in **incompatible blood transfusions**. The long period of time that had elapsed between these discoveries was because the MN types had little influence on transfusion incompatibilities. Actually, the system was discovered by matching human red blood cells to the serum from previously injected rabbits.

It is known that the red blood cells of humans carry a number of chemicals on their surface that act as antigens inside the body of another person or animal. Like the HLA antigens, these chemicals are basically protein and are under the direct control of genes. The body of an organism into which such antigens are injected responds to this "invasion" of foreign substances by manufacturing antibodies. The antibodies act specifically against the antigens and eliminate them from the system. In the case of antigens carried on the surface of red blood cells, the antibody causes a clumping together of the foreign cells as a part of the elimination process. By mixing red blood cells with serum carrying an antibody against a substance on the cell surfaces, an obvious clumping reaction can be observed.

The 1927 studies revealed that there are two antigenic substances on human red blood cells which stimulate antibody production by rabbits. These were labeled **antigens M** and **N** (Fig. 7.5). Injection of type M red cells stimulated rabbits to produce **anti-M antibodies**, while injection of type N caused **anti-N antibodies** to appear. It was discovered that some persons possessed only M antigen, some only N antigen, and others both M and N. The rabbit serum containing anti-M antibodies would clump the blood of M or MN persons, while serum containing anti-N antibodies would clump N or MN red blood cells. A rabbit could be stimulated to produce both types of antibodies by injection of MN red blood cells.

The genetic explanation for the system is based on the understanding that each antigen is under the control of a different form of the same gene. These two alleles are labeled L^{M} and L^{N} (*L* for Landsteiner). Each allele dictates the production of one of the antigens and neither allele is dominant to the other. Thus, a person who has the genotype $L^{M}L^{M}$ produces only the M antigen and is classified as type M. Likewise, the $L^{N}L^{N}$ gentoype is responsible for type N. Heterozygous individuals, $L^{M}L^{N}$, have a distinct phenotype that is a blend of the two antigens on the red blood cells and they are type MN.

It should be obvious that a single-factor cross involving heterozygous parents results in a simple modification of the 3:1 phenotypic ratio. Notice that each $L^{M}L^{N}$ parent still produces gametes carrying L^{M} and L^{N} genes with equal frequency (1/2): One-quarter of the offspring will still be $L^{M}L^{M}$ ($1/2 \times 1/2$), one-half will be $L^{M}L^{N}$ ($1/2 \times 1/2 + 1/2 \times 1/2$), and one-quarter $L^{N}L^{N}$ ($1/2 \times 1/2$). The modification is seen in the fact that each genotype results in a different phenotype. The phenotypic ratio is therefore the same as the genotypic ratio (1:2:1): One-quarter are type M, one-half are type MN, and one-quarter type N.

The situation becomes more complicated if two **codominant** pairs of genes are considered in a **dihybrid cross**. To consider such a cross, a characteristic which exhibits a similar phenotypic pattern is analyzed. Eye color in humans is often utilized as a model for dominance and recessiveness involving a single-gene pair. In fact, there are several gene pairs involved in the color determination, but there appears to be at least one pair of genes which can be labeled *B, b*. In the presence of certain other genes the *B* gene is responsible for the production of brown pigment granules in the iris, while the *b* allele is responsible for no pigment granules. The *B* is dominant to *b,* but the pair act like codominants, since a *BB* genotype produces more brown granules than a *Bb*. Thus, *BB* results in dark brown eyes, *Bb* gives light brown, and *bb* produces blue eyes.

If these two gene pairs are considered operating simultaneously in the same person, a dihybrid cross may be described ($L^{M}L^{N}$ *Bb* $\times$ $L^{M}L^{N}$ *Bb*). Once again, the

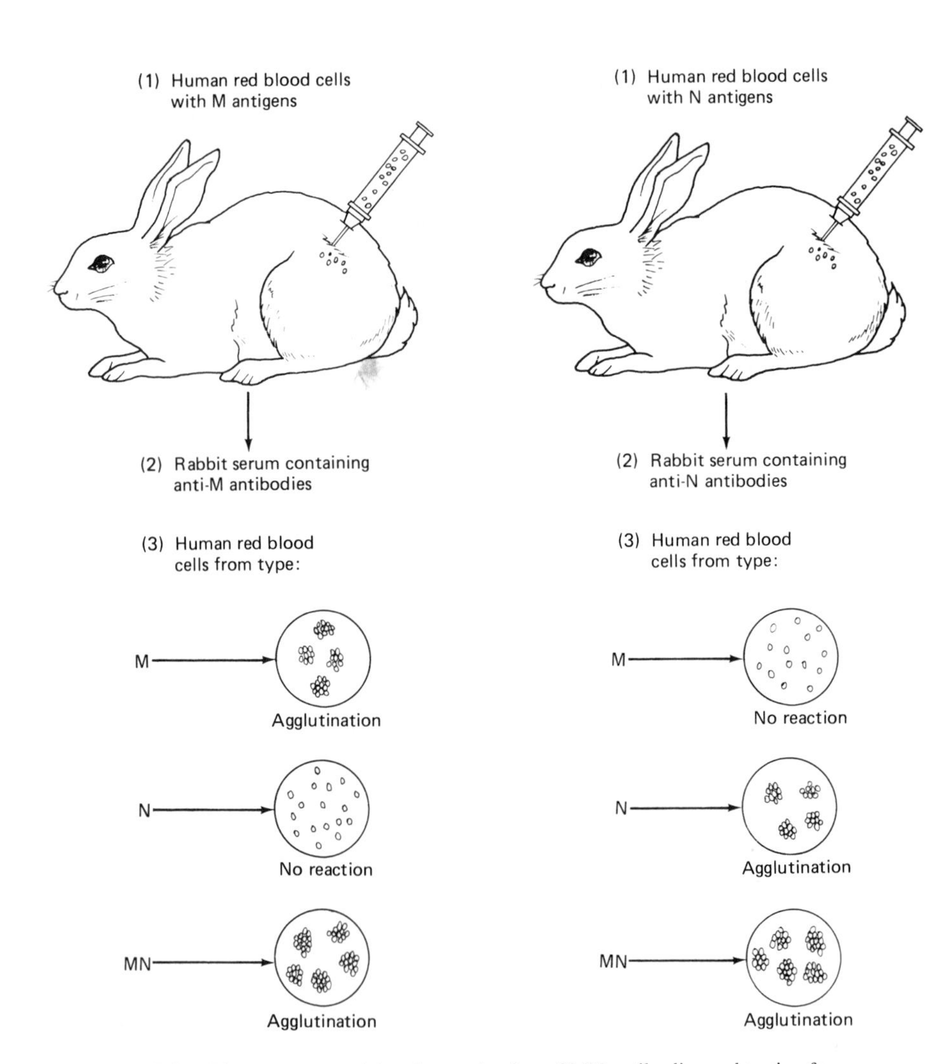

Figure 7.5. Diagram summarizing the production of MN antibodies and testing for MN blood tests. Immune antiserum formation is stimulated by injecting human M or N red blood cells into rabbits. The rabbit blood serum can subsequently be used to type human blood. Type M blood agglutinates when mixed with anti-M serum, type N agglutinates when mixed with anti-N serum, and type MN agglutinates with both.

phenotypic ratio is the same as the genotypic ratio. The two independent single-factor crosses that are taking place simultaneously can be diagrammed as follows:

$L^M L^N \times L^M L^N$	$Bb \times Bb$
↓	↓
1/4 M type blood	1/4 Dark brown eyes
1/2 MN type blood	1/2 Light brown eyes
1/4 N type blood	1/4 Blue eyes

When both blood type and eye color are considered simultaneously, there are nine different combinations possible. This is true because three different blood types are being combined with three different eye colors ($3 \times 3 = 9$). The different combinations possible and their chance of occurrence are as follows:

Genotypes	Frequency calculation	Blood type	Eye color
$L^M L^M$ *BB*	$1/4 \times 1/4 = 1/16$	M	Dark brown
$L^M L^M$ *Bb*	$1/4 \times 1/2 = 1/8$	M	Light brown
$L^M L^M$ *bb*	$1/4 \times 1/4 = 1/16$	M	Blue
$L^M L^N$ *BB*	$1/2 \times 1/4 = 1/8$	MN	Dark brown
$L^M L^N$ *Bb*	$1/2 \times 1/2 = 1/4$	MN	Light brown
$L^M L^N$ *bb*	$1/2 \times 1/4 = 1/8$	MN	Blue
$L^N L^N$ *BB*	$1/4 \times 1/4 = 1/16$	N	Dark brown
$L^N L^N$ *Bb*	$1/4 \times 1/2 = 1/8$	N	Light brown
$L^N L^N$ *bb*	$1/4 \times 1/4 = 1/16$	N	Blue

Each of the parents in the cross will produce four types of gametes ($L^M B$, $L^M b$, $L^N B$, $L^N b$). *Thus, a* **Punnett square** with 16 blocks could be constructed as before. There would be nine different phenotypes among the 16 blocks, giving the same results as in the preceding table.

The mechanism of gene transmission for these traits still operates in the characteristically predictable fashion that permitted Mendel to fathom the precise nature of the process. Now, however, complete dominance is lacking. This makes the situation seem more complex since there are more phenotypes, but the underlying principles of chance are the same. Many genes in humans show this effect; any time an inherited trait is investigated, it is essential to anticipate that a lack of dominance may modify the expected results.

Geneticists agree that given a means of measurement which is sufficiently refined, most **heterozygous organisms** could be distinguished from **homozygotes**. The use of precise measurements, such as those of chemical analysis, reveals many subtle differences that are not apparent to the naked eye. Therefore, the question of whether or not allelic genes exhibit the phenomenon of dominance and recessiveness depends on the level of analysis. What seemed obvious to Mendel's unaided eyes may not be so clear-cut when scrutinized with modern techniques. The original work in the science of heredity is still perfectly valid, but opinions must be revised concerning the absolute dominance of one gene over another.

The importance of techniques to identify heterozygotes among humans becomes obvious when defective or debilitating genes are investigated. If an analysis reveals that two parents carry a recessive gene for a lethal or seriously harmful genetic disease, decisions can be made by the parents in regard to future offspring. The risks involved in some cases will suggest alternate solutions, such as **abortion** or **adoption**.

An example of a human trait which has puzzled geneticists for many years is **thalassemia**, described in Chapter 4. At first glance the situation seems to fit into the category of traits in which dominance is lacking. Persons who do not have the disease are homozygous for the normal allele; persons who have the severe form of the anemia (thalassemia major) are homozygous for the anemia gene; and persons with the mild form of anemia (thalassemia minor) are heterozygotes.

However, closer examination revealed that the gene for hemoglobin was not defective, since any hemoglobin produced by the person with thalassemia minor was normal. A reduced amount of hemoglobin was manufactured and it was suspected that some sort of control gene was involved in a situation where one gene pair affects the manifestation of a separate gene pair (called **epistasis**). As now known, a mutation in an intron results in premature termination of transcription of the gene, seriously reducing the amount of β-chain production in the disease.

MULTIPLE ALLELES

When the mechanism of polymorphisms is considered, the possibility of numerous alleles for a single gene becomes apparent. It is interesting to recall that so-called recessive alleles can in fact be simply mutants of a gene that no longer produce functional products. The significance is that a mutation at any one of many points in the gene might result in a nonfunctional product. All such mutants might be classified as the same recessive allele. Only detailed base sequence analysis of the DNA would reveal the differences in such alleles.

On the other hand, some mutations result in alleles that produce functional products, as in the HLA system or MN blood types. In the case of another red blood cell antigen system the result is again codominance between alleles with two separate antigens under their control. The antigens in this case are classified as A and B, resulting in persons who are type A, B, or AB. This system differs from the MN system, however, in that there is a third form of the A-B gene that does not produce a functional product—neither **antigen A** nor **antigen B**.

The symbols used for the antigen-producing alleles are I^A (for A antigen) and I^B (for B antigen). The allele which produces neither antigen is clearly recessive to the other two and is symbolized with an *i*. It is important to note that here there are three alleles for the same gene. Of course, each person can carry only two alleles, regardless of what they are.

The ABO blood system was the first antigenic blood system discovered in 1900 by Landsteiner. Humans can be classified as type A (I^AI^A or I^Ai), type B (I^BI^B or I^Bi), type AB (I^AI^B), or type O (*ii*). Persons who are type A develop antibodies early in life against the B antigen. Likewise, persons who are type B produce anti-B antibodies, type O individuals develop both **anti-A** and **anti-B antibodies,** and type AB persons develop neither antibody. Such antibodies are considered to be naturally occurring because of their omnipresence, but they probably are induced during the first few weeks of life by exposure to similar antigens in the individual's environment.

The frequency of ABO blood types varies significantly from one human population to another (Table 7.1).

As a result of these antigens on the red blood cells and antibodies in the serum, attention must be given to blood types in whole blood transfusions. A blood transfusion was first utilized in 1818 but as a medical practice the procedure was considered unsafe until 1900 when the ABO mechanism was discovered. It is now known

TABLE 7.1. Approximate frequencies of ABO blood types in various populations (given as percentages).

Blood Type	Black Americans	White Americans	Navajo Indians	Japanese	British	Worldwide
O	48	45	76	30	50	32
A	27	42	24	39	39	22
B	21	9	0	22	9	39
AB	4	4	0	9	2	7

that the placement of incompatible red blood cells in a person's body during a transfusion will lead to the **clumping** of these foreign cells (Fig. 7.6). The clumped cells block vital capillaries, usually resulting in death. Donor red cells carrying type A antigen (type A or AB) would be clumped by the recipient's serum containing anti-A antibodies (type B or O). Donor red cells carrying type B antigen (type B or AB) would be clumped by the recipient's serum containing anti-B antibodies (type A or O).

The same principles of heredity can be applied to inheritance analysis of ABO blood types, except that there are now three alleles involved with a simultaneous codominance and dominance hierachy operating among the three.

Note that the principles determining inheritance of **multiple alleles,** such as ABO blood types, are the same as the basic rules utilized previously. Now the situation involves three forms of a gene instead of two. This results in more phenotypes and of course, a more difficult problem of analysis.

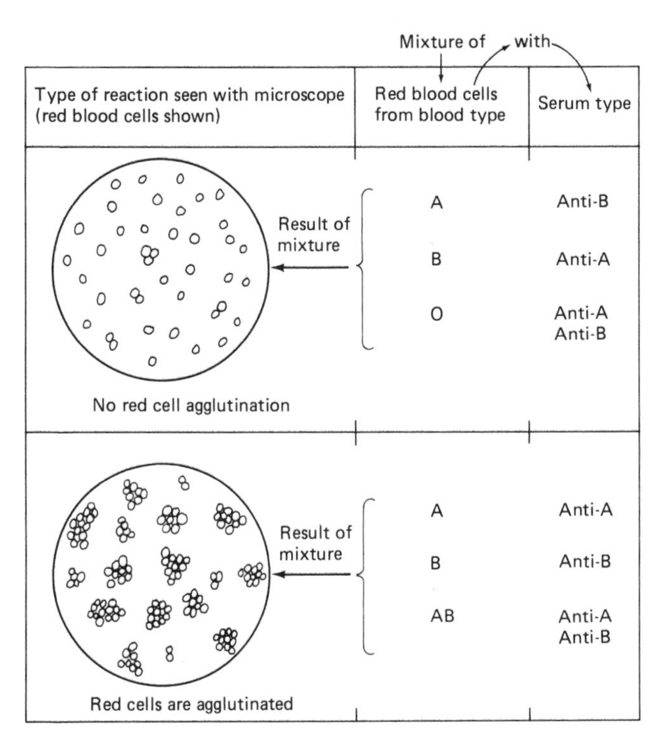

Figure 7.6. The various combinations of red blood cells and sera that may be used in determining blood types for the ABO system. Diagrams indicate the physical appearance of agglutination or nonagglutination of red blood cells. Anti-A serum occurs in blood types B and O. Anti-B serum occurs in blood types A and O.

Various types of crosses can be imagined involving blood types, but two that show the results and manner in which such crosses vary from typical Mendelian patterns are worth considering.

If two obvious monohybrid individuals (blood type AB) are mated, one could predict that the potential exists for them to have three types of children in a typical 1/4:1/2:1/4 ratio:

$$I^A I^B \times I^A I^B$$
$$\downarrow$$

1/4 $I^A I^A$	(Type A)
1/2 $I^A I^B$	(Type AB)
1/4 $I^B I^B$	(Type B)

Another type of mating involving two monohybrids with two different blood types (A and B), whose genotypes are not obvious from their phenotypes, is the only mating that has the potential of producing all four types of children:

$$I^A i \times I^B i$$
$$\downarrow$$

1/4 $I^A i$	(Type A)
1/4 $I^B i$	(Type B)
1/4 $I^A I^B$	(Type AB)
1/4 $i\ i$	(Type O)

Blood types have played a role in legal investigations, especially in identifying individuals involved in crimes who have left blood stains at the scene of a crime. Blood antigens can be identified from rather small, dried samples. As more antigens are discovered, blood types will become almost as distinctive as fingerprints in the identification of individuals.

The use of blood types in determining parentage has proved to be of some value in legal investigations. Some courts have accepted blood types as evidence in disputed parentage cases, although a pall of doubt was sometimes created for judge or jury if mutations were mentioned. The chance of a mutation from one blood type gene to another is so rare that it is practically meaningless, but it might be enough to provide reasonable doubt in the minds of some. In any event blood typing for disputed parentage situations is essentially only of value in excluding potential parents and not positively identifying them. Although blood typing is easier to perform than an analysis of HLA genes, HLA haplotypes are infinitely more revealing than red blood cell antigens in cases of disputed parentage. Combined together, the evidence is probably able to withstand any legal challenge of validity.

Rh BLOOD TYPES

An important human red blood cell antigen system presents a pattern that is difficult to analyze genetically. Although much research has been done, the genetic basis has not been clarified. One system proposed by Alexander Wiener is based on eight forms of the gene for Rh antigen in a complicated multiple allele pattern. Half of the alleles are symbolized by a form of *R* (Rh-positive alleles), while half are symbolized by an *r* (Rh-negative alleles).

The simplest plausible mechanism follows a system devised by Ronald Fisher and Robert Race. It is based on three separate pairs of alleles which are closely linked in a single group (*C-c, D-d, E-e*). Here capital letters indicate positive types and lower-case letters represent recessive negative types. The presence of the *D* allele in a genotype indicates the most common Rh-positive condition (*D* antigen), whereas *dd* indicates Rh negative (no *D* antigen on red blood cells). For the purposes of this text, it is possible to ignore the *C,c* and *E,e* alleles, and to assume that *D* is dominant to *d*.

As in the MN blood typing system, but unlike the ABO blood types, Rh antibodies do not occur naturally in blood serum unless there has been an exposure to Rh-positive red blood cells, which carry the Rh antigen. Actually, the Rh antigen was first discovered when human Rh-positive red blood cells were injected into **Rhesus monkeys** (thus, the Rh). Therefore, no transfusion problems could generally occur on the first transfusion. Even a mismatched transfusion to a **sensitized** person might not be expected to have similar severe consequences seen in ABO mismatches, since the red blood cells are **hemolyzed** or broken down, rather than clumped.

The disease aspect of the situation results when a sensitized woman is pregnant with an Rh-positive child. Note that only Rh-negative persons can develop antibodies to the Rh-positive red blood cells. Thus, a sensitized mother must be an Rh-negative woman who has received a transfusion with Rh-positive blood (unlikely) or who has had a prior pregnancy involving an Rh-positive child. In the latter case during childbirth some of the baby's Rh^+ red blood cells from the bleeding placenta may enter the mother's bloodstream and result in sensitization.

Notice that the requirements for the disease situation, called HDN (**hemolytic disease of the newborn**) or **erythroblastosis foetalis**, are a sensitized Rh^- mother (*dd*) and an Rh^+ fetus (*Dd*) (Fig. 7.7). To have an Rh^+ child, the father must also

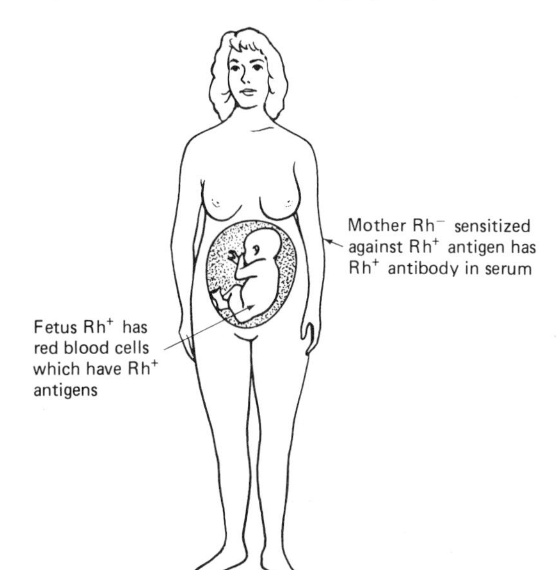

Figure 7.7. Hemolytic disease of the newborn (HDN). Conditions leading to HDN are shown (Rh^+ fetus, Rh^- mother who is sensitized against Rh^+ antigen). Rh^+ antibodies in serum will cross placenta to fetal bloodstream, causing destruction of fetal red blood cells.

be Rh$^+$. If he is homozygous for the dominant allele (*DD*), all children will be Rh$^+$. If he is heterozygous (*Dd*), there is a one-half chance that each child will be Rh$^+$.

Symptoms of HDN can be extremely serious for the fetus and infant. Often antibodies cross the placenta, resulting in such problems as the destruction by hemolysis of the red blood cells of the fetus (Fig. 7.8). In severe cases spontaneous abortions may result.

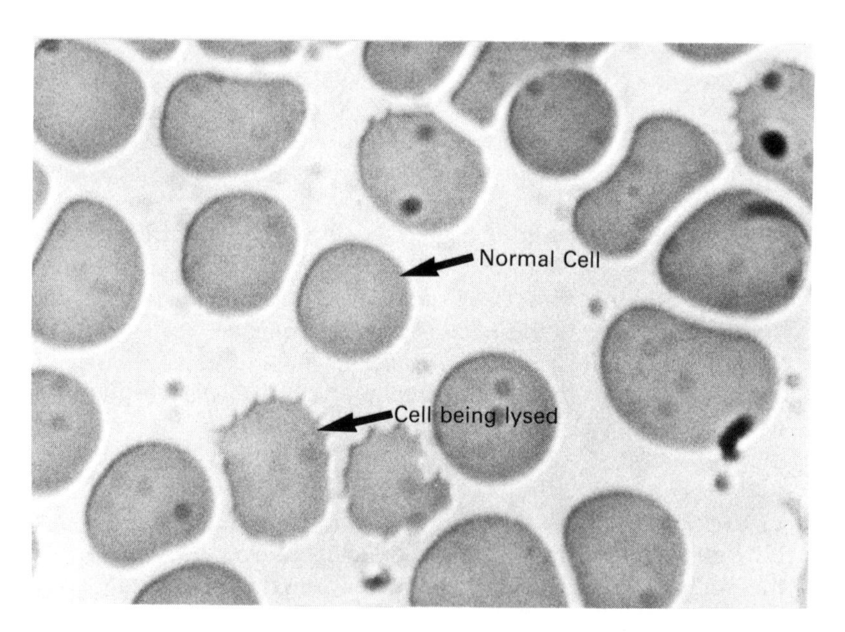

Figure 7.8. Red blood cells showing lysis due to Rh incompatibility. Light microscopic photograph shows Rh$^+$ human red blood cells that are being broken down because of exposure to Rh$^+$ antibodies.

An infant born with HDN is in great jeopardy because antibodies to its red blood cells have been passed on from the sensitized mother. The most obvious symptom is a severe anemia which continues to develop even after birth. Accompanying the anemia is a jaundiced condition resulting from an accumulation of **bilirubin** from the hemolyzed red blood cells. This yellow pigment is detrimental to brain development and can result in serious defects.

In the past the only therapeutic measures consisted of a complete removal of blood and transfusion with Rh-negative whole blood, coupled with exposure to bright fluorescent lights to destroy the bilirubin.

There is now an effective, although somewhat expensive, method of preventing the sensitization of women. If utilized for all Rh$^-$ women bearing Rh$^+$ fetuses, HDN could be completely eliminated. It is based on preventing the Rh$^+$ red blood cells from remaining in the woman's bloodstream long enough to stimulate antibody formation. An injection called **RhOGAM** must be administered within 72 hours of the birth of the Rh^+ fetus. If an abortion occurs, therapeutic or otherwise, the serum should also be used.

The RhOGAM serum is obtained from human sensitized blood containing Rh antibodies. Such serum can be obtained by injecting Rh^+ red blood cells into Rh^- males and later using them as serum donors or by utilizing the serum of already sensitized women. Its function is, of course the destruction of Rh^+ red blood cells that have entered the mother's body from the fetus during childbirth.

EPISTASIS

Another mechanism which influences the inheritance of some traits is an interaction between two separate pairs of genes. These two gene pairs are separate as to their location on the chromosomes and independent in their inheritance. A type of interaction between the pairs in which one gene has a masking or overshadowing effect on the other is known as **epistasis** and differs from dominance, since separate gene pairs are involved.

Human genetics does not contain a large number of situations that clearly demonstrate the phenomenon of epistasis. Probably many exist that have not been discovered. There are a few situations that easily fit the definition. Consideration of these situations not only helps to explain some complexities of human heredity, but also helps one to understand better the true nature of gene action.

Two men receiving blood transfusions after injuries in Bombay, India, in 1952 presented anomalous situations. Even though they were classified as type O, because their red blood cells did not carry the A or B antigens, the antibodies of their sera were not the same as those of O type persons.

Remember that the sera of type O persons carries both antibodies A and B. The blood of these men did not. It was discovered that a single precursor chemical was necessary for the production of both antigens A and B and, if it were lacking, neither A nor B substance could be produced. In its absence then there were no A or B antigens, as in type O persons. This precursor chemical also acts as an antigen and is considered to be under the control of a gene labeled *H*. The allele *H* is therefore responsible for the production of the precursor chemical and is dominant to allele *h,* which acts as a recessive mutant, unable to dictate the production of the precursor to the A and B antigens.

Usually persons have the *H* gene and therefore the precursor for antigens A and B is manufactured. When a person has the genotype *hh,* however, no precursor chemical is produced. Even though such a person might have the I^A or I^B genes, the A or B antigens would not be manufactured. The blood type resulting is classified as the **Bombay blood type**, due to its discovery in that city. It is like type O in the lack of A and B antigens, but it obviously is of different origin. Bombay type persons can be distinguished from type O persons, since they carry an antibody against the precursor chemical in their sera and type O persons do not.

Based on this dependence on a precursor, the *h* gene can be classified as being epistatic to the I^A and I^B genes. An interesting variation of a routine cross would be anticipated if persons heterozygous for the gene were mated:

$$\begin{array}{ccc} \text{Type A} & & \text{Type A} \\ I^A i\ \text{Hh} & \times & I^A i\ Hh \\ & \downarrow & \end{array}$$

$$\begin{aligned} 3/4\ I^A \times 3/4\ H &= 9/16 \text{ Type A} \\ 3/4\ I^A \times 1/4\ hh &= 3/16 \text{ Bombay type} \\ 1/4\ ii \times 3/4\ H &= 3/16 \text{ Type O} \\ 1/4\ ii \times 1/4\ hh &= 1/16 \text{ Bombay type} \end{aligned}$$

The familiar 9:3:3:1 ratio is modified so that a 9:4:3 ratio results. If extraordinary care were not exercised in the typing, the Bombay type would be identified as type O and a strange 9:7 ratio would result.

Another epistatic situation can be found in the production of **melanin pigment** in the body. Melanin pigment is responsible for all black or brown pigmentation in the body. A mutant form of a single gene is known to cause the condition called **albinism.** This trait is one originally proposed by Archibald Garrod as an inborn error of metabolism, and it has been shown to be based on the lack of an enzyme. Figure 8.1 on page 203 shows where the mutation and enzyme absence fit into the biochemical pathway involving phenylalanine and tyrosine.

Even though adequate amounts of a precursor chemical are available in a person's body, without the *A* gene no enzyme and therefore no melanin pigment is produced. As a result, regardless of the number of black skin genes, dark hair genes, or brown eye genes, if a person has the *aa* genotype, none of these traits can be exhibited, since they all require the lacking pigment to present their effect. Again, the recessive *a* gene is epistatic to these other pigmentation genes.

An interesting example of a gene which affects the phenotypic manifestation of another gene is the **secretor trait**. Although on first consideration it appears to be a case of epistasis, this is a difficult categorization. There is no interference or assistance in the production of a gene product, simply an action on the physical nature and location of the product.

The ABO blood type antigens occur on cell types other than red blood cells and in some cases may be present in the body fluids. Apparently the secretor gene, a simple dominant gene *(Se),* causes the formation of a water-soluble form of the antigens. Thus the A or B substances may be found readily in the saliva, as well as other such fluids. Secretors have the genotype *SeSe* or *Sese,* while nonsecretors (about 25% of the population) have the genotype *sese.* No particular advantage or disadvantage seems to be associated with the trait, but it is useful in identification procedures, as are the blood types.

SEX LINKAGE

When one considers that the X chromosome carries a number of genes responsible for significant human characteristics, it is obvious that such genes represent an important exception to classical Mendelian principles. This is true because females possess two blocks of X-chromosome genes, while males have only one. Even though one X chromosome is inactivated in each cell, the female's cells do not all have the same inactivated X chromosomes and the phenotype represents a mixture of the two blocks of genes. The male, on the other hand, truly represents a **monosomic** situation for the X-chromosome genes. Males are said to be **hemizygous** for these factors (since they cannot be homozygous or heterozygous). The male receives an X chromosome from his mother and a Y chromosome from his father. Thus, there is an obvious imbalance in the source of genes carried on the sex chromosomes.

Males carry only one X chromosome and the genes it contains are not matched by genes on the Y chromosome. Because of this, recessive genes on the X chromosome are exhibited more frequently among males than females. There is a greater opportunity among females for an X-linked gene to be overshadowed by a dominant allele. In the male there is only one of each X-linked gene and no chance for a recessive allele to be masked by a dominant allele. Genes carried on the X chromosome are said to be sex-linked because of an inheritance pattern which is associated with the sex of individuals. Such traits are seen more frequently among males and the pattern of inheritance follows a general line from grandfather to mother to son. Genes carried on chromosomes other than sex chromosomes (autosomes) are called autosomal genes.

G6PD Deficiency

As seen, the gene which controls the production of the enzyme glucose-6-phosphate dehydrogenase is physically located on the X chromosome. Therefore, it is a sex-linked gene and follows the pattern of inheritance typical for such traits. The anemia caused by exposure to such agents as primaquine, napthalene, sulfa drugs, and fava beans is sometimes called favism. Breathing air containing fava bean pollen can cause the anemic response. It is found most often among persons whose ancestors are from the Mediterranean area, but about 10% of American blacks show the trait. In Chapter 4 the Mediterranean area and some parts of Africa were noted as high malaria-risk areas.

The phenotype may be tested for by adding one of the offending agents to a blood sample and observing it to identify the possible breakdown (hemolysis) of red blood cells. A deficiency of G6PD does not have any unfavorable effects until a person eats fava beans or uses primaquine or sulfa drugs as medication. Without exposure to such agents, a person usually does not even know that the enzyme is lacking. Another agent that can precipitate anemia is napthalene, a popular component of mothballs. The defect is generally not fatal if the offending agent is removed from an affected person's environment.

The gene generally is found at a higher frequency among populations from malaria-risk areas. Even though this gene does not produce an aberrant hemoglobin, the hemolysis that occurs under some circumstances may be associated with a partial immunity to malaria. It is interesting that fava beans (broad beans) are popular as a food source in some of the same Mediterranean areas in which the gene is most prevalent. In these areas prolonged diets or the inhalation of pollen from the fields of plants has caused bouts of untreatable anemia before the source was recognized.

Obviously the treatment for the disease is simple. The offending agent needs to be identified in a person's environment and removed. Further problems are prevented by taking care to avoid exposure to such agents.

More men exhibit favism in a population than women. The mutated gene *(g)* is rare and recessive to the more prevalent dominant allele *(G)*. Because females carry two X chromosomes, a dominant gene will likely be present to overshadow the effects of a recessive gene, even though the dominant gene will be inactivated in many cells of the body. Males, on the other hand, are hemizygous and the enzyme will not be produced in the cells of an individual with a single *g* gene.

Because males are conceived by the fertilization of an egg carrying an X chromosome and by a sperm carrying a Y chromosome, the pattern of sex-linked inheritance can be worked out. Since all normal eggs carry an X chromosome, a mother passes along a block of sex-linked genes to all her children. On the other hand, half the sperm cells carry an X chromosome and half carry a Y. Thus, approximately half the children produced are female, receiving an X from each parent. The other half are males, since they get a Y chromosome from the sperm along with the X from the egg. This is not strictly true, since there is a better than one-half chance for the conception and birth of males.

If, for example, a male carries the recessive allele for favism, his genotype can be written X^gY. It is apparent that the *g* allele will not be passed on by the father to his sons, since only the Y chromosome is passed to sons. On the other hand, all his daughters will receive the *g* gene. If a man does not have the favism gene (X^GY), none of his daughters will have the disease, since they will all receive an X chromosome carrying the *G* allele from him.

Males who have favism have received the *g* allele from their mothers, since that is the source of their X chromosome. It should be apparent that half the sons of a heterozygous woman (X^GX^g) will have favism since there is a one-half chance of receiving the X chromosome carrying the *g* gene from their mother. If a woman has favism, it follows that all her sons will be lacking in the G6PD enzyme.

Such a sex-linked recessive allele may be passed from mother to son or mother to daughter, but a man may pass the gene along only to his daughters. A recessive sex-linked gene thus may be passed along from father to daughter, where it is often masked by a dominant allele. In the next generation the recessive gene may then be exhibited among the grandsons of the first man. This passage from grandfather to grandson will become more obvious in Chapter 9 when pedigrees are studied.

Colorblindness

Probably the most commonly encountered and familiar example of a sex-linked gene is red-green **colorblindness**. It seems that everyone knows someone who is colorblind and most colorblind persons are males. Because of its association with gender, it is easy to see why the term sex linkage is used, but keep in mind that the term, as it is used in this text, refers specifically to genes carried on the X chromosome. (Some texts include Y-linked genes under the same heading.) Some other gene effects that correlate with the sex of individuals, but which are not carried on the X chromosome, will be discussed later.

Red-green colorblindness is a sex-linked characteristic, but it is not the simple one-gene trait originally postulated. The physiology of color vision will not be pursued in depth, but a simple understanding will help to explain the mechanism of gene control.

Color vision in the human eye is due to light receptors called **cones** (because of the shape of the cells). These light receptors are normally of three types, based on the light-sensitive chemical contained in each type. There are red-, green-, and blue-sensitive cones and each kind contains a particular pigment that is "bleached" when exposed to the proper color of light. This stimulates an impulse to the brain which one learns to associate with the name of the color. Thus, red light bleaches the substance in red-sensitive cones, green light bleaches that in green-sensitive cones, and blue light affects only the blue-sensitive cones.

As in all proteinaceous substances produced by living cells, these pigments are under specific gene control. If a defect in the genes should occur, a protein might be produced which does not respond to the proper wavelength of light. It is rare to find the blue-sensitive pigment defective. More often the genes for either green- or red-sensitive pigments are defective, resulting in incomplete color vision.

If the substance for green vision is produced by a recessive mutation, a form of the light-sensitive pigment may be produced that responds to red light almost as effectively as it does to green. In fact, the person would be green-blind and could not effectively distinguish red from green.

Likewise, if the gene for red-sensitive pigment were mutated, the substance for red light might respond about as well to green, sending an impulse to the brain when either red or green light was viewed. Again, the person would find it difficult to distinguish red from green, since the eye would respond essentially the same to both. Technically such a person would be red-blind because the red pigment would be the defective one.

By definition, red-blind persons are said to suffer from **protanopia**, while green-blind persons have **deuteranopia**. Such persons are often classified as having the protan and deutan type of red-green colorblindness. In the United States about 8% of all males are red-green colorblind (about 5% deutan type, 3% protan type), while only about one-half of 1% of females are red-green colorblind.

One might expect that the genes for these pigments are **polymorphic** and the situation is even more complex than stated. Some persons show a partial red-green colorblindness because the pigments in the cones will respond to intense colors, but fail to detect differences in lighter or pastel shades. Thus, conditions known as **protanomaly** and **deuteranomaly** have been described.

For convenience in understanding the sex-linked basis of colorblindness, consider red-green colorblindness to be a single trait, controlled by a single gene. Remember, however, that this is not true. While deutan and protan types cause the same effect, the two types can be differentiated by proper tests. In spite of this, both genes are located on the X chromosome and it is convenient in working hypothetical problems to refer to colorblindness as a single sex-linked phenotypic effect. Predictions can generally be made in families using this method, but it is probably wise to identify the particular mutant involved in actual analyses.

As an example of the inheritance pattern seen in red-green colorblindness, assume that a man is colorblind and his wife is not. The genotype of a man who exhibits a phenotype for a sex-linked trait is immediately known because of his hemizygous situation. The man in this case carries the recessive gene on his single X chromosome (X^c). The woman, on the other hand, can be homozygous ($X^C X^C$) *or heterozygous* ($X^C X^c$). If the woman is homozygous for the dominant trait, she cannot produce any colorblind children, even though all daughters would receive a *c* allele from the father. If she is heterozygous, one-half the children would be expected to be colorblind, since all girls would receive the recessive allele from the man and all boys would receive a Y chromosome. Half the children would get a dominant gene from the mother and half would receive a recessive.

If the father were not colorblind, none of his daughters could be, since they would all receive the dominant allele for color vision on their X chromosome from him.

HOLANDRIC GENES

As stated, the Y chromosome appears to be of little significance to the genes that are essential to life. Although females get along well without any Y-linked genes, the significance of any such genes may only be realized in the environment of the male body. It seems obvious that there are potent male-determining genes which must be present and functioning at an early fetal stage to program the formation of testes. Once this has occurred and male hormones are produced, the other genes affiliated with secondary sex characteristics on other chromosomes can begin to function.

Because the Y chromsome is at least structurally similar to other chromosomes, there might well be genes affecting some characteristics which are not essential to life and found only in males. Such genes would be found wholly among males and thus labeled **holandric genes**. Genes for **histocompatibility antigens (H-Y)** have been discovered on the human Y chromosome and there is a strong likelihood that these and the sex-determining genes are one and the same. No other well-documented and confirmed holandric genes have been demonstrated.

One example of a proposed holandric gene is referred to as **hairy ear pinnae**. In this trait fairly long hair is observed to grow on the edges of the ear pinnae (Fig. 7.9). The trait seems to be passed along only from father to son, but so few cases have been found that the analysis is tentative. With productive new hybrid cell chromosome techniques, if any functional genes are there, they will be found.

There are no sound examples to demonstrate that genes other than male determiners exist on the Y chromosome. Some investigators anticipate finding Y-linked behavioral patterns and they are pursuing the study of the purported association between the XYY genome and crimes of violence.

To complicate further the situation, it appears that over a short region of the Y chromosome there is homology with the X chromosome, allowing synapsis to occur during meiosis. However, no good examples of homologous genes in human X-Y pairs have been demonstrated. It is likely that such an observation will only be

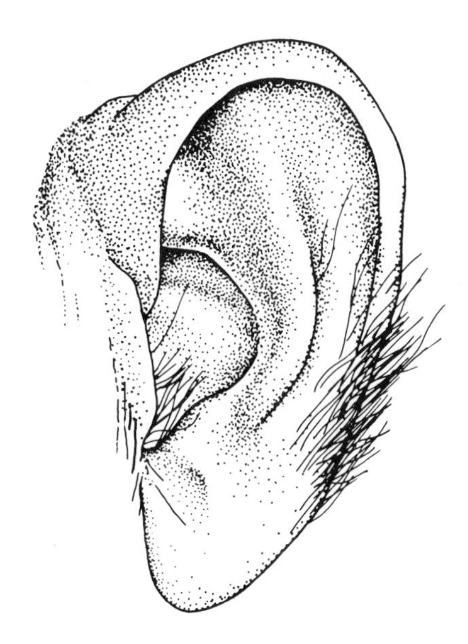

Figure 7.9. Hairy ear pinnae (hypertrichosis). Several inheritance patterns have indicated this to be a dominant holandric trait (carried on the Y chromosome). (Courtesy P. A. Moody and W. W. Norton & Co., Inc.)

made through clear-out cases of linkage with other genes in the X- and Y-linkage groups or through careful studies of gene products. This will surely be a difficult pattern to demonstrate but, as the human chromosome map becomes more detailed, homologies between the X and Y might be found.

SOMATIC CHROMOSOME-LINKED GENES SHOWING SEX DIFFERENCES

It must seem apparent that some human traits tend to be associated with gender but their genes are not physically located on the sex chromosomes. Because sex linkage is defined as applying only to genes carried by the X chromosome, the same term should not be used for genes otherwise located.

Sex-Influenced Genes

In the determination of their phenotypic effects some genes seem to respond differently to the environment of the cells of males and females. As a matter of fact, some genes give the appearance of acting completely opposite in the two sexes. In one sex two alleles are required to produce a given effect, while in the other sex only one of the same allele is required to be exhibited in the phenotype. This means that the allele is recessive in one sex and dominant in the other. Such traits are said to be **sex-influenced.**

The presence of certain genes in the environment of the male hormones seems to permit these genes to be exhibited as if they were dominant. Although considerably more research is needed to demonstrate the validity of the observations, two traits have been proposed as examples of genes dominant in males and recessive in females. Both **gout** and **pattern baldness** are much rarer in females than males, while clearly not carried by the X chromosome. An analysis of either situation presents a superficially simple task, but the problem becomes very confusing if one is not careful to analyze each sex separately.

Males need only one gene to develop pattern baldness (homozygous or heterozygous, thus a dominant trait), whereas females require two (homozygous for the baldness gene, which is now recessive). The father of a woman showing the trait would thus be bald. Unfortunately for the purpose of analyses, such traits probably also require the presence of other genes in the genome. The mode of inheritance is not understood well enough to attempt to determine whether or not such genes are fully penetrant under all conditions. The situation of analysis is like many other human traits, where it is difficult to think of only one gene as the sole determiner of a characteristic.

Sex-Limited Genes

Some genes will only exhibit their effects in one sex or the other, even though they may follow a classical Mendelian pattern and be transmitted from one generation to the next by either parent. It is obvious that there is no way to demonstrate genes for high milk production in cattle among males, yet the trait may be passed along by the bull to his offspring. Such an animal is clearly an asset in milk cattle breeding programs. Likewise, any of the genes responsible for secondary sex characteristics, such as **breast development** and **prostate gland formation,** are active in only one sex. Genes of this type, carried on the autosomes, result in the manifestation of traits that are called **sex-limited traits.**

MENDEL'S HERITAGE TO THE STUDY OF HUMAN GENETICS

The transmission of genes from generation to generation in humans is a chance process, but statistically a highly predictable one. An understanding of the use of simple mathematics to calculate in Mendelian fashion is the basis for all genetic analysis. Sophisticated understanding of gene action and the counseling of individuals regarding the inheritance of detrimental genes rely on this foundation.

Having said this, it is important to point out that it is not possible to relax in the knowledge that an understanding of Mendel's principles will provide a thorough knowledge of all aspects of human heredity. The basic patterns are there and often the classic ratios will serve to predict accurately. However, life is never simple and the fundamental principles are the beginning of understanding. They must serve as a start in all analyses, but one should anticipate other processes at work in the hereditary mechanism.

ADDITIONAL READING

AUERBACH, C., *The Science of Genetics*. New York: Harper & Row, Pub., 1961.

BARR, M. L., "Sex Chromatin and Phenotype in Man," *Science* (1959), 130, 679–85.

———, "The Sex Chromosomes in Evolution and in Medicine," *The Can. Med. Assoc.* (1966), 95, 1137–48. (Available in Birth Defects Reprint Series, The National Foundation—March of Dimes.)

BOYER, S. H., IV (ed.), *Papers on Human Genetics*. Englewood Cliffs, N.J.: Prentice-Hall, 1963.

DAVIDSON, R. G., "The Lyon Hypothesis," *The Journal of Pediatrics* (1964), 65, 765–75. (Available in Birth Defects Reprint Series, The National Foundation—March of Dimes.)

MANGE, A. P., and E. J. MANGE, *Genetics: Human Aspects*. Philadelphia: Saunders College/Holt, Rinehart, & Winston, 1980.

MCKUSICK, V. A., *Human Genetics* (2nd ed.). Englewood Cliffs, N.J.: Prentice-Hall, 1969.

———, *Mendelian Inheritance in Man: Catalogs of Autosomal Dominant, Autosomal Recessive, and X-Linked Phenotypes* (5th ed.). Baltimore: Johns Hopkins Univ. Press, 1978.

PAI, A, and H. MARCUS-ROBERTS, *Genetics: Its Concepts and Implications,* Englewood Cliffs, N.J.: Prentice-Hall, 1981.

RACE R. R., and R. SANGER, *Blood Groups in Man* (6th ed.). Oxford, England: Blackwell Scientific Publications, 1975.

REISFELD, R. A., and B. D. KAHAN, "Markers of Biological Individuality," *Scientific American* (1972), 226, No. 6 28–37.

ROTHWELL, N. V., *Human Genetics*. Englewood Cliffs, N.J.: Prentice-Hall, 1977.

Science (1981), 211, No. 4488. This March 20, 1981, issue is devoted to sexual dimorphism with several relevant papers.

VOLPE, E. P., *Man, Nature, and Society*. Dubuque, Ia.: Wm. C. Brown, 1975.

WINCHESTER, A. M., *Genetics: A Survey of the Principles of Heredity* (5th ed.). Boston: Houghton Mifflin Co., 1977.

———, *Heredity: An Introduction to Genetics* (College Outline Series) (3rd ed.). New York: Barnes & Noble, 1977.

REVIEW QUESTIONS

1. State reasons for the idea that chance is the basic underlying factor on which Mendel's principles are based. In what way is chance a factor in the genetic mechanism associated with sexual reproduction?
2. Explain how tossing a nickel and a penny at the same time could serve as a model of Mendel's principles of independent assortment. List the different results that could be obtained in such tosses. What is the chance for each type to occur? Assume that a nickel, penny, and dime are tossed at the same time. List the different possible results for this toss and tell the chance for each type. If you tossed three identical pennies and could not distinguish one from the other, what would be the chance of getting two heads and one tail?

3. Using the alleles for ear lobe shape (A, a) and PTC taste ability (P, p), set up a Punnett square for the cross

$$Aa\ Pp \times Aa\ pp$$

Does it make any difference whether the male gametes are lined up horizontally across the top as specified in the text, or vertically down the left side as directed for female gametes? How many blocks does your checkerboard have? If your answer is 16, would eight have been satisfactory? If your answer is eight, defend this number against those who utilize 16. Give the genotypic and phenotypic ratios for all classes of offspring represented in your chart.

4. Using the allele pairs Aa, and Bb, show graphically how the mathematical equation $3 \times 3 = 9$ (p. 170) represents the mechanism by which nine different genotypes may be produced in the cross $Aa\ Bb \times Aa\ Bb$. What do the 3's represent? Why is one 3 multiplied by the other? If you add a third allele pair, Cc, how many genotypes could be found among the offspring of $Aa\ Bb\ Cc \times Aa\ Bb\ Cc$?

5. *Note*: A trait has been discovered in humans which allows some persons to bend the terminal segment of one or both of their thumbs back almost to a right angle with the basal thumb segment. This hyperextensibility of the distal thumb segment is due to the homozygous presence of a simple recessive gene.

   If a man who cannot bend either of his thumbs back at a sharp angle is married to a woman who also does not show the hyperextensibility trait, what is the chance that their first child will show the thumb-bending ability? Both the man's and woman's fathers possess the hyperextensiblity trait.

6. *Note*: Another characteristic of human hands, although rather rare, has been shown to have a dominant genetical basis. This is called *crooked little fingers* because the last two segments of the "pinkies" are bent in rather obviously toward the ring fingers when the hands are placed flat on a table.

   If a man with crooked little fingers had a mother who did not show the trait, what was his father's phenotype? Can you give his mother's genotype with certainty? Can you give his father's genotype with certainty? Suppose the man has a sister who does not have crooked little fingers. Does this provide any helpful data for the analysis of the genotypes?

7. *Note: Brachydactly* is a trait found in humans which consists of shortened stubby fingers. It is rather uncommon but relatively easy to identify.

   If a man and a woman both exhibit the brachydactly trait and their firstborn child has fingers of normal length and shape would you predict that the trait is dominant or recessive? What are the genotypes of the two parents and the child? What is the chance that a second child will be brachydactylous? What is the chance that this second child will have normal fingers?

8. *Note:* The presence of hair on the back of any or all the fingers in the middle segment is an inherited trait which may be easily observed. While this *middigital hair* may be subject to loss due to environmental or occupational factors, the trait can still be identified by searching for hair follicles with a hand lens.

   After observing a large number of families having a child with middigital hair, and always finding that at least one parent showed the same characteristic, would you be able to propose a mechanism for the inheritance of the trait? If so, what would you propose the mode of inheritance to be? Based on this mode, determine the probable genotypes for the family that has one parent and a child with middigital hair and one parent without. Would it be possible for this same type of family (one parent and one child showing the trait and the other parent without the middigital hair) to have a child without middigital hair? What would be all the genotypes? What would be the probability for each type of child to be born?

9. *Note:* A trait known as *widow's peak* may be easily determined by examining a person's hairline across the forehead and looking for a pointed projection of hair.

   If a man and woman both have widow's peaks, but have a child without such a hairline, what are the probable genotypes of the family members? What is the probability that their next child will have a widow's peak? In looking through a photo album the woman notices that her father did not have a widow's peak. What information does this give about her mother?

10. If a man is heterozygous for the free or attached ear lobe alleles, has straight little fingers, and is not able to taste PTC, what is his genotype for these three traits? Suppose he is married to a woman with attached ear lobes who is heterozygous for PTC taste ability and heterozygous for crooked little fingers. What is the chance that their first child will have free ear lobes, crooked little fingers, and be able to taste PTC?

11. Using the following symbols, give the phenotypes of a man whose genotype is *Aa Tt mm Ss* and of a woman who is *aa tt Mm Ss*. What is the chance that they could have a child with the genotype *aa Tt mm SS?*

A	Free ear lobes	*T*	Taster of PTC
a	Attached ear lobes	*t*	Nontaster of PTC
M	Middigital hair present	*S*	Crooked little fingers
m	Middigital hair absent	*s*	Straight little fingers

12. Using the symbols in the preceding problem, consider the following cross:

$$Aa\ TT\ mm\ Ss \times aa\ tt\ Mm\ Ss$$

   What is the chance that a child will be born who has free ear lobes, is a nontaster of PTC, has no middigital hair, and has crooked little fingers?

13. Assume that a man has type 0, MN blood and a woman has type AB, N. Complete the following chart by placing a + in each blank where a reaction to an antiserum would occur (the serum containing the shown antibody would cause the person's red blood cells to clump). Place a 0 where no clumping reaction would occur.

Serum containing antibody:

	Anti-M	*Anti-N*	*Anti-A*	*Anti-B*
Man	_____	_____	_____	_____
Woman	_____	_____	_____	_____

   What phenotypes of children could these two produce? What would be the chance for each type?

14. Assume that the three alleles controlling ABO blood types are symbolized I^A, I^B, and i. I^A and I^B are codominants for antigens A and B, while i allele is recessive to both I^A and I^B, resulting in no antigen production. What are the possible blood types of children in the following matings, and what is the chance for each type to occur?

   (a) AB × O **(b)** $I^Ai \times I^Bi$ **(c)** $I^Bi \times$ AB

15. **(a)** If a woman of type O blood has a child who is type B, what would be the possible blood types of the father?

(b) If a woman has type B blood and has a child who is type AB, can any of three men with the following blood types be excluded as the possible father?

(1) O (2) A (3) AB

(c) By adding MN blood types to part (b), can the choice be narrowed of the possible father?

Mother: Type B, N
Child: Type AB, N
Possible Father: (1) O, N (2) A, MN (3) AB, M

16. Assume that four children are born in a hospital at about the same time to four different sets of parents. If their identification tags were lost, could you assign these four children to the proper sets of parents?

Children: (1) Type A (2) Type B (3) Type AB (4) Type O

Parents:	Mother		Father	Child?
(a)	A	x	O	_____
(b)	O	x	O	_____
(c)	AB	x	O	_____
(d)	A	x	AB	_____

17. Recall that a trait in humans which is due to a single dominant gene is the ability to taste PTC. The alleles can be symbolized with *P* for PTC taste ability and *p* for inability to taste PTC. Assume that, under the conditions of the cross, dark brown eyes are due to the homozygous presence of the one allele (*BB*), blue eyes are due to the homozygous presence of the other allele (*bb*), and the heterozygous condition results in light brown eyes (*Bb*).

If a couple have the gentoypes *PpBB* (father) and *PpBb* (mother), what phenotypes of children can they produce and what is the chance for each type? What are the phenotypes of the two parents?

18. Piebald spotting is a condition seen in some persons who have patches of hair and skin on their bodies from which pigment is totally lacking. The piebald trait is due to a single dominant gene (*S*). Normally pigmented persons have the genotype *ss*. Albinism, the total lack of pigment in skin, hair, and eyes, is due to a homozygous recessive condition for a single, rare gene (*aa*). Normally pigmented persons carry the *A* allele. Note that the *aa* has the capability of masking the presence of the *S* gene.

If a man with the genotype *AaSs* is mated to a woman with the same genotype (*AaSs*), what phenotypes would be possible among their offspring? What chance would exist for each phenotype?

19. Assume that the dominant gene *Se* (secretor) is responsible for the presence of soluble ABO blood antigens in the saliva. The recessive form (*se*) does not result in soluble blood antigens. In the cross I^AI^B *Sese* × I^Ai *Sese*

(a) What chance is there for a child to be born who will have type A antigens secreted in the saliva?

(b) What is the chance for a child with type B antigens secreted in the saliva?

(c) What is the chance that a child will be a secretor for both A and B antigens?

20. Assume that an Rh-negative woman is married to an Rh-positive man and their first child is Rh negative (considering only the *D* locus). Give the genotypes for both parents and child. What is the chance that their next child will be Rh positive? If the second child is Rh positive, is there a chance that they are at risk for a case of hemolytic disease of the newborn in subsequent children? If so, can you suggest a method of prevention?

21. Assume that a man has been identified as having type O blood on the basis of red blood cell coagulation by antisera. On further analysis it is observed that his blood serum does not contain either anti-A or anti-B antibodies. What blood type is indicated by the serum analysis? Speculate about what is causing this anomaly. Can you suggest a blood test that will verify his correct blood type?

22. Assume that eye color is controlled by a single pair of genes that operates as indicated in Question 17 and that the ABO blood types are controlled by three alleles as described in Question 14. Consider the *D* gene to be dominant to *d* and to result in Rh-positive blood type. If a couple have the genotypes $dd\ I^{A}I^{B}\ Bb \times Dd\ ii\ bb$, what are their phenotypes? What types of children can they produce in regard to blood types and eye color? What is the chance for each type?

23. An antigen on the red blood cells labeled Xg^{a} has been found to be sex-linked. Persons having this antigen are classified as Xg^{a} positive, while those lacking the antigen are Xg^{a} negative. The allele for the presence of the antigen (dominant) is symbolized Xg^{a} and the allele for its absence is *Xg*.

In the following matings what types of children would be possible in regard to the sex of the child and the Xg blood type? What is the probability for each type to be born?

	Mother		*Father*
(a)	$Xg^{a}Xg$	×	*Xg* ____
(b)	*Xg Xg*	×	Xg^{a} ____
(c)	$Xg^{a}Xg$	×	Xg^{a} ____

24. Assume that the dominant allele for G6PD production is symbolized *G* and the recessive allele for lack of the enzyme is indicated by *g*. In which of the following families would a possibility of favism (anemia resulting from eating fava beans) exist? What is the chance for a child to show the sensitivity in each case? Include the sex of the children involved in each family.

	Mother		*Father*		*Mother*		*Father*
(a)	*Gg*	×	*G*____	**(c)**	*GG*	×	*g*____
(b)	*gg*	×	*G*____	**(d)**	*Gg*	×	*g*____

25. Although it has not been proved with certainty, rare cases of hypertrichosis of the ears seem to be holandric. The trait is dominant and the phenotypic expression consists of long hairs growing on the pinnae of the ears.

If a man showing ear pinnae hypertrichosis is married to a woman without long hairs growing on the ear pinnae, what chance do they have for children showing the trait? Specify the sex of the children in your answer.

26. Persons' hands may be classified into two general categories, based on the relative length of index and ring fingers. In one type the index finger is longer than the ring finger, in the other group the index finger is about the same length, or slightly shorter. It appears that the long index finger is dominant in women and recessive in men (sex-influenced trait).

Assume that a man having long index fingers is married to a woman who has long index fingers. The woman's mother had short index fingers.

(a) What is the chance that this couple's daughter will have long index fingers?

(b) What is the chance that their son will have short index fingers?

CHAPTER SUMMARY

1. Mendel reported the results of studies of traits that showed complete dominance. Such traits can be conveniently studied by utilizing Mendel's logic and original system of symbols. Thus, capital letters are often used to indicate dominant genes and lowercase letters, the recessive forms.
2. The mathematical law of independent simultaneous events can be used to calculate the probability that two gene forms in independent gametes will occur together in a single zygote. Parents homozygous for a particular gene (*AA* or *aa*) can supply only one form of the gene (*A* or *a*), whereas heterozygous parents (*Aa*) have one-half chance of passing along one allele and one-half chance of contributing the other (1/2 *A* and 1/2 *a*).
3. Two heterozygous parents can produce *AA* offspring (1/4 chance), *Aa* offspring (1/2 chance), and *aa* offspring (1/4 chance). A heterozygous parent and a homozygous recessive parent can produce heterozygous offspring (1/2 chance) or homozygous recessive offspring (1/2 chance). A heterozygous parent mated with a homozygous dominant parent can produce homozygous dominant offspring (1/2 chance) or heterozygous offspring (1/2 chance). Two like homozygous parents produce offspring of the same parental type (*AA* or *aa*). A homozygous dominant parent and a homozygous recessive parent can produce only heterozygous offspring (*Aa*).
4. Some trivial or innocuous traits in humans can be traced genetically within families. Convenient examples of single-gene pairs showing dominance and recessiveness are ear lobe shape (free versus attached), PTC taste ability, middigital hair, facial freckles, and widow's peak.
5. Not all genes exhibit their effects with the constancy predicted by Mendel. In some cases a gene is not fully penetrant and the phenotype is affected in only a certain proportion of the cases where it is present. The expressivity of a gene in some cases varies from one individual to another, resulting in differing phenotypes among those carrying the gene.
6. Some allelic pairs do not show a full dominance-recessiveness situation. Heterozygous individuals show a different phenotype than homozygous dominants. In many cases the heterozygotes are intermediate in phenotype and the alleles are considered to be codominants.
7. In some human traits there are multiple alleles, where more than two forms of a gene exist. Even though only two alleles normally can be present at one time in an individual, the analysis is complicated by the number of alleles available. Because several combinations of alleles are possible, more than the typical two phenotypes usually occur.
8. Another exception to the typical Mendelian pattern is seen in epistasis where an entirely separate gene pair exerts an effect on the manifestation of the other pair.
9. Human blood type antigens show some classic exceptions to Mendelian inheritance patterns. The MN blood types show codominance. The ABO blood type are an example of multiple alleles (three alleles and four phenotypes) as well as codominance for two alleles.
10. A good example of epistasis is in the ABO system, where the gene for a precursor to antigens A and B controls the production of those factors. The homozygous presence of recessive genes for albinism prevents other pigmentation genes from showing their effects.
11. A complicated system of genes is responsible for human Rh blood types. The Rh antigen-antibody system is of importance in cases where an Rh-negative mother carries an Rh-positive fetus. The Rh-positive red blood cells can sensitize the mother and

her antibodies may cause hemolytic disease of the newborn in subsequent Rh-positive fetuses.

12. Genes carried by the X chromosome are said to be sex-linked since they follow a pattern of inheritance associated with gender. Males have only one X chromosome and are much more likely than females to exhibit such recessive sex-linked traits as favism and colorblindness.

13. The Y chromosomes seems to be devoid of significant genes. There are H-Y antigens produced by genes which are thought to be the same as those causing testes development. Based on some meager evidence, hairy ear pinnae have been reported to be due to holandric genes.

14. Autosomal genes may react differently in the two sexes. Some genes, such as the one producing pattern baldness, seem to be dominant in one sex and recessive in the other. Such genes are said to be sex-influenced. Some autosomal genes are classified as sex-limited, since they only show their phenotypic effects in one sex or the other, not both.

15. By using the mathematical law of independent simultaneous events, it is possible to analyze and predict the inheritance pattern of more than one gene pair simultaneously. It is a simple matter to consider each gene pair separately and then calculate the product of the independent crosses occurring simultaneously.

16. A device sometimes used to solve genetics problems is the Punnett square or "checkerboard method." In this graphic process the gametes of one parent are lined up horizontally across the top of the chart, while the gametes of the other parent are lined up vertically down the left side. The number of blocks formed is dependent on the number of gamete types and each contains a genotype with a chance of occurrence equal to its relative proportion within the whole checkerboard.

CHAPTER

8

GENETIC DISEASES

Treasure your exceptions! When there are none, the work gets so dull that no one cares to carry it further.

William Bateson, The Methods and Scope of Genetics (1908)

It is important to understand that gene-caused defects or disabilities can be analyzed and predicted, as innocuous traits, according to Mendelian principles. A study of the inheritance of ear lobe shape or PTC taste ability may be interesting, but some single-gene-based diseases may involve life or death numerical predictions. As serious in many instances as the chromosomal defects studied in Chapter 6, **genetic diseases** have become a major focus of attention.

A disease can be thought of as any deviation from a normal, healthy bodily condition. Once any normal function is impaired through either the interference of external agents or the failure of some intrinsic factor, the well-attuned and balanced equilibrium with the environment is disrupted. This leads to disability in functioning and sometimes painful responses. More than 2000 different diseases can be ascribed to genes that operate according to typical hereditary patterns. The mechanisms of genetic action are not understood thoroughly for all diseases, but definite hereditary patterns have been identified. In some cases detailed biochemical analyses have been carried out.

Much information about genetic diseases is available from many sources. Because some genes are responsible for medically recognized defects that follow basic hereditary patterns, it is possible to analyze family pedigrees and make predictions in many different situations. Once the basic patterns are recognized, it is not difficult to extend one's knowledge to many other specific disease situations.

It is important to note that even though a disease may be determined by genes which are inherited in Mendelian fashion, this does not necessarily mean that the defect is **congenital**. The term congenital implies that the defect is present at birth. Some defective genes are not manifested until sometime after birth. It is also true that some congenital defects are not genetically determined. Some may result from environmentally stimulated aberrancies.

DISEASES CAUSED BY RECESSIVE GENES

The fact that certain human abnormalities appear to be due to a homozygous combination of recessive genes quickly leads to the conclusion that an essential gene product is lacking. Clearly, such situations can result from mutant alleles that produce aberrant and ineffective forms of a normal product. Because proteins are the main gene products, enzymes or structural proteins are the major components whose defective manufacture results in such diseases.

Galactosemia

Milk is a most important food source in infants' nutrition and is normally digested and utilized by the overwhelming majority of babies. According to a Massachusetts screening program covering the years 1962 to 1973, approximately one child in 120,000 had **galactosemia**, the disease caused by the inability to digest a milk sugar called galactose. The children's cells lack the ability to produce an enzyme called **galactose-1-phosphate uridyl transferase**, causing a build-up of galactose phosphate in the tissues. The overabundance of this substance causes nausea and vomiting and shortly leads to malnutrition. The accumulated substance results in severe mental retardation and is commonly fatal within one year, if no therapy is utilized.

The lack of one enzyme obviously leads to the conclusion that a single pair of recessive alleles is responsible for the disease. This recessive gene hypothesis is further supported by the observation that infants having the disease are born of two normal parents. Thus, the causative gene behaves like a Mendelian recessive gene and serves as an effective model for the study of gene-caused diseases. It is a good example of the interaction of heredity and environment and demonstrates clearly how knowledge of a mechanism can suggest means of alleviating disease symptoms, as well as providing information for prospective parents.

Two obvious mechanisms for overcoming the lack of the enzyme in galactosemia might be proposed: (1) the mutant gene might be replaced by a normally functioning dominant gene for galactose-1-phosphate uridyl transferase; (2) the enzyme might be injected directly into the body. Unfortunately, neither is yet technically feasible. Because there is no known mechanism for replacing or repairing defective human genes, the first option is improbable: the second is not workable because it is practically impossible to inject protein molecules into the body and insure their dispersal to all the required sites of need.

Without a presently available means of overcoming the genetic cause of the disease, medical science is forced to a form of therapy. In this case the solution is relatively simple: Remove galactose from the person's diet. This dietary control has been utilized in many cases to prevent death and allow normal mental development. Finding a substitute for milk that is nutritious, palatable, and free of galactose is expensive, but it is effective.

Those who have brothers and sisters who are galactosemics might wish to know whether or not they carry the gene, since they would have a one-half chance of passing it on to offspring. In such cases a reasonably easy analysis will answer the question. Although classical Mendelian recessive genes are masked in the heterozygous condition, in this situation the reduced production of the gene product can be measured. While the heterozygous individual is normal in being able to digest galactose, there is only about one-half the amount of galactose-1-phosphate uridyl transferase present in the body. This enzyme activity level is detectable in the red blood cells as well as other tissues, allowing an assay to be made.

Phenylketonuria (PKU)

In another metabolic disorder it has been shown that a mutant gene is responsible for the lack of an important enzyme. The normal dominant allele is responsible for an enzyme that catalyzes the conversion of one amino acid called phenylalanine into another amino acid called tyrosine. Phenylalanine enters the body through a normal diet of protein-containing food. As a result of the enzyme's absence, phenylalanine build ups in the tissues, with other products like phenylpyruvic acid. There is a large amount of such materials excreted in the urine, but their excess accumulation in the body of an infant causes abnormal brain development and may result in varying degrees of mental retardation. Profoundly lower I.Q.'s have been documented in cases of PKU and victims of the disease have been labeled *phenylpyruvic idiots* in the past.

The primary location of the enzyme **phenylalanine hydroxylase** is the liver, and it has been difficult to distinguish heterozygous from homozygous normal individuals. However, a rather sophisticated procedure of measuring the ability to break down phenylalanine has been devised and can be used to identify heterozygotes.

Fortunately, because of the excretion of the phenylalanine in the urine and the accumulation of excess phenylalanine in the blood early in life, it is a relatively simple matter to test for the homozygous recessive condition in infants. In 1962 Massachusetts became the first state to mandate that all infants be tested for the condition. Since then most states have instituted requirements for similar screening programs. The incidence of PKU in the Massachusetts screening program has proven to be about one in 15,000 live births.

The question of treatment is similar to the galactosemia situation. Although PKU is not a fatal disease, reduced life spans may result. The most serious symptom is the mental retardation caused by the impairment of normal brain development in the first few months after birth. Again, a true cure is not yet available and may never be, but control is possible by greatly limiting the dietary intake of phenylalanine.

Phenylalanine is an amino acid commonly found in many proteins. Thus, it is a challenge to prepare food that is free of this component. In the early years of treatment it was feared that persons with PKU might be forced to utilize an unpleasant and unattractive phenylalanine-low diet as long as they lived. It was decided about 6 years of such a diet commencing immediately after birth would be sufficient to prevent brain damage. There are indications, however, that some caution about too much intake of phenylalanine throughout life may be well advised. Close monitoring of the phenylalanine level of the body is important, especially during formative years. A diet of low-protein foods, supplemented by an amino acid mixture low in phenylalanine, can be rather unpalatable, as well as expensive, but it is an essential part of the treatment. There is evidence that properly monitored individuals who are maintained on the phenylalanine-free diet for a satisfactory period have I.Q.'s that compare closely to those of their disease-free brothers and sisters.

As those females who have the homozygous recessive genotype for PKU reach reproductive age, a new problem is encountered. If they become pregnant, it is essential for them to return to the phenylalanine-free diet to prevent abnormalities of the fetus, especially brain damage, growth retardation, and spontaneous abortions. If they continue to eat a normal diet, still being unable to change phenylalanine to tyrosine, the high levels of phenylalanine and phenylpyruvic acid in their bodies will be transferred to the fetal environment and cause developmental abnormalities.

The abbreviated biochemical pathway in Figure 8.1 shows some of the chemical changes that are mediated by gene-controlled enzymes in humans. Two amino acids, phenylalanine and tyrosine, taken in by normal diets are shown. In addition some of the various products of their metabolism are given. A number of key points that are under enzyme control are demonstrated. If blocks at these points occur as a result of a lacking enzyme, there may be an excess accumulation of some compounds. For example, when **phenylalanine hydroxylase** is lacking (1), phenylalanine and

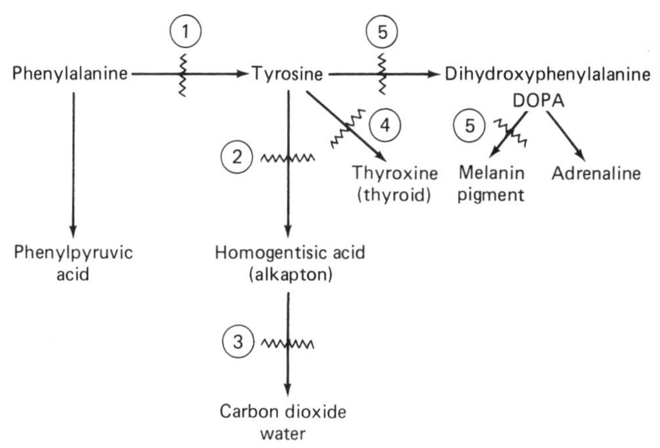

Figure 8.1. Schematic biochemical pathway of metabolism of amino acids phenylalanine and tyrosine in humans. Some points in the pathway where recessive mutations have resulted in lacking enzymes are shown. The resulting genetic diseases are the following: (1) Phenylketonuria (PKU) caused by excess build-up of phenylalanine and phenylpyruvic acid; (2) Tyrosinosis, caused by build-up of excess tyrosine; (3) Alkaptonuria, caused by build-up of excess alkapton; (4) Cretinism, caused by lack of growth hormone, thyroxine; (5) Albinism, caused by lack of melanin pigment or its precursor.

phenylpyruvic acid accumulate, causing PKU. When **tyrosinase** is lacking (5), the precursor of melanin is lacking and albinism results. Note that another gene product for the conversion of DOPA to melanin has the same phenotypic effect. Albinism therefore may be caused by more than one pair of recessive genes.

It is interesting that this pathway contains the genes originally studied by Garrod in 1901, as mentioned in Chapter 1. Due to the lack of **homogentisic acid oxidase**, alkapton (homogentisic acid) accumulates in the body and causes the rather mild disease **alkaptonuria.**

The fact that more than one autosomal gene for albinism exists presents an interesting situation. In the past such genes were called *duplicate genes*, but a careful look at their molecular basis reveals that only the phenotype is duplicated. Two albinos might produce a normally pigmented offspring, even though the gene is recessive and both parents are obviously homozygous. One might be an albino because the first enzyme blockage was responsible (tyrosinase). The other might be an albino because the block in the pathway occurred after DOPA (the exact step has not been determined and there are several possible intermediaries that might be involved). Thus, the first parent would have the dominant gene for the second enzyme, while the second parent would have the dominant gene for the first enzyme. The first could be called a **tyrosinase negative albino** and the second, **tyrosinase positive albino**.

Using conventional symbols to diagram the cross, it might look like this:

A_1 = Normal enzyme
a_1 = Lack of tyrosinase
A_2 = Normal enzyme
a_2 = Lack of second enzyme (DOPA to melanin)

First parent $a_1a_1\ A_2A_2$ × *Second parent* $A_1A_1\ a_2a_2$

↓

$A_1a_1\ A_2a_2$
(Normal pigmentation)

Tay-Sachs Disease

One of the most severely disabling genetic diseases in children also causes tremendous anguish for the parents. In **Tay-Sachs disease** a newborn infant appears to be perfectly normal and healthy at birth, but by about 6 months the victim begins to lose motor abilities through the loss of nervous control of muscles. The disease is invariably fatal, usually at about 3 or 4 years of age. Death follows an almost total loss of nervous functions, including sight and controlled movement.

It is mildly surprising to learn that this disease is based on the lack of one enzyme—**hexoseaminidase A**. As a result of the missing enzyme, fatty deposits gradually build up in the nerve cells, accounting for the loss of function. Unfortunately, no method of cure or therapy has been found, in spite of concentrated efforts.

It is possible to detect heterozygous carriers who could be potential parents. Furthermore, through **amniocentesis** an afflicted fetus can be detected in a pregnant woman in time to permit the option of **abortion**. Amniocentesis will be examined in detail later; for now simply define it as a means of obtaining and analyzing fetal cells from the fluid surrounding the fetus in a pregnant woman. Not all fetal genetic defects can be detected by amniocentesis, but chromosomal defects can, as described in Chapter 6, as well as some obvious enzyme defects such as Tay-Sachs disease.

The gene for Tay-Sachs disease occurs at higher frequency among persons of northern European Jewish descent (Ashkenazi Jews) than among other populations. About one in 30 persons of such ancestry is a carrier of the recessive allele. The risk is high enough among such persons to warrant an analysis, because of the serious consequences when an afflicted child is born.

Cystic Fibrosis

An example of a very serious disease that is caused by the homzygous presence of recessive alleles is **cystic fibrosis** of the pancreas. The disease is thought to occur in about one in 2500 live births in Caucasian populations. Most geneticists classified the gene as lethal until the advent of detection techniques and therapy in the 1960s; until recent years, a patient seldom survived beyond age 10. It was estimated in 1970 that 5% survived beyond age 17 and there is constant improvement.

Major symptoms are pancreatic insufficiency resulting in a lack of some digestive enzymes and a clogging of the organs of the body with thick mucus. The chief cause of death has been chronic pulmonary disease resulting from airway obstruction. Symptoms are frequently complicated by the reduced tolerance to infectious diseases.

In the past there was considerable difficulty in identifying the disease, because so many symptoms resembled those associated with other known defects. Children with the disease show a spectrum of symptoms in varying degrees that are related to the excess fluids. Deaths often have resulted from pneumonia, intestinal obstructions, pancreatic failure, malnutrition, and a general sensitivity to infectious diseases.

In spite of the problem of identification based on symptoms, it has been discovered that cystic fibrosis victims excrete an excessive amount of salt in their sweat. Stimulation of perspiration and a salt analysis appears to be an accurate and convenient way to diagnose the disease and is now routinely done where the disease is suspected in infants.

The pattern of inheritance in diseases like cystic fibrosis is essentially similar to that of genes that do not cause defects. Because the disease is based on an autosomal recessive allele, Mendelian techniques can be used to study it in families. When cystic fibrosis occurs in a child, one can assume that both parents are heterozygous carriers of the recessive allele. The symbols *C* and *c* might be selected for the controlling alleles. Thus the family's genotypes would be

$$Cc \times Cc$$
$$\downarrow$$
$$cc$$

Furthermore, after such an analysis, one could predict that with subsequent children the chance for cystic fibrosis would be 1/4, for heterozygous carriers of the gene, 1/2, and for homozygous dominant individuals (normal), 1/4.

Although research has proceeded with the assumption that cystic fibrotic individuals were homozygous for a recessive allele, there are some unresolved questions. Cystic fibrosis does not follow the typical pattern seen in recessive gene-based diseases, in that lack of a single enzyme has not been discovered. The symptoms do not seem to be explainable by such a mechanism, although future research may tie the recognized effects to a single product.

Xeroderma Pigmentosum

A well-known type of cancer, **xeroderma pigmentosum,** is caused by a single recessive gene pair. It is manifested by very heavy freckling with open sores and severe skin cancer in areas exposed to sunlight. This is also a disease of children and usually results in death before adulthood.

In xeroderma pigmentosum an enzyme which can normally repair defects in DNA caused by **ultraviolet light** is replaced by a defective form. Ultraviolet rays in sunlight then apparently cause unrepairable gene defects that release the control on cell division, resulting in skin cancer. From the description of **oncogenes** in Chapter 6, recall that oncogenes may be turned on by mutations that occur in **promoter genes** adjacent to them in the chromosomes. In addition there is now evidence that mutations in the oncogenes themselves may stimulate them to action.

This situation indicates that there is much more mutation occurring under normal environmental conditions than has been estimated in the past. A natural repair mechanism normally prevents problems by correcting such defects.

Hemophilia

A very rare example of a sex-linked recessive gene, and one that has much more severe consequences than previously studied sex-linked genes, is found in the disease **hemophilia**. It is sometimes referred to as the "bleeder's disease," because the symptoms are based on the inability of the blood to clot properly. Thus, a small wound can be lethal. The most common ailment for persons suffering from hemophilia consists of extremely painful **joint bleeds** that develop when tissue within the limb joints is injured during relatively normal movements.

There are several types of known hemophilia, but the type which is clearly due to a recessive allele carried on the X chromosome is **hemophilia A**. Blood clotting is a complex process involving several chemical and cellular components produced in the human body. The clotting system consists of a series of factors that contribute to the process in a sequential stepwise fashion. The clot itself is composed of **fibrin**, an insoluble substance formed when **fibrinogen** in the blood is acted on by the enzyme **thrombin**. To produce thrombin to catalyze the reaction, several factors must act on the chemical **prothrombin** which is released when blood platelets are broken open. The blood platelets are fragile cellular particles that breakup when they contact the roughened regions of a wound.

One of the factors needed to convert prothrombin to thrombin is **Factor VIII** or **antihemophilic globulin (AHG)**, a protein under X-linked gene control. If the gene for the production of Factor VIII has been mutated so that an inactive form of the compound is produced, prothrombin is not converted to thrombin and blood clotting does not occur in the normal short period required. This lack of functional Factor VIII is the basis for hemophilia A.

Fortunately, the mutant form of the gene is extremely rare and only about one in 7000 males has the disease. It is interesting that long before the genetic mechanism was discovered, and especially before sex linkage was recognized, there were indications that associations with the inheritance pattern had been noticed. The Hebrew Talmud, or Book of Laws, made the provision that the sons of a woman or those of her sisters were not required to undergo circumcision, if an earlier son had died due to excessive bleeding in the ritual operation. Her brother's sons were not included in the proscription. More recently, but still prior to a knowledge of sex linkage, hemophilia was described in 1803 by John Otto. It was noted that only males were subject to the disease, but that females in the same families sometimes passed on the trait to their sons.

Medical science has made it possible for afflicted individuals to inject themselves with Factor VIII obtained from the blood of normal individuals. In past years, the majority of hemophiliacs did not survive to reproduce, whereas today most can survive to normal old age, although the treatments are very expensive.

Although the disease is rare, an interesting occurrence of the trait probably had a profound influence on world history. In the latter half of the nineteenth century Queen Victoria of Great Britain carried a mutant gene for Factor VIII heterozygously. It appears that the mutation actually arose in one of the gametes giving rise to Queen Victoria, since the disease is not found among her ancestors. She transmitted the allele to three or four of her nine children, and two of her granddaughters received this recessive gene from their mothers. These two women subsequently married into other European royal families. One of them transmitted the allele to the Spanish Royal family, where two afflicted males died as a result of fairly moderate injuries sustained in automobile accidents. However, the gene had no real effect on the political fortunes of the Spanish monarchy.

Queen Victoria's other affected granddaughter Alexandra married Czar Nicholas II of Russia and bore a hemophiliac son, Alexis (Fig. 8.2). Although many factors led to the Russian revolution, there is no doubt that much of the corruption in the Czar's court centered about the self-styled monk Rasputin, who exercised an inordinate amount of power in the government (Fig. 8.3). Alexis's parents, especially Alexandra, believed that Rasputin had almost magical powers to relieve the painful seizures that their son experienced. If it had not been for the influence of Rasputin, some historians think that the Czar might have survived the attack on his regime and prevented the communist overthrow. Whether true or not, it is fascinating to consider that a single recessive gene could be the source of such a scenario.

Today it is possible to examine the blood of a female in a sophisticated analysis to determine whether or not she is a carrier of the hemophilia gene. Such a test is seldom deemed necessary, unless the trait has occurred elsewhere in her family. The analysis would have been appropriate among Queen Victoria's daughters, since she had a son who was a hemophiliac and the daughters had a one-half chance of being carriers. By testing her daughters, those who were carriers could have been identified. In any event, the present British royal family does not carry the allele since the family arose from King Edward VII, a nonhemophiliac son of Victoria.

Today women who have been recognized as heterozygous carriers may opt to have no children. There is no prenatal test to identify the hemophiliac fetus, but again the sex of the fetus is ascertainable. Some may opt to abort all male fetuses, knowing that there is a 50% chance of aborting a nonhemophiliac son. As in the fragile X syndrome, half the boys would receive the "normal" X chromosome and would not exhibit the defect; however, half the male offspring could be expected to suffer from the affliction.

Figure 8.2. Czar Nicholas II and his wife Alexandra with their son Alexis. Alexandra was a carrier of the recessive gene for hemophilia, as was her grandmother, Queen Victoria. Alexis suffered from the disease, since he was hemizygous for the allele.

Duchenne Muscular Dystrophy

The disease **muscular dystrophy** is known to be caused by at least three different genes. One gene responsible for the condition is an autosomal dominant and another is an autosomal recessive gene. The most frequently encountered type, however, is caused by a sex-linked recessive gene. This type is called **Duchenne muscular dystrophy** (after its discoverer, Guillaume Duchenne), and approximately three in 10,000 males manifests the disease (Fig. 8.4)

The symptoms consist of a deterioration of the muscles which begins during childhood, sometimes at birth. Leg muscles are soft and flabby and young boys exhibit a clumsy, uncoordinated ability to move. The patients sooner or later must resort to the use of a wheelchair and death results by age 20 in most cases.

Hemizygous males with the disease do not normally pass on the defective gene

Figure 8.3. Grigori Efimovich Rasputin was very influential in the palace of Czar Nicholas II and has been accused of responsibility for many corrupt government appointments. He derived his influence with the royal family from his reputed power to control the attacks of hemophilia in young Alexis.

through reproduction; the reservoir for the X-linked recessive allele is among heterozygous female carriers in the population. Because no means of identifying such carriers has been discovered, a decision not to have children by such women is impossible until a defective son has been identified (possibly 5 or 6 years after birth when the symptoms appear and other sons have already been born).

Likewise, no means of prenatal identification for fetuses has been devised, so that therapeutic abortion is not a possibility. There is hope, and some promise, that a defective or lacking gene product may be identified. If this discovery is made, it may be possible not only to identify gene carriers, but also to propose methods of prevention or therapy.

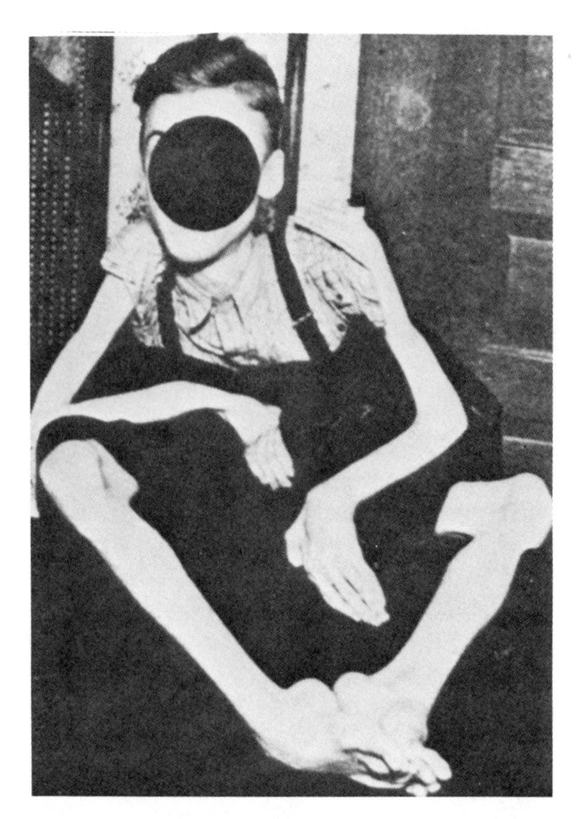

Figure 8.4. Duchenne muscular dystrophy. This teenaged boy shows the phenotypic effects of the sex-linked recessive gene for muscular dystrophy. For several years the muscles have been deteriorating, a condition that will eventually lead to a premature death. (Reprinted by permission from HEREDITY, EVOLUTION AND HUMANKIND by A. M. Winchester, Copyright © 1976 by West Publishing Company. All rights reserved.)

Lesch-Nyhan Syndrome

A sex-linked genetic disease was discovered at Johns Hopkins Hospital in 1962 and has been named for the two researchers who analyzed its basis, Michael Lesch and William L. Nyhan. The **Lesch-Nyhan syndrome** provides an interesting example of the effect of a gene on the production of a particular enzyme, **hypoxanthine-guanine phosphoribosyl transferase** or **HGPRT.** The function of the enzyme is to "recycle" metabolically degraded guanine to a form that can be utilized in DNA and RNA. Such a chemical pathway has been labeled by biochemists as a **salvage pathway**.

When the enzyme is absent or nonfunctional due to a mutant gene, there are excess purines accumulated in the body and these are in turn converted to unusually high amounts of **uric acid**. This uric acid results in gout, kidney stones, and kidney failure. The overproduction of uric acid also causes increases in enzymes that enhance the production of purines, a condition that heightens the problem. There are apparent mental problems and muscle spasticity in the disease. Most serious of all is the effect on the behavior of the individual.

Although apparently normal at birth and for several months thereafter, the child with Lesch-Nyhan syndrome becomes largely hostile and exhibits compulsive self-destructive behavior at age 2 or 3 years. A common symptom is the chewing of the child's own fingers and lips. Extreme measures are required to protect the patient from very serious injury. The syndrome is found only in males, since it is due to an X-linked gene. The frequency of the defect has been estimated at about one per 380,000 births. Because there is a missing enzyme, the disease can be discovered prenatally. By removing fetal cells from a pregnant woman in the process of amniocentesis, Lesch-Nyhan fetuses can be identified from cultured cells. In addition, because a missing enzyme is the primary feature, heterozygous females show a mosaic condition as a result of randomly inactivated X chromosomes throughout the body. Thus, where there is some reason to suspect the presence of the mutant gene, it is possible to identify a carrier mother and an affected fetus. Under such circumstances, the option of abortion is a possibility. Such analyses are warranted where the gene is known to occur in a family or where prior Lesch-Nyhan births have occurred.

In 1982 the HGPRT gene was isolated experimentally and, through the techniques of genetic engineering, was implanted in bacteria and in cultured mouse cells where it was transcribed and dictated the production of the enzyme. Such techniques will lead to the detailed analysis of the gene and may even eventually lead to therapy or cure.

DISEASES CAUSED BY DOMINANT GENES

In general there are fewer human diseases caused by dominant than by recessive alleles. The mechanism underlying their action is more difficult to understand, but progress is being made. Frequently it appears that dominant alleles interfere with structural components of the body, where formation of the normal functional product is blocked by an alternate product. Because mutations most often result in the change from a dominant to a recessive form, more abnormalities might be anticipated in the recessive category.

Important also to the greater incidence of recessively controlled diseases is that selection against a dominant trait is extremely effective when the trait is lethal or otherwise interferes with survival and reproduction. In contrast recessive diseases, such as cystic fibrosis or galactosemia, in untreated cases may be lethal, but this selection does not profoundly affect their already low frequencies. Such rare recessive traits are maintained in the population by heterozygote reproduction, but a dominant gene that was lethal or prevented reproduction would quickly be eliminated. The phenomenon of selection will be considered in more detail in Chapter 11.

Often dominant alleles that result in diseases exhibit variable effects on the phenotype. Where they produce their severe results in some individuals and not others, some disease-causing dominant genes are able to survive, similar to masked recessive genes. Thus, they are maintained in the gene pool and continue to be transmitted within a population at a low frequency (recall in Chapter 7 the phenomena of variable **penetrance** and **expressivity**).

Huntington Disease

In 1872, long before Mendel's principles had been rediscovered, George Huntington described the symptoms of a particularly disabling nervous disorder. Its cause remained a mystery for 80 years and today almost all that is known is that it is caused by a dominant gene. The disease was called **Huntington chorea** (*chorea:* Greek "dance"), based on the jerky involuntary motions that are one of the obvious symptoms. It is more commonly known now as Huntington disease. From the first signs of a weaving, uncoordinated walk, a steady deterioration of nervous control and mental ability occurs until death ensues in about 10 or 11 years.

Probably the most famous person to suffer with Huntington disease was folk singer Woody Guthrie (Fig. 8.5). His life and music have been popularized through films and television and many people know that his death in 1967 was due to this inherited disease. His wife Marjorie was very active during her life on the Committee to Combat Huntington Disease, a foundation that supports research and provides information about the disease.

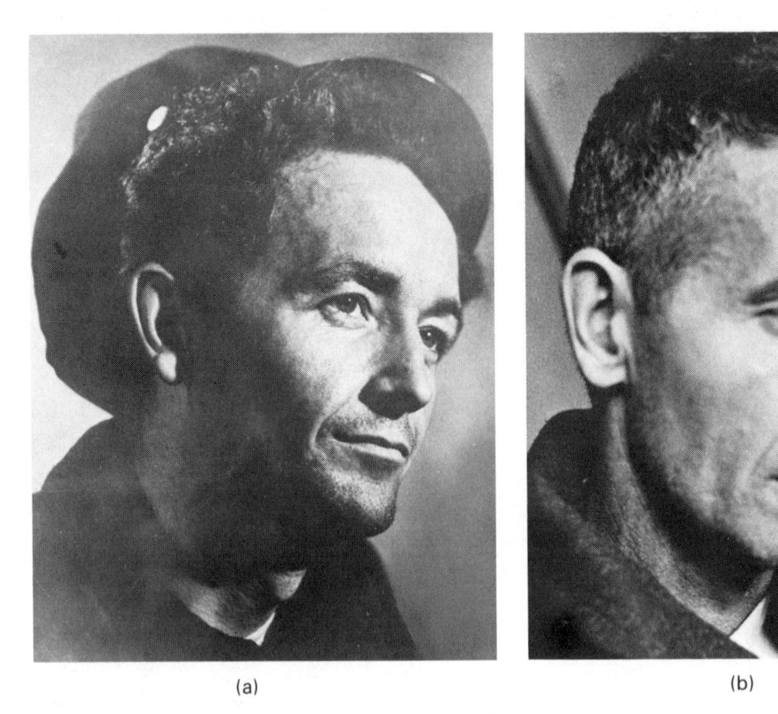

(a) (b)

Figure 8.5. Woody Guthrie (1912–1967). A renowned folk singer and composer until he was struck by the effects of Huntington disease, Guthrie is an example of the devastating nature of the disease. Dominantly inherited, the disease is often manifested after already having children and often at the height of a person's career. His mother also suffered and died with the affliction. Photograph (a) was taken before the onset of symptoms; photograph (b) shows Guthrie after the effects of the disease were obvious. (Courtesy Marjorie Guthrie)

The incidence of the disease is estimated at about one in 10,000 and all such persons are heterozygous. It is not known whether or not a person homozygous for the dominant gene would be able to survive as a fetus. The most unusual feature of the disease is its late manifestation. The average age of first appearance of the symptoms is about 35 to 40 years. The manifestation of the disease has occurred beyond age 65 and as early as the young teens. Most persons carrying the gene will have exhibited the symptoms by the time they are 50.

An obvious problem in controlling the transmission of the gene from one generation to the next is that most carriers have produced their families by the time the disease is diagnosed. Thus, even though the causative allele is dominant and the disease is fatal it has not been effectively selected against in reproduction.

The children of a parent with Huntington disease have a 50% chance of manifesting the disease themselves at some time in life. The genotypes involved may be diagrammed as follows:

Affected parent		*Nonaffected parent*
Hh	×	*hh*
	↓	
	1/2 *Hh*	(May manifest disease)
	1/2 *hh*	(Disease-free)

In recent years research has shown that some transmitter chemicals are depleted. These changes can be duplicated in laboratory rats by injecting **quinolinic acid**, a compound that is a natural breakdown product of tryptophane. It is not known whether or not quinolinic acid is present in abundance in persons with Huntington disease. If this proves to be the case, an effective treatment may be discovered for this mysterious and destructive condition.

In late 1983, James Gusella, a Massachusetts General Hospital geneticist, announced that a technique for identifying the presence of the gene responsible for Huntington disease had been devised. The technique was experimental and not available for widespread use until testing was completed. The analysis is based on the identification of a DNA base sequence on chromosome number 4 peculiar to Huntington disease victims. By synthesizing a molecular probe that recognizes the 17,000 nucleotide sequence, potential victims can be recognized with great accuracy. This opens the door to identification of fetuses carrying the gene and it implies that similar techniques may also be used for such diseases as cystic fibrosis.

Profound Deafness

Several different genes are involved in causing various forms of deafness. Dominant and recessive genes have been identified and some of the different alleles seem to be responsible for essentially the same phenotypic effect.

Profound childhood deafness, an example of a dominantly inherited congenital defect, follows a typical pattern of transmission from parent to offspring, where one or both parents manifest the problem. The genetic disease occurs with a frequency of about one person per 10,000 births for the dominant form.

Because persons possessing the defect are unable to hear from birth, they do not develop the ability to speak without specialized training. No cure for the problem is yet available, and no means of correcting the situation to allow normal hearing seems possible. At this point the organic nature of the problem is not well understood. It is probably a fundamental structural defect which arises in fairly early embryogenic stages of the individual.

Marfan Syndrome

Another example of a dominant trait whose effects are puzzling in terms of understanding the mode of action is Marfan disease. The symptoms of this abnormality of body structure were described by A. B. Marfan in 1896, before the rediscovery of Mendel's principles.

The expression of disease symptoms is variable, but bone, muscles, and connective tissues are affected. As a result of unusual bone growth, victims of the defect have long, slender arms, legs, and digits. Secondary symptoms involving vision and a tendency to fatal aortal rupture are sometimes associated with the syndrome.

Because an 11-year-old supposed relative of Abraham Lincoln was diagnosed as having Marfan syndrome in the 1960s, it has been proposed that Lincoln himself was afflicted with the disease (Fig. 8.6). The family connection provides weak evi-

Figure 8.6. Abraham Lincoln (1809–1865), sixteenth President of the United States. Lincoln was diagnosed in the 1960s as having Marfan syndrome, a genetic condition caused by a single autosomal gene. The syndrome is characterized by long limb and digit bones and a gaunt, awkward stature. However, the proposal was refuted in 1982 by a careful analysis of the facts. (Reprinted from *The American Conscience* by Saul Sigelschiffer copyright 1973, by permission of the publisher, Horizon Press, New York)

dence, since some 300 years and nine generations separate the boy with the syndrome and Lincoln. No other cases have been reported in the family line, an unusual situation where a dominant gene is involved.

John Lattimer in 1982 provided a convincing refutation of the suggestion that Lincoln suffered with the Marfan condition.* The diagnosis from such a distant time serves as a warning that genetic analysis may be difficult unless a careful firsthand survey is possible. The fact that Lincoln was tall, had long limbs, and experienced visual problems, can hardly provide unique or adequate diagnostic features for a genetic disease. Unfortunately, such pronouncements about a famous person are often utilized by the media as facts and become widely known.

Porphyria

A portion of the hemoglobin molecule is a submolecule called porphyrin. Under normal circumstances, hemoglobin is constructed according to a specific sequence of steps requiring enzymes for each change. Several genes are known to interfere with the normal sequence of events in such a way as to cause an excess accumulation and excretion of porphyrin. This component part of heme results in the excretion of urine that turns bright red on exposure to light and air.

Both dominant and recessive alleles are known to cause the same general effect. Exactly how this occurs is not known, but the enzyme **aminolevulinic acid synthetase** has been shown to occur in over-abundance in the disease.

Besides abdominal pains, nausea, headaches, and other physical symptoms, porphyria is often accompanied by bouts of emotional disturbance and behavioral changes. The use of drugs to relieve pain may even worsen these emotional problems. Porphyria is difficult to diagnose; this response to drugs is one of the chief clues.

Although the disease is very rare, it occurs with a frequency of about one in 250 persons among South African whites. It has been postulated in a much delayed diagnosis that King George III of Britain may have had porphyria. Some historians even speculate it was during a period of emotional disturbance that he imposed the famous Stamp Act. (This action led directly to overt rebellion of the American colonies and eventual independence from British rule.) One should be extremely cautious about such a diagnosis for King George III, or any other famous person.

Retinoblastoma

A type of cancer that is characterized by the development of malignant tumors in the retina of one or both eyes is usually caused by a dominant gene on an autosome. It was seen in Chapter 6 that retinoblastoma may sometimes be caused by a deletion in chromosome 13. The disease has a frequency of one in 25,000 children. It is usually first diagnosed around the age of 30 months when vision problems appear, but the cancerous growth begins about one year of age. The condition is responsible for 1% of the total number of cancer deaths in infancy and is the cause of 5% of childhood blindness cases.

Retinoblastoma is therefore a disease of children and it is fatal if untreated.

* John K. Lattimer, "Abraham Lincoln Did Not Have the Marfan Syndrome," *N. Y. State Jour. of Med.*, Vol. 81, Nov. 1981, 1805–13.

The usual treatment, which is life-saving, is to remove surgically both eyes to prevent the spread of the malignancy. In some cases microirradiation of the tumors has preserved partial eyesight, although the chance of success for this procedure has not been too good. The condition is obviously severely disabling and extremely sad for affected families.

The diseases described in this chapter are only a sampling of the more commonly recognized medical conditions that are attributed to genes. These simply provide a spectrum of the various modes of action and levels of phenotype manifestation. More genetic diseases are constantly being added to the list as their bases are recognized.

Those diseases already identified are becoming more well known as research into modes of action progresses. As they are studied intensively, the chances increase that treatments will be discovered that modify or alleviate the symptoms. The future holds promise that medical science will provide effective answers for situations that once appeared hopeless. That the cause of disease is inherited as a part of a genome does not signify that humans are incapable of overcoming its disabling results.

ADDITIONAL READING

BRADY, R. O., "Inherited Metabolic Diseases of the Nervous System," *Science* (1976), 193, 733–39.

CROW, J. F., "Genetics and Medicine," in Brink, R. A. (ed.), *Heritage from Mendel.* Madison, Wisc.: Univ. of Wisconsin Press, 1967. (Reprinted in Baer, A. S. (ed.), *Heredity and Society* (2nd ed.). New York: Macmillan, 1977.)

DEAN, G., "The Porphyrias," *British Medical Bulletin,* (1969), 25, 48–51. (Reprinted in Levine, L., *Papers on Genetics.* St. Louis, Mo.: C. V. Mosby, 1971.)

JOLLY, E., *The Invisible Chain: Diseases Passed on by Inheritance.* Chicago: Nelson-Hall, 1972.

MILUNSKY, A., *Know Your Genes.* Boston: Houghton-Mifflin Co., 1977.

STANBURY, J. B., J. B. WYNGAARDEN, and D. S. FREDRICKSON, *The Metabolic Basis of Inherited Disease* (4th ed.). New York: McGraw-Hill, 1978.

THOMPSON, J. S., and M. W. THOMPSON, *Genetics in Medicine* (3rd ed.). Philadelphia: Saunders, 1980.

REVIEW QUESTIONS

1. Suppose that a blood sample taken from an infant is found to be lacking the normally present enzyme galactose-1-phosphate uridyl transferase. Would you propose any treatment for the infant? What symptoms of disease would you look for? What is the genotype of the infant and its two parents? What is the chance that a child will be born to these parents with the same genotype as the first child? What chance is there for a child who produces the enzyme in adequate amounts?
2. It is reported to the normal parents of a newborn infant that the child has maple syrup urine disease, so called because of the odor of the urine. The diagnosis is possible because of high levels of the amino acids leucine, isoleucine, and valine in the blood and urine. The condition results in poor development of a child, followed by an early death.

(a) Speculate about the genetic basis of the disease.
(b) Suggest a method of treating a child who has the disease.
(c) What is the chance that the couple's next child will have the disease?

3. In screening newborn infants for PKU, the Guthrie test is routinely used. In this test a drop of an infant's blood is absorbed on a paper disk and placed in a special growth medium with bacterial cells. If the bacteria grow, the infant is presumed to have PKU. No bacterial growth indicates that the infant does not have PKU.

   (a) What is deficient in the special growth medium? (b) Why does the growth of bacteria indicate PKU? (c) How might a screening test for galactosemia differ from the type of analysis in the Guthrie test?

4. Why is the fetus in jeopardy in the body of a pregnant woman who is homozygous for the recessive gene for phenylketonuria? Why is the mother's body tolerant of the situation and the fetus is not?

5. A mutation that results in an achondroplastic dwarf has been observed to occur at a rate of 1 per 100,000. This condition is due to a defect of cartilage formation in the arms and legs which causes very short limbs to form. The head and body of such an individual are essentially of normal size. Using as a model the genetic basis of other developmental structural traits, such as polydactyly or Marfan syndrome, speculate about the mode of inheritance of achondroplasia. If a worldwide survey were possible, what sort of information would help to verify or negate your speculative analysis?

6. Suppose that a hypothetical wide-ranging survey provides the following information about two structural defects resulting from dominant genes:
   (a) Achondroplastic dwarfism: By surveying a large number of families with such dwarfs as parents, it is found that where one parent is a dwarf, about one-half the resulting children show the trait, with almost exactly the same symptoms as their dwarf parent.
   (b) Polydactyly: In a similar type of survey of families where one parent possesses extra fingers or toes it is found that about 20% of the children possess the trait. Among those who do, sometimes in the same family, some have six toes on each foot with five fingers on each hand, some have six toes on each foot with six fingers on one hand (right or left) and five fingers on the other hand, while others have six digits on hands and feet alike.

   Explain the basis for the different patterns seen in the inheritance of these two traits.

7. Retinoblastoma is a rare type of cancer that is first manifested in a child's eyes (usually before 6 years). If the affected eye or eyes are removed, the disease can be terminated. Otherwise, the cancer spreads and is fatal. It is caused by a single dominant gene, and the gene is about 80% penetrant.

   If a person who had retinoblastoma as a child, due to the heterozygous presence of the dominant gene, is married to a person who is homozygous for the recessive normal genes, what is the chance that their child will manifest retinoblastoma?

8. A rare disease of humans exists which does not reveal itself until a child has reached the age of about 6 years. At this time the child's muscles swell, most obviously those of the legs. Later the child becomes weak and wastes away; as a teenager, he is virtually "skin and bones." In almost every case the disease is lethal prior to maturity. The name given to the condition is pseudohypertrophic muscular dystrophy or Duchenne muscular dystrophy and is dependent on a sex-linked recessive gene (*d*).
   (a) What are the genotypes of a family (mother, father, son) in which a boy develops Duchenne muscular dystrophy?
   (b) What is the chance that, if these parents have another son, he will also develop the disease?

9. (a) Why do you think that geneticists have assumed that no female would ever be found with Duchenne muscular dystrophy?
   (b) In fact, a case of the disease has been identified in a female. Recalling the material in Chapter 6 on translocations, show how this could occur. Describe a possible karyotype for such a female.
10. Suppose that a woman with a headache and a cold self-administers aspirin in moderately heavy doses over a period of 2 days. In another day or two the woman's urine turns very dark in color; she becomes weak with various body pains. A medical examination reveals hemolysis of her red blood cells and an anemic condition. Consultation with a physician reveals that her father developed a severe anemia as a soldier during World War II when treated with primaquine for malaria. She also recalls that her mother's father (maternal grandfather) complained of similar symptoms for a short period each winter when his wool clothes were taken out of mothball storage.
   (a) Speculate about what might be the cause of her anemic illness.
   (b) Suggest ways to prevent its recurrence.
   (c) The woman has a daughter and a son by the same father (who shows no symptoms of the disease). What is the chance that each child has inherited the same condition?
11. (a) If the mother of a son with Lesch-Nyhan syndrome were analyzed by cloning cells from several parts of her body, and the tissue culture cells were examined for HGPRT, what would you expect to find?
   (b) What would you tell the parents about the chance for another child to be born with the disease?
   (c) Could you suggest a means of preventing such a birth if they wished to have another child?

CHAPTER SUMMARY

1. A genetic disease is a deviation from a normal healthy bodily condition that is caused by a gene.
2. Some genetic diseases are caused by autosomal homozygous recessive genes. Examples are galactosemia, phenylketonuria, Tay-Sachs disease, cystic fibrosis, and xeroderma pigmentosum. In some cases the defective or missing gene product, an enzyme, has been identified. In others, the cause is still not clear.
3. There are recessive genes on the X chromosome (sex-linked) that cause genetic diseases. They are primarily diseases of males. Examples of this type are hemophilia, Duchenne muscular dystrophy, and Lesch-Nyhan syndrome. Defective or missing enzymes have in some examples been identified.
4. Queen Victoria of Great Britain carried a mutant allele for hemophilia which was passed on through several of her children to other royal families in Europe, including those of Spain, Russia, and Prussia. The most famous of those affected was probably Alexis, the son of the last Russian Czar, Nicholas II.
5. Some genetic diseases are caused by autosomal dominant genes. Examples of this type are Huntington disease, profound childhood deafness, Marfan syndrome, porphyria, and retinoblastoma. Dominant genes often show a mode of action that affects structural components, but overabundance of gene products is seen in some cases.

C H A P T E R

9

FAMILY PEDIGREES AND ANALYSIS OF HEREDITARY PATTERNS

The Pedigree of Honey
Does not concern the Bee—
A Clover, anytime, to him
Is Aristocracy.

Emily Dickinson, Poem No. 1627 (Version II), The Complete Poems of Emily Dickinson. Ed. by Thomas H. Johnson. Little, Brown and Co. 1960. (1884)

Even though the principles discovered first in pea plants apply to humans as well as experimental organisms, studies in human genetics seem to be more difficult because of exceptions to those fundamental principles. More important to an analysis of human heredity than deviations from straight Mendelian inheritance patterns, however, are the problems encountered in methods of study. As seen, experimental laboratory organisms can be crossed according to prearranged plans, without regard to any ethical principles. Furthermore, large numbers of offspring can be produced, permitting statistically valid samples to be produced from a single cross.

Analysis of humans must be based on the small family units that are in existence. This means that one must attempt to trace characteristics through as many generations and members of a family as possible.

Because it is often necessary to follow the inheritance of a characteristic through several generations and lines of a family, a systematic and convenient means of recording and tracing the data is needed. To follow a trait through a family and find correlations that explain the mechanism, a family diagram or **pedigree chart** is used. The chart produced is analogous to a map and allows one to follow the transmission of a gene through the members of a family from generation to generation. This diagram can be used to analyze the genotypes of persons in the family and then make reasonably accurate numerical predictions for inheritance of given traits among the offspring.

One feature of human pedigree analysis that is similar to the technique of Mendel is that it is most convenient to follow one trait at a time. Various modifications of pedigree chart construction have been devised which permit several traits to be followed simultaneously in the same diagram, but care must be exercised to avoid confusion. It is often better to utilize two separate diagrams.

The symbols and method of arrangement are by no means universal, but there is a commonality of overall appearance and usage of conventions in all pedigree charts. If one is familiar with one system, the slight variations from one technique to another are generally so obvious that the new system can be figured out by simply perusing a pedigree.

Other than analyzing for the sake of understanding the mode of inheritance of a trait, pedigree analysis is used for practical reasons. Probably the most important use of human pedigrees is in the increasingly important area of **genetic counseling**. To provide reliable information regarding the chances for seriously detrimental genes to be passed on to offspring, it is essential that a thorough analysis be made. Without complete and accurate data, decisions of profound importance to potential parents cannot be made rationally.

CONSTRUCTION OF PEDIGREE CHARTS

Essentially, three distinctive types of features must be shown in a functionally useful pedigree chart: Individual persons must be indicated as distinct entities, the familial relationship among the various persons is required, and it is necessary to be able to identify the **phenotype** of each person at a glance. Simplicity of design is a prerequisite, since one should be able to make such a diagram easily, as well as identify readily the various persons in the family.

To show the individuals in the family, it is most common to use **circles** to represent females and **squares** to represent males (○ and □). The most convenient method of diagraming an individual who possesses a given trait is to draw a solid symbol (● and ■); the hollow or unshaded symbol represents those without the trait. Of course these symbols simply provide information about phenotypes, and some workers like to show genotypes as well. If the trait under scrutiny is due to a recessive gene (*a*), the solid symbol would designate homozygous recessive persons (*aa*) who show the trait. It would be impossible to tell whether a hollow symbol represented a homozygous dominant (*AA*) or a heterozygous individual (*Aa*). Sometimes various techniques are used to distinguish homozygous dominant individuals in such a pedigree from heterozygous persons. The hollow symbol may be used for the homozygous person, while heterozygotes may be indicated by some device such as a spot in the symbol (⊙ and ⊡), by stippling (◌ and ⬚), or by half-filled symbols (◐ and ◧).

Because it is often not known whether persons are homozygous or heterozygous when the pedigree chart is constructed, no attempt is made to distinguish them. Furthermore, some are not ascertainable and it would be misleading to leave a hollow symbol for them, with known heterozygotes marked distinctly. At any rate, the pedigree is designed to present in compact, easily followed form a summary of the family in terms of the phenotypes of a particular trait. It is probably an improper use to embellish one with more details. Most pedigrees do not include complete genetic information; they do, however, provide the phenotypic data from which genetic analyses and predictions can be made.

The relationships that exist among persons in a family are shown basically by three types of straight lines. Two of these lines are horizontal (within a single generation) and the third is a vertical line (connecting two different generations). One horizontal line is a **marriage** or **mating** line connecting two parents; it always connects a male and a female symbol together (□—○).

The other type of horizontal line is a **sibling line** which connects the brothers and sisters (siblings or sibs) from a single family unit (○□○○). Usually an attempt is made to keep the children arranged in the proper temporal sequence from left to right. Thus, in this example the first child born to the marriage was a daughter, then a son, followed by two daughters. Sometimes it is necessary to diagram the sibling line with the children in the wrong sequential order so that a person can be located in such a position to indicate another mating.

When there are twins in a family, it is necessary to show them born at the same time. Identical twins arise from the same zygote and are often indicated by a single, forked attachment line (○○). They are always of the same sex. Fraternal twins come from two different eggs fertilized simultaneously and are shown as two children born at the same time (□○).

The vertical lines in a pedigree simply show the issue of the next generation from two parents. A simple pedigree might then appear as follows:

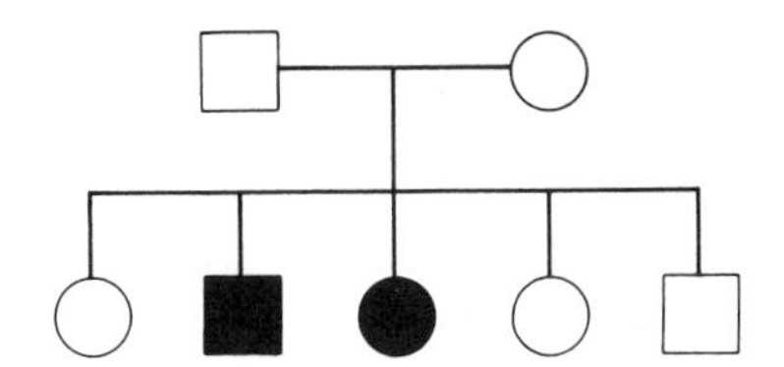

Another convention applied to most pedigree charts is a numbering system that allows one to refer specifically to certain individuals in the chart. The generations are numbered with Roman numerals, starting at the top. The individual members in the chart are numbered with Arabic numerals from left to right in each generation. All persons are numbered sequentially, left to right, regardless of their relationship. Although numbers usually do not appear on the chart, if written on the preceding pedigree, it would look like this:

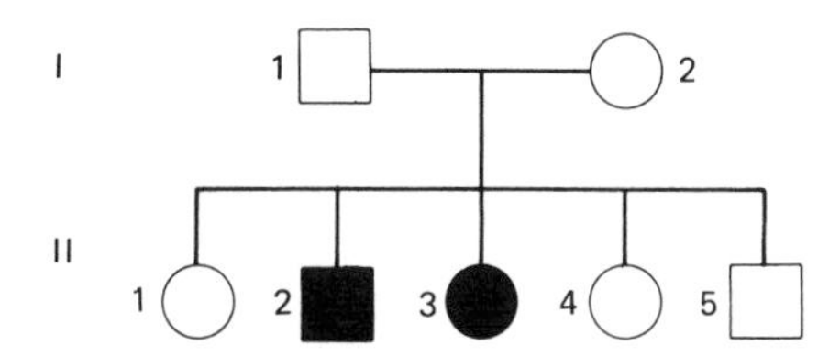

If it is assumed that the solid symbols indicate albinism, one can see quickly that individual II-2 is an albino son and II-3 is an albino daughter.

Sometimes in reconstructing a family's history the sex of an offspring will be unknown for one reason or another. In such a case a third type of symbol, a **diamond** (◇), is used. Another device sometimes utilized to compress the data is to use one symbol for several individuals all showing the same characteristic. This is done simply by writing the number of persons in the symbol (◇3, [3], or (3)).

An even more puzzling pattern may be encountered. When one parent is unknown or irrelevant to the analysis, only one parent is shown:

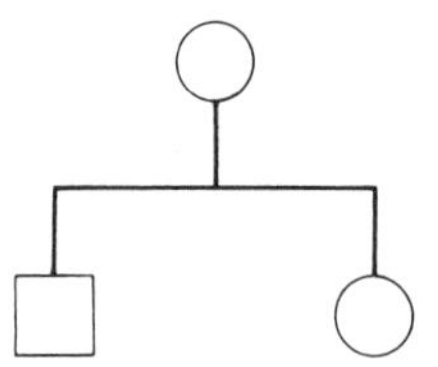

Because pedigrees are commonly used in counseling situations where a detrimental trait is being analyzed, it is often based on one person in the family who was first identified with the trait. Such an individual is called the **proband** or **propositus** and often indicated on the pedigree with an arrow:

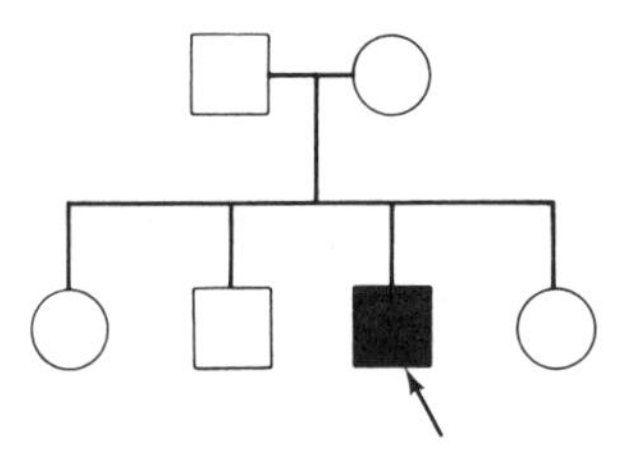

When there is doubt about whether or not a person possessed a genetic trait, it has been the custom to include a question mark within the symbol. If a person in the pedigree has died, notation is sometimes made, with the year of the person's death given.

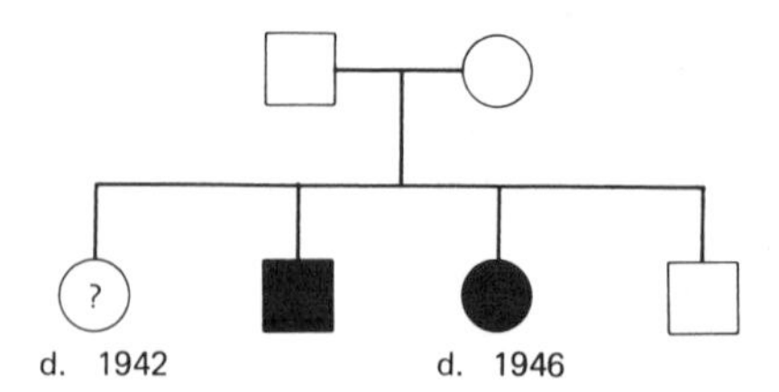

ANALYSIS OF PEDIGREES

Because a pedigree chart in most instances records only physical evidence concerning characteristics, it is necessary to study them to determine the genetic basis for the trait in question. In some cases the explanation is obvious through the use of simple logic. In other situations it is elusive and sometimes an answer is impossible to derive from the data presented.

Simple Recessive Traits

Probably the simplest and most convincing phenotypic pattern which is encountered in a pedigree is seen when neither parent shows a trait, but a child is produced who does. The case of two normally pigmented parents who have an albino daughter is an example:

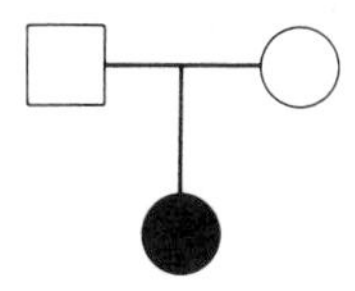

Despite any knowledge about the inheritance of the trait, one would be wise to use Mendel's example and test to determine if the characteristic is based on a single-gene pair showing dominance and recessiveness. It makes no sense to start out looking for a more complex explanation in the absence of other evidence. If the hypothesis is correct, there is only one way to explain the mode of action of this gene for **albinism**: It must be controlled by a **recessive** allele. If it were a dominant trait, one of the parents would be albino. Naturally, assurance must be tempered somewhat because of the rare possibility that a mutation has occurred, or that the gene is not fully penetrant in the parents.

To analyze the individuals in the family, based on this explanation, one would have to assume that the daughter was homozygous recessive for the albinism allele (*aa*). Because she got one of these genes from each parent, they would both have to be heterozygous for the normal-albino alleles (*Aa*).

Such information may be used in counseling situations to make predictions for future children. Knowledge of chance in monohybrid crosses must be utilized. It is obvious that each subsequent child born to the family would have a 1/4 chance of being albino. There is a 3/4 chance of being normally pigmented for each child: 1/2 heterozygous (*Aa*) and 1/4 homozygous dominant (*AA*).

Notice that among normally pigmented children (disregarding for the moment the albinos) 2/3 are heterozygous and 1/3 are homozygous, since only three choices remain when one is not considered (2:1). While consideration of only a segment of the children (normally pigmented ones) may seem unwarranted, it becomes important later in pedigree analysis. Remember also that each child is a separate, independent event in the family and the 3/4:1/4 ratio applies to each, regardless how many prior children have been normal or albino.

This simple family pattern illustrates the method of employment of pedigree charts. It should not be difficult to construct a chart for a family using this information. This simple pedigree was first used to **determine the mode of inheritance** of the trait. Once accomplished, it was possible to **analyze the family** and assign a genotype for each person. Finally, based on this analysis, one could **make mathematical predictions** about the statistical chances for future children.

Simple Dominant Traits

The mode of inheritance of a dominant allele is not so easily recognized as a recessive. Of course, if the dominant trait is being followed as solid symbols in the chart, the hollow symbols represent homozygous recessive individuals and they can be followed as easily as the recessives in the last example. However, it is worth looking at a pattern involving a **dominant** allele simply to anticipate the sort of problems one might encounter.

If one recalls that **PTC taste ability** is based on a **dominant** gene, a pedigree involving the trait can be envisioned:

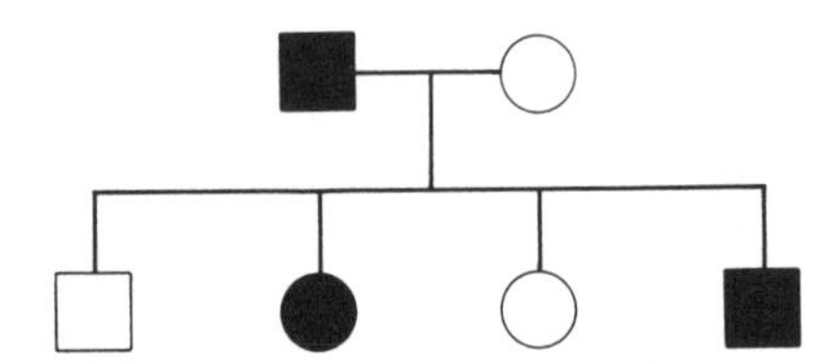

The shaded symbols represent PTC tasters; there would be more than one interpretation possible, if the mode of inheritance were not known. Actually, taste ability is due to a dominant allele and the father is heterozygous (*Aa*). He has passed on the dominant gene to two children (II-2 and II-4). However, there are several explanations that would fit this pedigree.

A plausible second alternative is the interpretation of the allele as recessive, in which case the father would be homozygous recessive (*aa*) and the motor heterozygous (*Aa*). However, the mode of inheritance for taste ability is known and therefore this is incorrect. Usually it is necessary to collect a large amount of information from several families to verify the transmission of a dominantly inherited trait. The major supportive evidence would be a repeated pattern of offspring exhibiting a trait only when shown by one or both parents.

A simple situation does reveal that a gene is dominant. In the following family it is apparent that the trait in question is determined by a dominant gene. It could not be due to a recessive gene.

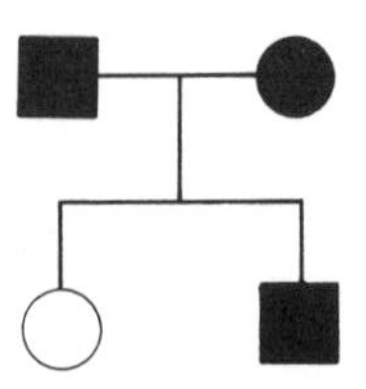

Sometimes it is difficult to be sure of a dominant pattern. Remember that where a dominant allele exists, a recessive allele is also found in the population. A recessive can be recognized much more easily in the same pedigree. Even though the dominant allele is more difficult to follow, if it were a very rare trait, one might choose to trace it in a pedigree. An example of such a trait is **polydactylism**, an extra digit on hands or feet (six fingers or toes). A pedigree involving such a trait could conceivably look like the following chart:

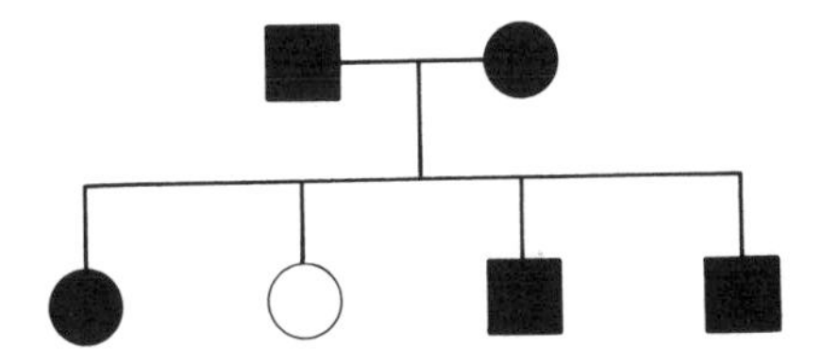

Because the dominant trait is unusual, it is indicated with shaded symbols; by noticing female II-2, one can observe that the normal five-digit situation is recessive. By proving that five digits are due to a recessive gene, it is possible to show that polydactylism is dominant. The fact that dominant traits are often quite rare will be explored in more depth in Chapter 11.

Another dominant allele pedigree might look like this:

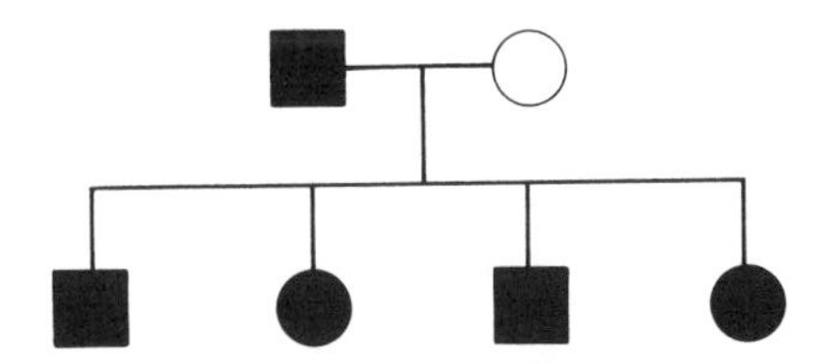

Four children is not a sufficient number to make a strong case, but the possibility exists that the father is homozygous for the dominant allele. While there is no reason to suspect that he carries a recessive gene, it would be unwise to attempt a prediction with any assurance for either possibility. Sometimes it is impossible to determine genotypes completely.

More Complex Traits

Obviously, characteristics which are not controlled by simple dominant or recessive traits present more challenging problems of analysis. Just the mechanism of diagraming a pedigree is much more difficult. The typical system of symbols is often not adequate to prepare a simple chart.

Examples of the sort of problem one encounters are found in cases of codominant inheritance or in traits controlled by multiple alleles. Without going into the detailed problems of analysis or the features of each type of pattern, the following two pedigree charts will give an idea of this sort of system.

First, consider a trait that is based on a codominant gene pair. Because each genotype results in a different phenotype, a system must be devised that gives three different symbols. For example, consider sickle-cell anemia: Remember that persons can have the severe form of anemia (*ss*), the *mixed* hemoglobin state called sickle cell trait (*Ss*), or normal hemoglobin A (*SS*). For the purpose of this pedigree, two different forms of crosshatching (▨ = S hemoglobin, ▧ = A hemoglobin) will be used with both in the same symbol representing the heterozygotes (▩ = S and

A hemoglobins). It is obvious that two heterozygotes may give birth to all three types of children:

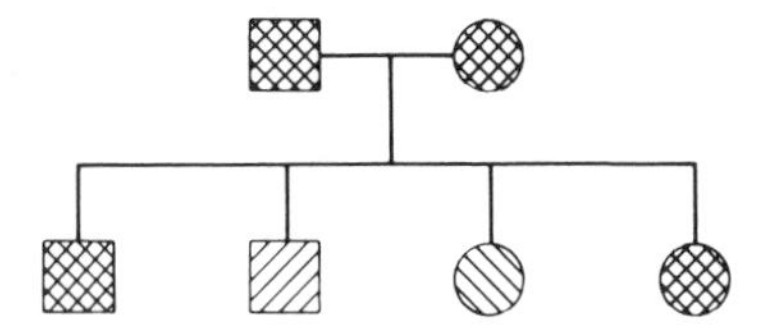

This method could be cumbersome and it is sometimes easier to write identifying letters in each symbol.

A trait that presents a more difficult problem for pedigree analysis is a multiple allele situation, such as ABO blood types. Here not only is codominance encountered between two genes, but there is also a third allele recessive to the other two. Thus, there are four phenotypes based on six potential genotypes. In this case it would be easier to write the letters for each phenotype in every individual's symbol.

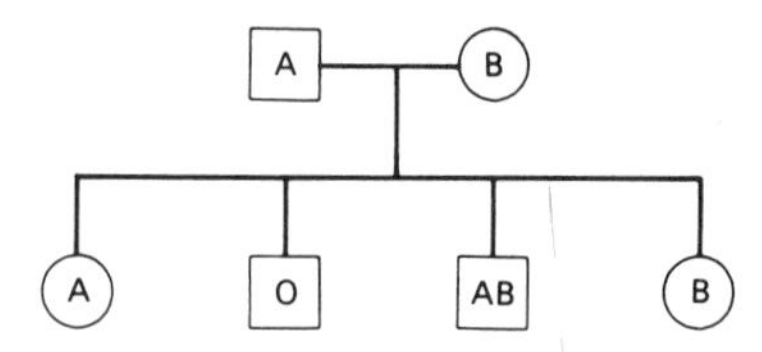

From this pedigree, the parents are heterozygous ($I^{A}i$ and $I^{B}i$). The children represent four different genotypes: $I^{A}i$, ii, $I^{A}I^{B}$, and $I^{B}i$.

Sex-Linked Recessive Traits

As seen, **sex-linked traits** follow a line from father to carrier daughter to grandson. More extensive information is necessary to determine that such a pattern exists. A pedigree chart for a sex-linked trait as **colorblindness** might look like this:

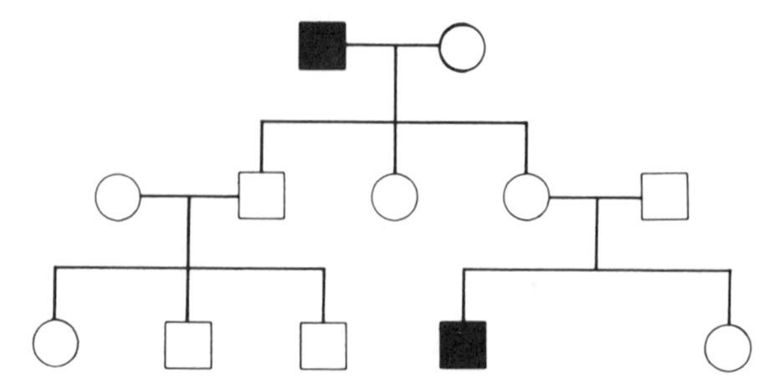

Notice that the chart shows two sets of grandchildren: One set arises from the son (II-2) in the family and the other arises from a daughter (II-4). The trait might be recessive since one finds a child (III-4) showing the characteristic, when neither parent does. However, it has been passed along from female to son, not from male to any of his children. The pedigree might not be totally convincing if the trait were not known to be a sex-linked trait. It might be necessary to seek more information to be confident about an analysis. One should be able to predict that there is a one-half chance for other colorblind sons to the marriage of II-4 and II-5, while the chance of a colorblind daughter is zero. Good clues to the involvement of a sex-linked recessive trait are that a trait skips generations and seems to be passed on from grandfather through mother to grandson.

PREDICTIONS BASED ON PEDIGREE ANALYSIS

One of the major uses of pedigree charts has been the analysis of an inheritance pattern for a trait within a family for genetic counseling. Where a seriously detrimental trait is involved, it is critical to be able to provide a reasonably valid estimate of the chance of gene transmission to offspring. Sometimes life or death situations are at stake, not to mention the financial hardship or extreme mental anguish that might be involved.

It is important to be able to analyze and understand the mechanism of inheritance for a trait in a family. Following this, it is necessary to be able to estimate with accuracy the **genotypes** of involved individuals. Once such knowledge has been gained, it is a simple matter to make predictions based on Mendelian calculations.

Rare recessive traits have a higher chance of appearing in the homozygous condition among the products of close **intermarriage** within a family. It is obvious that two members of the same family have a higher chance of carrying the same mutant allele than two members of the population at large. Many laws and taboos have been formulated to prevent familial intermarriage in societies over the years. Such action, stated or not, is based on experience gained from such marriages involving the appearance of rare and detrimental recessive traits.

Groups that control and authorize marriages in various societies have devised rules to prevent close familial intermarriages. Some churches exercise authority in this regard, as do state laws. Generally the degree of kindred is specified in the permission for marriage by such rules. Commonly in the United States no two persons more closely related than second cousins are given license to marry (e.g., children of first cousins), although some states permit first-cousin marriages.

The following pedigree, exhibiting a family with first-cousin matings, provides a hypothetical example of a situation where counseling for albinism might be important:

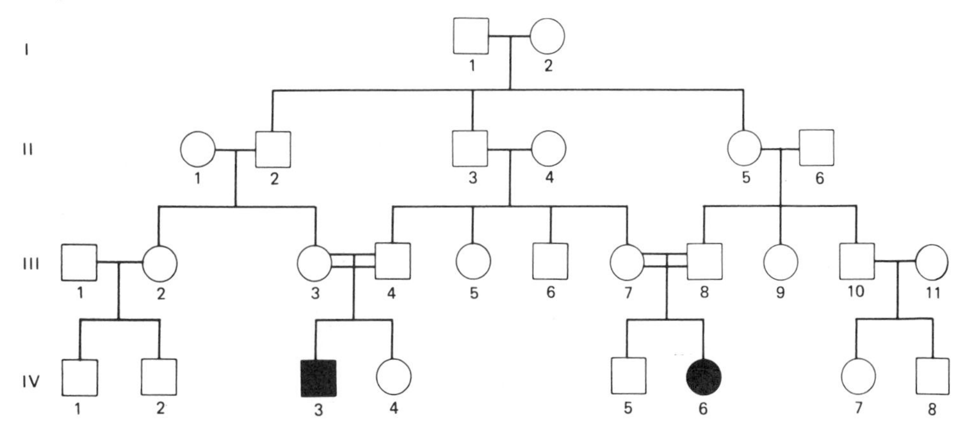

First, it should be pointed out that albinism is quite rare, occurring among only one in 20,000 persons. Thus, it can be assumed that persons marrying into the family from the general population do not carry the rare gene (there would be a chance of about one in 72). One could be wrong in making such an assumption, but the chance of being right is greater.

One should be able to determine quickly that the trait is due to a recessive allele. With such information, persons IV-3 and IV-6 are homozygous recessive (*aa*) and their parents (III-3 $\times$ III-4 and III-7 $\times$ III-8) are heterozygous (*Aa*). Notice that both albinos come from **first-cousin matings,** indicating strongly that one of the first parents in the family (I-1 or I-2) carried the recessive allele (*Aa*). Another convention sometimes used in diagraming pedigrees is the double mating line indicating **consanguineous matings** (between close relatives) (Note its use in preceding pedigree).

Now that an analysis has been made, one can make predictions in a counseling situation. Suppose that persons IV-1 and IV-4 wish to get married but recognize that albinism, a hereditary trait, exists in their family. They might like to know what the chances are that their child would be an albino. Intuitively, one would probably assume that first-cousin matings in a family carrying the trait run a higher than normal risk of producing children with the defect. In this case a stronger statement should be made, since one of the proposed parents has a brother who is an albino.

Actually, it is possible to assign a numerical value to such a chance. The chance that each parent carries the gene must be determined and such independent chances must be multiplied by the probability that the gene could be passed along by each parent to a potential offspring.

Each proposed parent is normally pigmented, but both might be carrying the recessive gene. The only way they could have an albino child would be if they were both heterozygotes. Of course if each parent is a heterozygote, the chance of having an albino child is one-fourth.

To determine the chance that the male IV-1 is a **carrier** of the *a* gene, one must look first at his parents. His father (III-1) and his grandmother (II-1) both married into the family from the general population and it can be assumed that neither of them carries the rare gene (*AA*). His mother, though, is uncertain. It is apparent that his grandfather (II-2) was a carrier (*Aa*), since his daughter (III-3) gave birth to an albino son (IV-3). This means that there is a one-half chance that the woman in question (III-2) is a carrier (*Aa*):

$$\begin{array}{c} AA \times Aa \\ \downarrow \\ 1/2\ AA \\ 1/2\ Aa \end{array}$$

The logic behind this is based on II-1 being *AA* and II-2 being *Aa*. Thus, there is also a one-half chance that she is not a carrier (*AA*); however, the chance that IV-1 *is* a carrier is still in question.

If III-2 is a carrier (*Aa*), which has a probability of one-half, there is a one-half chance that the gene has been passed on to IV-1 (again, $AA \times Aa = 1/2\,Aa$). Thus, there are two independent steps involving one-half chance each in the chain. Therefore, the law of simultaneous independent events says that there is a one-quarter chance ($1/2 \times 1/2$) that IV-1 is a carrier.

The chance of IV-4 being a carrier is somewhat different. First, both of her parents were carriers (*Aa*) since she has an albino brother. A monohybrid cross provides a one-half chance that an offspring will be a heterozygote. However, notice that the female is not *aa* because she is not an albino. This means that she can be either *AA* (1/4) or *Aa* (1/2). In reality, with one choice ruled out, she has a one-third chance of being *AA* and a two-thirds chance of being *Aa*. Because of the sole interest in the chance of being a carrier, one can say that there is a two-thirds chance of carrying the albinism gene.

Now a calculation can be made based on three independent individuals (IV-1 = 1/4, IV-4 = 2/3, and 1/4 chance for the hypothetical child they might produce). The analysis was started with several "ifs" and each one of these now has a tentative predictive number assigned to it. The chance they might have an albino child would be $1/4 \times 2/3 \times 1/4 = 2/48 = 1/24$. This may not sound like a high statistical chance but it is much higher than the 1/20,000 chance occurrence of albinos in the population.

Remember the earlier intuitive prediction that there would be higher risk than normal in such a first-cousin mating. Suppose that IV-4 had requested information about the risk of albino children if she married IV-5. Now an intuitive prediction would have led to the conclusion that the risk would be greater in this case, since all four potential grandparents would be from the same family (none from the outside population). Furthermore, both potential parents (IV-4 and IV-5) have siblings who are albinos.

Mathematically, the greater risk prediction would prove to be correct. As seen, the chance that IV-4 is *Aa* is two-thirds. The same line of reasoning will show that the chance for IV-5 to be a carrier is also two-thirds. The calculation based on three independent chances in this case (IV-4 = 2/3, IV-5 = 2/3, 1/4 chance for albino *if* both parents are *Aa*) results in a chance of 4/36 or 1/9. This is almost three times as great a risk as in the proposed IV-1 × IV-4 mating, so obviously the intuitive predictions were logically sound.

The ultimate decision to have children in situations such as these resides with the potential parents. They must decide whether or not the risk of a 1/24 or 1/9 probability is too high when weighed against the incapacity or hardship caused by the condition. Notice that if they do have an albino child, the predictions must change. The chance that the parents are both carriers has been eliminated (1/4 and 2/3). These points in the pedigree no longer represent "ifs." It is definite that they are both *Aa*. Thereafter, one would say that there is a one-fourth chance for each subsequent child to be an albino.

One might intuitively predict that second-cousin matings in the family would have a smaller chance of producing albino children than these first-cousin matings. To prove such a statement numerically, suppose that IV-4 had planned to marry IV-8.

In tracing IV-8's lineage, it is found that the first certain carrier in his ancestry is his grandmother (II-5). She had a one-half chance of passing on the gene to her son (III-10) and he had a one-half chance of passing on the gene to the potential father (IV-8). Therefore, the chance that IV-8 is a carrier is one-fourth ($1/2 \times 1/2$).

The chance for an albino child from this second-cousin marriage (IV-4 × IV-8) is one in 24 (2/3 × 1/4 × 1/4 = 2/48 = 1/24). The chances are the same as the first-cousin mating considered first (IV-2 × IV-4). The risk is clearly less than the second cross analyzed (IV-4 × IV-5). Usually risks are lower in second-cousin matings, but some second-cousin marriages could have a higher probability than first-cousin marriages. Such predictions depend on evidence which indicates the presence or absence of the *a* gene. It is essential in all such analyses to have the most complete information included in the pedigree.

QUEEN VICTORIA PEDIGREE

The Queen Victoria pedigree, the most famous and most often published pedigree, is shown in Figure 9.1. It is interesting because it traces a very rare sex-linked trait, **hemophilia**. The pedigree has been studied frequently because it shows the flow of this rare trait through European royalty. Because it is rare (probably about 1 per

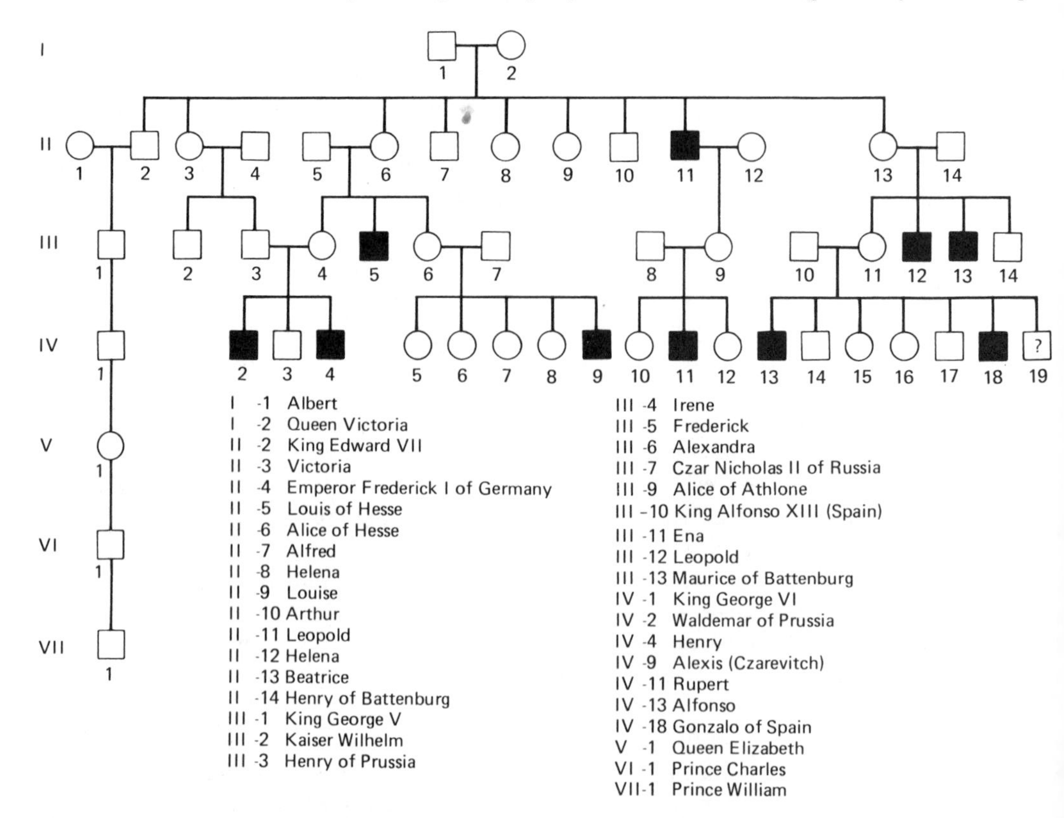

Figure 9.1. The Queen Victoria pedigree of hemophilia (incomplete). The mutant gene for this rare sex-linked recessive trait apparently arose in one of the gametes that gave rise to Queen Victoria herself. She passed it on to at least three of her children and it was carried to other European royal families by marriage.

10,000 males), a high proportion (about 30%) of those who have the gene received it as a new mutation. As seen in Chapter 8, Victoria apparently received the gene as a new mutation in one of the gametes that produced her.

The pedigree is incomplete: There are so many individuals in the many branches of the family that it is difficult to include them all. It is evident that members of the present British royal family are free from the *h* allele as they are descended from Edward VII who was free of the disease.

A reality of the disease and constant hazard is also part of the story, since Rupert (IV-11), Alfonso (IV-13), and Gonzalo (IV-18) were all killed as a result of uncontrollable bleeding from fairly minor injuries in automobile accidents.

In the absence of experimental procedures it is possible to determine the genetic basis for some human traits using family pedigree charts. One can analyze the passage of a gene through several generations of a family and estimate numerically what the chances are for a given genotype in specific individuals. It is even possible in certain instances to be sure of a person's genotype. Once the mode of inheritance and the genotypes, or the probability of certain genotypes, is understood, predictions for the chance of the trait's occurrence in future generations can be made. Thus, the pedigree chart becomes an indispensable tool to the human geneticist.

Pedigrees in real situations are not always as easily constructed or analyzed as some of the contrived examples used here, but with care it is possible to record a large amount of worthwhile information in a small space. By the use of logic, much solid information can be provided for those who want or need to know about the action of a human gene.

ADDITIONAL READING

FUHRMANN, W., and F. VOGEL, *Genetic Counseling.* English translation by S. Kurth. New York: Springer-Verlag, 1969.

MCKUSICK, V. A., *Human Genetics* (2nd ed.). Englewood Cliffs, N.J.: Prentice-Hall, 1969.

STINE, G. J., *Biosocial Genetics.* New York: Macmillan, 1977.

THOMPSON, J. S., and M. W. THOMPSON, *Genetics in Medicine* (3rd ed.). Philadelphia: Saunders, 1980.

REVIEW QUESTIONS

1. (a) Neither grandmother nor grandfather on the father's side manifests ocular albinism (the nearly complete absence of pigmentation from eyes). Grandmother on the mother's side does not manifest ocular albinism and neither does the mother. However, the mother's father (grandfather) does have the defect. The mother and father have four children: Their first son does not manifest ocular albinism and neither do the second and third children, both girls. Their fourth child, a boy, does have the trait.

Diagram a pedigree for the family.

(b) Based on this pedigree, what is the most logical genetic mode of inheritance of ocular albinism?

(c) Give the genotypes of each person in this pedigree as completely as possible.

2. The following pedigree is for galactosemia, a disease due to the lack of an enzyme which converts a milk sugar (galactose) to glucose. Without the enzyme, galactose accumulates in an infant's body, causing liver damage and mental retardation soon after birth.

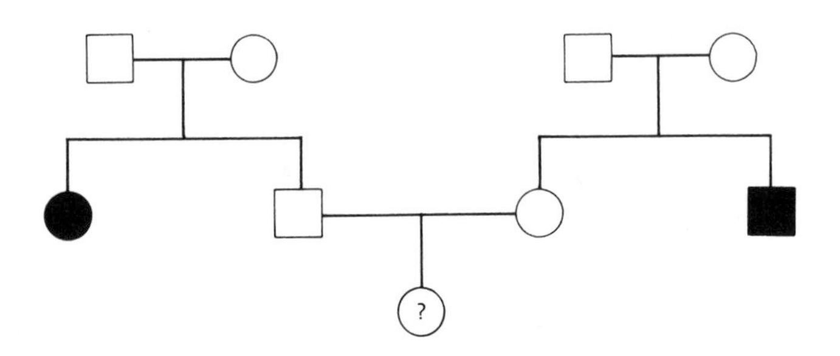

(a) What is the mode of inheritance of galactosemia?

(b) What is the chance that newborn female III-1 will have the disease?

(c) Can you suggest a way to prevent liver and brain damage in an infant?

(d) Can you suggest a possible means of identifying a potential parent as a heterozygous carrier of the galactosemia allele prior to marriage or conception?

3. What is the most likely mode of inheritance of the trait in the following pedigree? Is it possible that another mode of inheritance can be applied correctly for the information given? Explain.

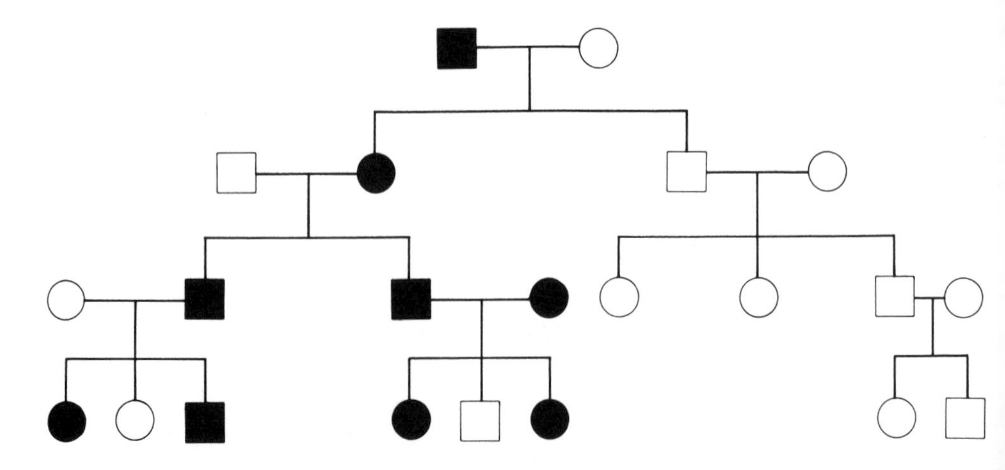

4. The following pedigree is for a very rare trait, nystagmus, a constant involuntary cyclical movement of the eyeballs.

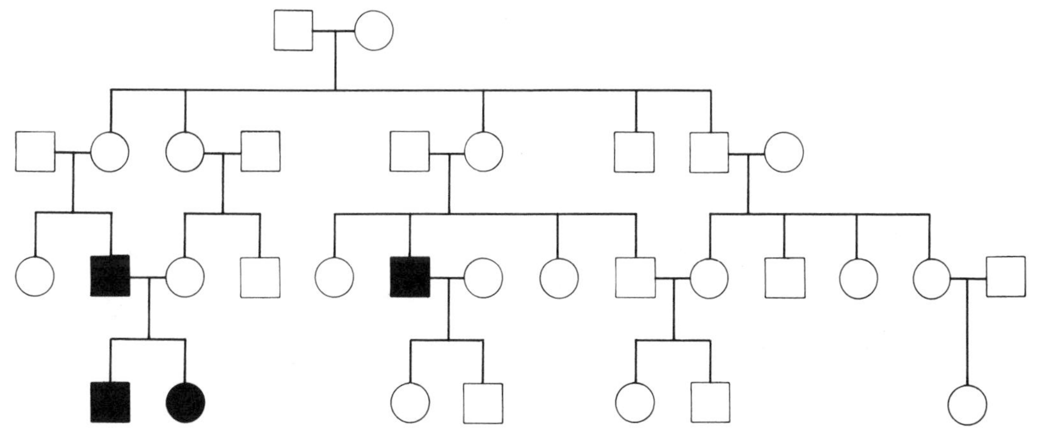

(a) What is the mode of inheritance of nystagmus, as indicated by the pedigree?

(b) What are the genotypes of the three females in the second generation (II-2, II-3, II-6) and females III-3 and III-10?

(c) In a marriage of III-1 and III-4 what is the chance for a child with nystagmus? Is the risk different for male and female children?

(d) If IV-1 married IV-3, what would be their risk for sons with nystagmus? For daughters with nystagmus?

5. Give the genotypes for the hemophilia alleles as completely as possible for the following persons in the Queen Victoria royal pedigree (Fig. 9.1):

Queen Victoria	Alice of Athlone
Queen Elizabeth	Alice of Hesse
Prince Charles	Waldemar of Prussia
Alexandra	King Alfonso XIII (Spain)
Czar Nicholas II	Gonzalo of Spain
Czarevitch Alexis	Maurice of Battenburg
Beatrice (daughter of Victoria)	Henry of Prussia
Irene (granddaughter of Victoria)	

6. In the example of a pedigree of a sex-linked trait given on page 226, what is the chance that II-3 is a carrier of the gene? What is the chance that individuals III-1 and III-5 are carriers?

7. In the pedigree for albinism given on page 227, what is the chance that individuals II-2, II-5, III-4, III-6, and III-9 are carriers of the recessive gene? If IV-2 marries IV-7, what is the chance that their child will be albino?

8. Go back to the Chapter 7 Review Questions on page 194 and construct pedigree charts for Questions 5 and 6. Are the problems easier to solve using this method?

9. Construct pedigree charts of the families described in Questions 15(a) and 15(b) at the end of Chapter 7 (p. 195). Can you devise a way of using a single chart to diagram 15(c)? Use the same sort of device to diagram a pedigree for the family described in Question 19.

10. Prepare a pedigree chart for the situation described in Review Question 10, Chapter 8 (p. 218). Provide as much detail as possible from the information given. Is the problem easier to understand or explain using a pedigree chart?

CHAPTER SUMMARY

1. A systematic and convenient means of recording and tracing genetic information within families is the pedigree chart. Although symbols are not universal, a common style is used by all human geneticists and knowledge of a few principles permits the construction and use of pedigrees. Such charts are necessary in many situations for genetic counselors.
2. Pedigree charts are composed of circles (females) and squares (males) connected by a series of horizontal and vertical lines that indicate relationships. Solid symbols indicate that the person represented by a symbol shows a trait in question.
3. In many pedigrees, if enough information is provided, the mode of inheritance can be determined. Recessive and codominant traits show a distinctive pattern. Sex linkage can often be identified and there are sometimes clues that a trait is controlled by a dominant gene.
4. It is often possible to determine the genotypes of family members, based on information provided by the pedigree. Following such analysis, mathematical predictions can be made concerning future children. Once such patterns are established, genetic counseling can be provided.

CHAPTER

10

POLYGENES AND QUANTITATIVE INHERITANCE

You cannot conceive the many without the one.
Plato, Dialogues *(Ca. 400* B.C.*)*

All traits studied so far have been shown to result in clear-cut, easily recognized genetic ratios from various crosses. This is because the characteristics selected for consideration have exhibited **sharply defined phenotypes.** Of course, Mendel likewise selected the traits in pea plants which he used to demonstrate the underlying mechanisms of genetics. Although some minor deviations from his principles have been seen and examples involving multiple alleles have been studied, these patterns still involve clear-cut, distinct phenotypes.

In 1910 Nilsson-Ehle in Sweden recognized that not all traits under the control of genes show distinct, easily categorized phenotypes. Some obviously inherited characteristics do not seem to fit a pattern which produces clear-cut ratios in controlled crosses. It was noticed that some traits result in a spectrum of variation ranging from one phenotypic extreme to the other. Such traits are referred to as showing **continuous variation,** which distinguishes them from the classic examples of discontinuous phenotypes so far studied.

CHARACTERISTICS OF QUANTITATIVE INHERITANCE

Continuously variable traits do not fit typical Mendelian descriptions and ratios. Furthermore, one cannot conduct experiments that produce distinct categories fitting the classic ratios. Remember that, even though the ratios are not apparent, the basic Mendelian mechanisms are still operating. In fact, the ratios are simply not recognizable, because the traits are under the control of more than one pair of genes, each exercising a **cumulative effect** on the phenotype. Because there is more than one pair of genes operating to produce the phenotype, they are sometimes called **polygenes** (poly = many).

Because a whole spectrum of gradations in phenotype from one extreme to the other is obtained, it has proved appropriate to describe the characteristics with some unit of measure, such as intensity of color, height in millimeters, weight in pounds, or another highly variable classification. The study of such characteristics has thus come to be known as **quantitative genetics,** to distinguish it from the more familiar qualitative type of study discussed up to this point..

IDENTIFICATION OF POLYGENIC INHERITANCE PATTERNS

Nilsson-Ehle's discovery of quantitative inheritance was made with his studies of wheat kernel color. He observed that there is a complete spectrum of shades of red, ranging from white to the darkest red, in the seeds of cultivated wheat. It was found that an intermediate shade of red resulted in the offspring from a cross of a white-kerneled plant with one having the darkest red seeds. This is what might be expected if a single-gene pair were controlling the characteristic and there was no dominance. However, when two resultant F_1 plants having intermediate red kernels were crossed to each other, Nilsson-Ehle noticed that about one-sixteenth of the offspring were white, instead of the anticipated one-fourth. The figure 1/16 indicated that two pairs of genes were operating, since it was well known that double recessives (*aabb*) occurred to this extent in dihybrid crosses.

From this clue Nilsson-Ehle was able to predict that two independent pairs of genes were operating to control a single characteristic. He postulated that there were two pairs of **codominant genes,** and that the genes acted in cumulative fashion. It is important to an understanding of the basic mechanism to assume that these genes are physically independent and follow the laws of chance as in the other traits studied.

If each separate gene is envisioned as contributing to the color in the dominant form, while adding nothing in the recessive form, one can construct the same sort of hypothesis as Nilsson-Ehle. Remembering the classic results of two-factor problems, the various genotypes can be categorized into five basic types:

1. No red color genes *aabb*
2. One red color gene *Aabb, aaBb*
3. Two red color genes *AaBb, AAbb, aaBB*
4. Three red color genes *AABb, AaBB*
5. Four red color genes *AABB*

These would account for five different seed colors, showing a darker red intensity as the number of red color genes (capital letters) increases.

Nilsson-Ehle's experiments can be summarized in routine fashion by using the classical notation system that has been used for other traits:

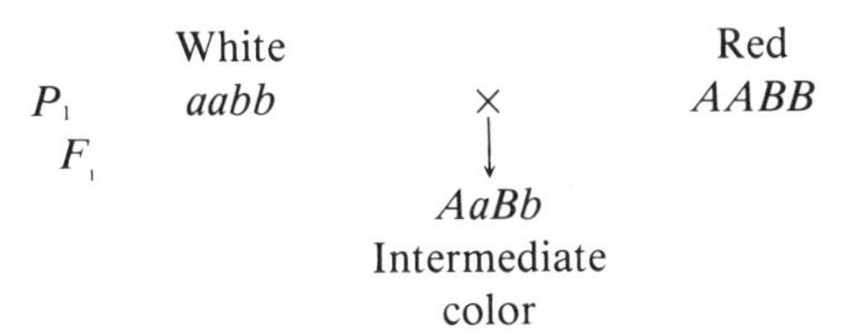

A cross of two heterozygous intermediate red-kernel plants would produce a typical dihybrid F_2 generation in regard to gene distribution. However, without dominance and with the two gene pairs affecting the same characteristic, a distinct modification of the 9:3:3:1 ratio is seen. The results of such a cross are shown in the following Punnett square:

Intermediate $AaBb$ × Intermediate $AaBb$

	AB	*Ab*	*aB*	*ab*
AB	*AABB* 4	*AABb* 3	*AaBB* 3	*AaBb* 2
Ab	*AABb* 3	*AAbb* 2	*AaBb* 2	*Aabb* 1
aB	*AaBB* 3	*AaBb* 2	*aaBB* 2	*aaBb* 1
ab	*AaBb* 2	*Aabb* 1	*aaBb* 1	*aabb* 0

The Punnett square is used in this case to show the numbers of uppercase letters in each genotype. Keeping in mind that these uppercase letters represent genes for red color, and that they have a cumulative effect, it may be seen that five types of offspring are produced. The F_2 results may be summarized: 1/16 have four color genes and are the darkest red, 4/16 (1/4) have three and are dark red, 6/16 (3/8) have two and are intermediate in color, 4/16 (1/4) have one and are light red, while 1/16 have none and are white.

BINOMIAL TECHNIQUE AND POLYGENIC TRAITS

While the Punnett square may be useful to visualize the results of a two-factor quantitative trait, there is a mathematical technique which is simpler to interpret. As seen in earlier sections, when more than two traits are analyzed simultaneously, the Punnett square becomes almost unmanageable.

Using the two-factor polygenic system of wheat kernel color as an example, one can utilize a fundamental maneuver that serves as the basis for many statistical predictions. The procedure is known as the **expansion of a binomial;** "binomial" is used because there are two forms of each gene (*A* or *a*, *B* or *b*, etc.). It may also be applied to other genetic situations that are based on two alleles for each trait, but there has been no need to utilize it in the simpler situations studied.

In the cross shown as a Punnett square, there are four different genes involved and each can be either a red color-producing gene or a no color-producing gene. Mathematical abstractions can be used to represent the chance for a particular gene to be present in an individual. Thus, x can symbolize the chance for a capital letter gene and y can symbolize the chance for a lowercase letter gene.

The symbol $x + y$ then represents the total opportunity for color genes at each position (either *A* or *a*, *B* or *b*, etc.). When four positions (loci) are involved, each one being occupied by a capital or lowercase letter, the distribution of results is based on the product of the independent probabilities (that is, position one × position two × position three × position four). This is equal to $(x + y) \cdot (x + y) \cdot (x + y) \cdot (x + y) = (x + y)^4$

There is a simple, although somewhat cumbersome, means of expanding the binomial. It is based on a progression which assumes that the first item in the expansion is always the first character (x) expanded to the full power. In the expansion $(x + y)^4$ the first term would then be x^4.

To find subsequent terms in the expansion, the **exponent** of each x term must be multiplied by its **characteristic** number and then that product divided by the **number of the term** in the sequence. Notice that the number of the term in the sequence can be found simply by counting from left to right. One can determine the exponent of each term by calculating that the exponent of x decreases by one in amount as one shifts from one term to the next in the sequence. Likewise, the exponent of y increases one in amount as one progresses along the sequence. All that is required to be calculated for the completion of the expansion is the characteristic number for each term.

To follow the example of the four-gene (two-pair) situation, the following pattern of exponents is derived for $(x + y)^4$:

$$x^4 + x^3y + x^2y^2 + xy^3 + y^4$$

Note that the first term is the first chance expanded to the full power of four. In each subsequent term x decreases by one and y increases by one in exponent. Represented of course are the five classes of red color. If x represents the capital letter genes, then x^4 (four capital letters) is the darkest red and y^4 (four lowercase letters) is the white.

To determine the chance for each shade of color in the spectrum, first calculate the characteristic number for each term in the sequence. Because the characteristic number of the first term in the sequence is always 1, multiply 1 (characteristic number) × 4 (exponent of x in the first term) and divide by 1 (the number of the term in the sequence). This yields 4, the characteristic number of the second term in the sequence.

To determine the third characteristic number, multiply 4 (characteristic) × 3 (exponent of x) and divide by 2 (second term in the sequence), resulting in 6 for the characteristic number of the third term.

When the sequence has been completed, the full binomial expansion looks like this:

$$(x + y)^4 = x^4 + 4x^3y + 6x^2y^2 + 4xy^3 + y^4$$

To utilize the expansion, insert the numerical chance for capital letter genes *(x)* and lowercase letter genes *(y)* in the formula and obtain numerical values for the chances of each color shade in the spectrum from a given cross.

If the cross is between two heterozygous parents *AaBb* × *AaBb,* then the chance at each of the four positions for an uppercase or a lowercase letter gene is one-half (1/4 *AA,* 1/4 *Aa,* 1/4 *aA,* 1/4 *aa* and 1/4 *BB,* 1/4 *Bb,* 1/4 *bB,* 1/4, *bb*):

	Position 1	Position 2		Position 3	Position 4
	A	*A*	1/4 chance each	*B*	*B*
	A	*a*		*B*	*b*
	a	*A*		*b*	*B*
	a	*a*		*b*	*b*
Chance for capital letter	1/2	1/2		1/2	1/2

Thus $x = 1/2$ and $y = 1/2$. The darkest red offspring will be found among 1/16 of the offspring of such a cross ($x^4 = (1/2)^4$); white offspring likewise have a 1/16 chance ($y^4 = (1/2)^4$).

The most commonly found type would be the intermediate red color with a 3/8 chance of occurrence ($6x^2y^2 = 6 \cdot (1/2)^2 \cdot (1/2)^2$). If graphed, a simple **normal distribution curve** would result:

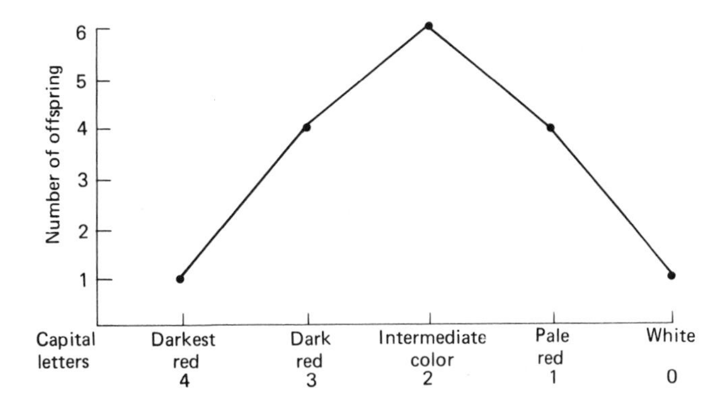

It is sometimes the practice to describe a population by giving the **average** or **mean** size, color, or other parameter. With a small population (or family), as shown in the normal curve, the mean is two (intermediate color). Another descriptive term occasionally used is the **median.** The median is a point on the curve at which half the population lies above in value and half lies below. In a perfect binomial curve as diagramed, and in the next graph showing $(x + y)^8$, the median and mean are at exactly the same point (two and four capital letters, respectively).

An additional factor often used in describing a population, the **variance,** takes into account the number of individuals and the amount of deviation from the mean of these individuals. In calculating variance the plus or minus deviations from the mean are squared and summed. This sum of squares is then divided by the number of individuals in the population. The variance can then be used as a comparative value to describe how closely the population is clustered around the mean. It gives a rough estimate of how accurately the mean represents the group. Likewise, the variance indicates how much diversity or heterogeneity exists within the group for a particular trait.

If the number of gene pairs controlling a trait in cumulative fashion were four, instead of two as found in Nilsson-Ehle's wheat studies, more phenotypic classes would exist (nine instead of five). If $(x + y)^8$ is expanded and the results graphed, with the larger number of points a smoother normal curve is the apparent result. Now there would be only 1/256 chance for an offspring to look like the homozygous grandparents (1/256 chance for *AABBCCDD* or 1/256 chance for *aabbccdd*).

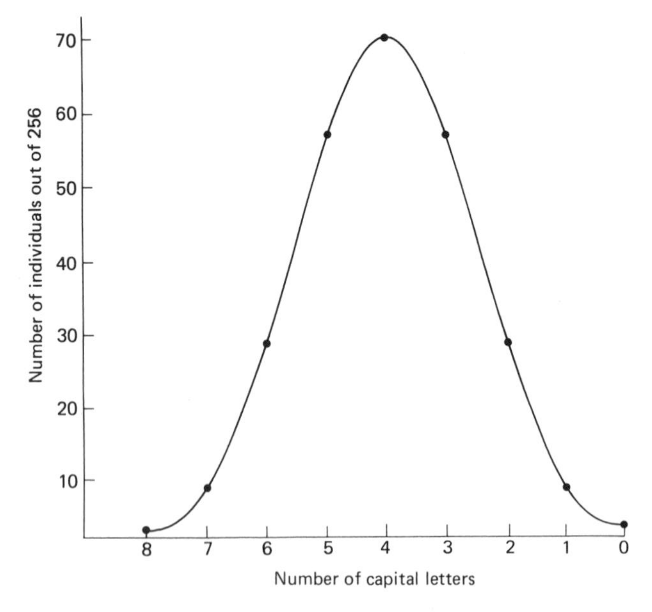

It should be apparent that the method of determining the number of gene pairs involved in controlling a quantitative trait is based on a cross of two individuals showing the intermediate phenotype. In humans this means finding as many marriages of persons showing such phenotypes as possible, because of the limited number of offspring. Once the proper crosses have been made or identified, it is necessary to estimate the proportion of the extreme phenotypes produced in the spectrum of offspring. Even under the best conditions, in a simple situation, this process can be difficult. In complex situations it can be virtually impossible.

Table 10.1 indicates the proportion of offspring expected to be homozygous

TABLE 10.1. Heterozygous Crosses Involving Quantitative Genes.			
Number of hetero-zygous gene pairs in parents	Binomial expansion involved in hetero-zygous crosses	Proportion of offspring having all capital letter genes or all lowercase letter genes	Number of classes of offspring
1 *(Aa)*	$(x + y)^2$	1/4	3
2 *(AaBb)*	$(x + y)^4$	1/16	5
3 *(AaBbCc)*	$(x + y)^6$	1/64	7
4 *(AaBbCcDd)*	$(x + y)^8$	1/256	9
5 *(AaBbCcDdEe)*	$(x + y)^{10}$	1/1024	11
6 *(AaBbCcDdEeFf)*	$(x + y)^{12}$	1/4096	13
n *(AaBbCc . . . ??)*	$(x + y)^{2n}$	$1/(2)^{2n}$	$2n + 1$

for a quantitative gene among the offspring of heterozygous crosses. Each line in the chart represents an increasing number of controlling gene pairs. As the number of controlling gene pairs that determine the trait increases, the probability of encountering the extremes in the spectrum decreases and the number of classes increases.

HERITABILITY

To describe the interaction of environmental factors and heredity on the phenotype of an organism, geneticists often utilize a quantity known as **heritability.** Heritability is expressed as a number derived from a mathematical formula. Unfortunately, it is a complex idea, difficult to work with, and often unrealistic where human genetics is involved.

One can develop a rather superficial idea of heritability if it is remembered that the variance in a population measures the deviation from the mean present among the collective members of the sample. If perfect binomial distributions as shown in the preceding two graphs are found the variance would all be due to the genes alone. However, in practically all traits, some variations would be due to environmental factors and would produce distortions from expected results. This means in effect the **total variance** (V_{tot}) would be the sum of **genetic variance** (V_{gen}) and **environmental variance** (V_{env}):

$$V_{tot} = V_{gen} + V_{env}$$

To express how much variation is due to genetic factors as a portion of the total variation, one might write a somewhat crude formula for heritability:

$$\text{Heritability} = \frac{V_{gen}}{V_{gen} + V_{env}}$$

or

$$\text{Her} = \frac{V_{gen}}{V_{tot}}$$

Heritability is sometimes abbreviated h^2, because of the squaring process used to derive the variance.

It is important, especially in human genetics, to note that a heritability figure calculated in this way can only be applied correctly to the population from which it is derived and only under given conditions of measurement. Controlling environmental factors and properly assessing the genetic component of many human traits is extremely difficult, if not impossible. Great caution must be exercised in applying the heritability concept to human studies.

HUMAN POLYGENIC TRAITS

It is evident that many human hereditary traits fit a quantitative pattern. Thus, some traits one might like to study and understand often present bewildering problems due to their control by several gene pairs.

The problems noticed before in studying human inheritance patterns, such as small families and the inability to control crosses, present even greater difficulties in quantitative traits. Traits that are apparently based on polygenic systems can be identified, but in no case has it been possible to provide unequivocal analyses of the number of genes and their basis of action.

Complexities of Polygenic Traits in Humans

Within small human family units it is virtually impossible to ascertain the true mechanism of action of quantitative traits. Because there are many human polygenic traits, many inheritance patterns cannot be thoroughly understood. This is in spite of the fact that these difficult characteristics are based on the same underlying fundamental pattern of Mendelian genetics as single-gene determined traits.

To complicate further the ability to understand human polygenic systems, quantitative traits are generally subject to profound effects of **environmental circumstances.** Slight differences in the physical or chemical surroundings may cause variations in the phenotype which are greater than those due to the genes involved. Even if a reasonable estimate of heritability is possible, its application may be restricted to a single, well-defined population or segment of a population.

The problem is compounded still further if the different genes that affect the characteristic are physically located on the same chromosome. The results of **linkage** will confuse the ratios to such an extent that the genetic basis is likely to be unrecognized entirely.

It is no wonder that some human traits defy genetic analysis. One must be cautious with any assumptions about hereditary patterns or predictions involving polygenic traits. However, in an attempt to understand mechanisms, apparent cases of quantitative phenotypes in admittedly oversimplified fashion can be examined and a reasonable impression of the action of an important category of genes can be gained.

Human Skin Color

Human **skin color** seems to fit the pattern of a quantitative trait and it is most often cited as an example of such a characteristic. However, the pattern is not clearly understood and the number of genes involved has not been definitely established. Likewise, the exact mode of action of such genes is not known. While one can speculate knowledgeably, the true mechanism remains somewhat vague.

Gertrude and Charles Davenport postulated in 1913 that two gene pairs were involved in the determination of amount of **melanin** or black pigment in human skin. While the Davenports' idea that skin color is a polygenic trait is not seriously questioned, it is obvious that the full story of pigmentation heredity is not adequately explained by their simple model. The implication is, as seen in wheat kernel color, that all hybrids between black and whites can be conveniently categorized into five skin color classes.

Even allowing for great variation caused by environmental factors, such as tanning due to sunlight exposure, it is not possible to simplify the situation to this extent. The shades of color caused by amount of melanin pigment vary continuously throughout the population. Any grouping into categories is necessarily artificial.

The useful ideas that are provided by the Davenports' research explain that skin color determination involves more than one gene pair, and that the genes show a codominant, cumulative pattern.

The Davenports' work was done under ideal, naturally controlled conditions in Jamaica and Bermuda, where marriages among blacks and whites were fairly common and the ancestry of the parents could be traced fairly clearly. The model devised to explain resultant skin colors among offspring is highly instructive in understanding inheritance of varying amounts of black color pigment (melanin) in human skin.

To investigate the phenotypes of individuals studied, the Davenports utilized colored paper disks which were spun on tops to blend the white, black, red, and yellow colors together in an attempt to match skin colors (Fig. 10.1). The major pigments which were varied in the proportionate blends were, of course, black and white. In matings between black parents (about 80% black pigment in the blend) and white parents (about 20% black pigment), the offspring showed an intermediate skin color (about 45% black pigment). Such data verified the codominant nature of the genes and led to the thesis that one allele of a pair is responsible for producing melanin pigment, while the other member of the gene pair does not contribute to melanin production. Obviously, even so-called white persons produce a certain amount of melanin pigment, so that a basic amount is produced by all individuals except albinos, and the black skin color genes add more than this basic amount.

It is important again to stress the strong influence of environment. Melanin pigment is produced in increasing amounts in skin that is exposed repeatedly to the ultraviolet rays of sunlight, in both blacks and whites alike. Attempts were made to alleviate this factor by observing areas of the skin that are not subject to such repeated exposure, but even this technique does not eliminate environmental differences completely.

The real test of the multiple gene hypothesis is in observing the results of matings

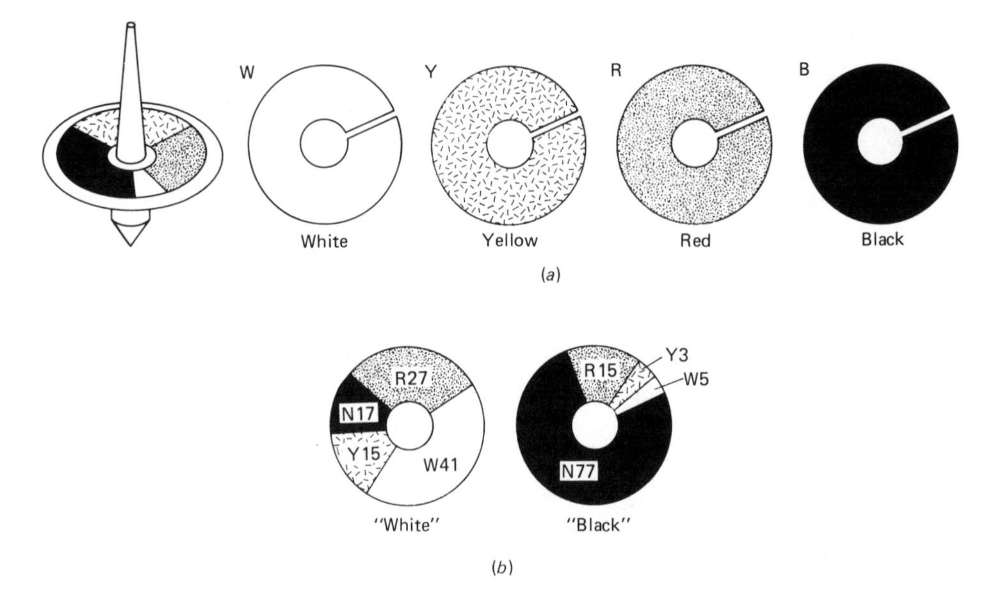

Figure 10.1. The tops utilized by Gertrude and Charles Davenport to investigate the inheritance of human skin color. The spinning tops matched various skin colors. Overlapping colored paper disks (a) seemed to blend together to a uniform color when the tops were spun. Two examples of the tops with colored disks set to match white and black skin are shown in (b) with the percentage of each color given. (With permission from *Natural History,* Vol. 26, No. 1; Copyright the American Museum of Natural History, 1926.)

between two offspring (intermediate skin color) of the black-white marriages. In spite of the useful data collected in the Davenport study, only six families showing the desired marriages were observed with 32 grandchildren from matings of intermediate parents. These offspring could be classified into a number of different skin color categories, ranging between the extremes of the grandparents, with some lighter and some darker than their parents. The number of examples is too small to serve as the basis for an explanation, but it verifies the cumulative, multiple nature of the inheritance pattern.

Because the Davenports found about 1/16 of the offspring fit the white description, as Nilsson-Ehle found in wheat kernel color, they thought two pairs of genes might be operating. Based on this interpretation, they grouped the offspring into five categories of skin colors paralleling the five possible genotypes.

The genes involved can be labeled A,a and $B,b,$ with the capital letters symbolizing melanin-producing alleles and the lowercase letters symbolizing melanin-nonproducing genes. Such separate genes would be equal and cumulative in their effect. However, the allelic pairs would show Mendelian segregation and the nonhomologous genes would show Mendelian independent assortment. There would be five phenotypes based on the number of melanin-producing genes involved:

Number of melanin-producing genes	Possible genotypes	Skin color phenotype
4	*AABB*	Black
3	*AABb, AaBB*	Dark
2	*AaBb, AAbb, aaBB*	Intermediate
1	*Aabb, aaBb*	Light
0	*aabb*	White

It should be apparent that the most frequent genotype from such crosses is the intermediate, two-melanin allele individual, since there are three ways such individuals can occur. The one or three melanin gene individuals can occur two ways, and there is only one way for the zero- or four-allele individuals to occur. This is exactly the same system used to describe Nilsson-Ehle's work with wheat kernel color.

As seen, the binomial expansion may be used to describe the probabilities of various polygene combinations without the use of a confusing Punnett square. The results of the Davenport analysis involving children of marriages between the intermediate color individuals should follow the predictions of $(x + y)^4$. With only 32 children, of course, no definitive interpretations can be made. Because the categorization into five color groups was essentially artificial, the results did not contradict their predictions.

The binomial expansion is much more important as a technique of analysis in quantitative inheritance when more than two pairs of genes are involved. It becomes significant in recent investigations of skin color, where it is assumed that the Davenport hypothesis was an oversimplification. Modern interpretations of skin color genetics, such as those of Curt Stern, implicate four to six pairs of genes in the control of pigment production.

The complexities of such a pattern, may be realized if the simplest of the modern proposals—four pairs of genes—is considered. If four pairs of alleles are involved, there would be eight genes determining skin color in humans. If two individuals with the genotype *AaBbCcDd* (each having one black *AABBCCDD* parent and one white *aabbccdd* parent) were mated, the possible offspring and the probability of their occurrence could be described with the binomial expansion $(x + y)^8$. With the chance for the capital letter, melanin-producing genes (*A, B, C, D*) equal to one-half in the cross, the mathematical analysis would be relatively simple for the nine predicted types of offspring:

$$(x + y)^8 = x^8 + 8\,x^7y + 28\,x^6y^2 + 56\,x^5y^3 + 70\,x^4y^4 + 56\,x^3y^5 + 28\,x^2y^6 + 8\,xy^7 + y^8$$

Thus, there would be $70\,(1/2)^4\,(1/2)^4$ or 35/128 chance for a child to have the same intermediate skin color as the two parents. There would be only $(1/2)^8$ or 1/256 probability that a child would be as black $(x)^8$ or as white $(y)^8$ as the original grandparents. The graph for such a distribution is shown on page 240.

Table 10.1 indicates that if five pairs of genes are involved, there are 11 categories of skin color and 1/1024 probability of the extremes in crosses between true hybrid parents. With six pairs of genes, the chance for homozygous white or black offspring is 1/4096 with 13 different skin color shades.

In the late 1950s studies were conducted which attempted to refine the analysis of skin color genetics through the use of **reflectance spectrophotometry.** Even when such modern instrumentation is used, however, definitive answers are impossible. Such techniques provide convincing evidence that there is a continuous spectrum of variation in melanin pigment amounts. The theory that skin color is controlled by polygenes is supported by the results of research by G. A. Harrison, but specific statements about number of genes or their exact mode of action simply cannot be made. General predictions about skin color can, however, be made with reasonable reliability in individual families or in given populations. With the knowledge that the trait follows a quantitative pattern involving about five pairs of genes, it is possible to analyze in fundamental studies with some confidence. There are still some interesting twists that interfere with clear patterns, but more accurate analyses may someday be available.

It is unfortunate that such an interesting hereditary trait, and one that is apparently incidental to human survival and function, should be viewed with such emotion in a social context. The categorization of individuals into various social strata almost solely on the basis of the innocuous presence or absence of melanin pigment in the skin should be deplored in modern culture.

Other Human Quantitative Traits

After studying Mendel's data on height in pea plants, one might be tempted to survey **human stature** in search of a simple genetic control mechanism. A quick look at **height variation** among humans should discourage such an attempt, however, since it is apparent that the trait is a continuously variable one. The height characteristic surely follows a genetically controlled pattern, and its association with families to one degree or another is obvious.

If an adequate system of measurement is used, practically any random group of persons can be arranged according to height in a broad spectrum. If measured in inches, one might anticipate that a small population of male students would fall between the extremes of about 62 to 76 inches. Depending on the preciseness of measurement, the persons could be grouped into a series of categories representing a continuous spectrum of variation. The two extremes of 62 inches and 76 inches would probably represent less than 1% of the population, while about 40% would likely be found in the 67- to 71-inch categories.

If the heights were graphed, the so-called normal distribution curve would be recognized, which can be demonstrated mathematically by the binomial expansion. Thus, it is obvious that such a collection of data leads logically to the conclusion that height in humans is a polygenic trait. This is generally considered to be the mechanism which is responsible for genetically determining stature, but no one is prepared to speculate on the number of polygenes involved or their mode of action. The effects of environmental factors are probably too great to permit an accurate analysis in a group of persons. Remember that variation in environmental factors would possibly result in a normal distribution of heights, even if there were no genetic control involved.

It is likely that, even without the effects of environment, the spectrum of heights would show a continuous intergrading series which would probably reflect a large number of genes affecting the phenotype. While the data indicate a quantitative situation, there is probably little hope of sorting out the complete picture of genetic control of height in humans.

If something so obvious and easily measured as height virtually defies genetic analysis, consider the possibilities involved in classifying **intelligence** as a polygenic trait. Probably most geneticists would not object to classifying intelligence as a quantitative characteristic strongly affected by environment. Most studies have indicated a definite genetic component to human intelligence, but there can be little hope of counting genes or specifying their mode of action. Intelligence is virtually undefinable and must be considered to have a large number of components.

During the years since Alfred Binet's original work in 1905, the psychological and educational communities have attempted to reduce the measure of intelligence to a single score determined by examination (either written or verbal). The exams attempt to evaluate the innate ability of an individual to achieve in areas associated with success in a societal context. Thus the test should be able to predict with regard to ability to handle normal sorts of educational tasks.

Primary abilities have been identified in such areas as spatial visualization, memorization, and inductive reasoning, but it has been traditionally the practice to provide one single score representing a measure of intelligence. The score, arrived at by dividing the **mental age** by the **chronological age,** is known as the **intelligence quotient (I.Q.)**. The number is multiplied by 100 for convenience, so that someone whose mental age and chronological age are the same has an I.Q. of 100 and is considered average. The results are adjusted so that the spectrum of data fits a normal curve, as would be predicted by pairs of genes incorporated into a binomial expansion. As a result, the complex situation can be simplified to speculate that a number of polygenes for intelligence control the trait measured by I.Q. tests.

Undoubtedly, there is some measure of truth to such an analysis, but it is unwise to place too simple an explanation on the inheritance mechanism for intelligence and then attempt to apply the assumptions to some practical educational procedure. To assume first that I.Q. data should follow a normal distribution, then adjust results so that they do, followed by the logic that such binomial distribution indicates a polygenic pattern, is a procedure that requires a critical careful study.

To utilize a single number for lifelong evaluative procedures, while neglecting such unmeasured relevant qualities as motivation, intellectual environment, sociological environment, and nutritional environment, seems questionable, if not totally unwarranted. It is probably not reasonable to reject totally the use of the I.Q., but caution must be employed in its use, especially where genetic associations are involved. Probably a combination of significant components of intelligence are measured by the I.Q., but surely not all components of what may be classified as intelligence. The I.Q. is the only significant attempt that has been utilized in this important evaluation and, as such, there is much information that has been compiled about it.

Based on its importance as an indicator of potential in intellectual endeavors, the I.Q. has been studied with regard to its relationship to genetics and environment. Several studies have indicated that the heritability of I.Q. is in the vicinity of 40 to 80%. This indicates that up to 80% of the total variance in I.Q. measurements is due to genetic factors. Notice that such figures, based on averages for an overall populational analysis, are not necessarily applicable to any single given individual, and possibly not transferable from one population to another. Numerical data are derived from comparisons of individuals, compiled into larger masses of figures, and finally averaged to provide meaningful information.

Based on evidence that has been accumulated, probably any prudent geneticist would agree that there is clearly a genetic component to the I.Q. Likewise, it is impossible to view the data without admitting the importance of environmental factors in this human trait. In fact, many workers who have looked closely at the research assume that there is a third component involved in calculating the heritability and that is a mutually reinforcing combination of the two factors (gene-environment covariance). The real dispute over inheritance of the qualities that control human performance on I.Q. tests involves the responsibility of either genes or environment for variation.

When heritability is analyzed, the comparative figures arrived at seem to depend on the data chosen for analysis, as well as the investigator doing the analysis. Christopher Jencks of Harvard University has estimated the heritability of I.Q. to be 0.45 (with environment accounting for 0.35 and gene-environment covariance equal to 0.20). At the high end of estimates for heritability of I.Q., Arthur Jensen of the University of California (Berkeley) and the late Cyril Burt in England presented figures of 0.80 and 0.87, respectively.

In general, a knowledge of the level of control of I.Q. by genes is not essential to the broad picture of inheritance of intelligence. The qualities measured by the I.Q. are inherited polygenically and are affected by environmental factors. Both genes and environment seem to exert substantial effects on the final measurements and are significant to an understanding of the I.Q.

Because any heritability figure is based on the analysis of a population, the figure is meaningless in its application to a single individual. The figure does not mean that 40%, 60%, or 80% of a person's I.Q. is based on genes. Heritability describes the amount of variance in average I.Q. that is due to heredity in a population.

It is important to note that heritability figures are usually obtained from research on a given population. Thus, one must be cautious about applying such figures to an entirely separate population or even combining figures from several populations and then reapplying them to the individual groups. The validity of such practices in questionable. In England there appears to be more cultural (environmental) uniformity than in the United States. Thus, if an English population were used as a source of data, a lower V_{env} factor might be expected in the heritability equation. The actual genetic component might be no greater, but in England the heritability would be relatively higher than for the U.S. population. In fact the only difference might be the environmental background of the two populations.

Generally, English data have provided higher heritability estimates than U.S. data. This, coupled with a serious question about the validity of Cyril Burt's research and its subsequent rejection by many authors, has led to lower estimates of heritability of I.Q. in recent years. The figures indicating that the heritability of I.Q. is in the range of 0.80 are highly debatable for general application. A figure of about 0.60 is more commonly cited, with a word of caution for anyone who wishes to apply the figure to different situations.

The control of other less obvious, and possibly less significant, human characteristics has been attributed to polygenes. The sort of trait that is placed in this category is, of course, one that can be shown to follow familial inheritance patterns, but without clear-cut phenotypic categories that exist in definite ratios. Continuous variation is generally found over a spectrum, with the most common type an intermediate in the scale. In short the distribution of such traits follows a normal curve.

Two other examples of probable quantitative traits are **fingerprint patterns** and **length of life**. Without delving into the details of analysis, the method of counting fingerprint ridges on human hands results in a single figure for all 10 fingers from 0 to 300 (Fig. 10.2).

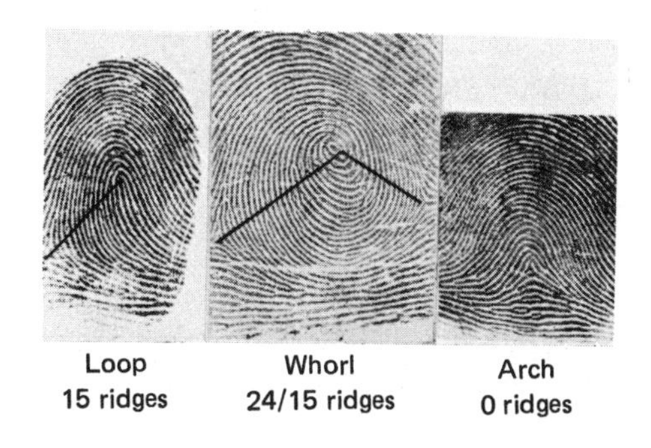

Figure 10.2. The technique used in making fingerprint ridge counts is illustrated in these photographs. A line is drawn from each triradius (note triangular area formed by intersecting arrays of ridges) to the core or center of pattern. There are three general patterns: loops with one triradius, whorls with two triradii, and arches with no triradii and thus a ridge count of zero. Polygenic hereditary patterns for total ridge counts (sum of all the fingers) have been observed in family studies. (Courtesy Dr. Milton Alter and *Medicine.*)

When a population is studied, the distribution of ridge counts follows a normal curve with the median number at about 130. The ridge counts have been shown to correlate well with familial relationship and the trait is therefore considered polygenic. It is essentially an unimportant characteristic, reflecting an embryological development during the first 3 or 4 months after conception. The ridge patterns have been shown to respond to such chromosomal abnormalities as Down syndrome, Klinefelter syndrome, and Turner syndrome. Obviously these aberrant genomes interfere with normal development. One of the key indicators of this abnormal development is the flow of embryonic skin cells, which is reflected in handprints and footprints.

Longevity in humans seems to show a hereditary correlation, since the length of life of ancestors of 90- and 100-year-old persons is statistically longer than persons in the general population. Furthermore, a normal distribution of longevity is observed, indicating a quantitative pattern. Another note of caution should be sounded in this context, since persons having above average age may be living in the same sort of environment as their ancestors. Family tradition, such as diet, atmosphere, and habits, among many other factors, might well contribute more to longevity than the genes carried by the family.

TWIN STUDIES IN HUMAN GENETICS

Human **twins** are useful in analyzing the genetic basis of any trait, but they are especially valuable in a study of quantitative inheritance. With polygenic traits environmental factors have been shown to play such a significant role that it is almost impossible to distinguish the role of genes in determining the phenotypes. In fact, it is often difficult to determine whether or not there is a genetic component.

Under such circumstances, methods must be devised of separating these two factors—heredity and environment—to conduct meaningful studies. It is impossible to estimate heritability unless one can evaluate how much of the total variance results from genetics and environment. It is easy to study persons with different hereditary backgrounds, and sometimes possible to maintain a reasonably constant environment for different individuals. However, to find individuals with the same genotypes in different environments is a rare situation. Even persons with the same genotype in the same environment could provide a valuable aspect for the study of human genetics.

It has been known for many years that **identical** or **monozygotic** twins are unusually valuable in understanding inheritance patterns. Such twins are sometimes referred to as monozygotic, because they come from a single zygote. One conception involving an egg and a sperm gives rise to two separate individuals because the embryo separates at an early stage into two viable groups of cells which become two separate embryos. The real asset to such twins is that they both have the **same genotype.** Naturally, they are always of the **same sex.**

Dizygotic or **fraternal** twins result from two separate, essentially simultaneous conceptions involving two different eggs and sperm. Actually, such twins are no more or less closely related than brothers and sisters and they are of different sexes about half the time. Conceivably their environments are more similar than typical brothers' and sisters', since they are carried in utero at the same time and place, while siblings are usually separated from each other in time by a year or more. Most often they spend their childhood together in reasonably similar environments. In the last section the matter of inheritance of I.Q. and the estimation of its heritability was explored briefly. Most studies supporting such estimations have been based at least in part on twin studies.

An early study of separated identical twins in the United States was published by Newman, Freeman, and Holzinger in 1937. They located 19 pairs of identical twins separated from each other since birth. Their environments were evaluated for various types of differences, and a statistical test was run to see how well the I.Q. differences correlated with environmental differences.

Although the number of individuals studied by Newman and coworkers is small to build a case, their data showed average I.Q. differences of 5.9 points between 50 identical twin pairs raised together, 9.9 points between 52 pairs of nonidentical twins reared together, and 8.2 points for the 19 identical twin pairs separated from birth. The average difference between pairs of siblings raised together was found to be 9.8 I.Q. points for 47 pairs of the same sex. A relevant study has shown that the average difference in I.Q., as measured on the same individuals with different test administrations, is about 5 points.

Three other more recent twin studies of a similar nature were separately conducted in the 1950s and 1960s by Cyril Burt and James Shields in England, and Niels Juel-Nielsen in Denmark. Definite correlations with both environmental factors and genetic background have been shown for I.Q., with the genetic factors appearing to be somewhat stronger. As mentioned, the work of Burt has come under considerable criticism as being fabricated in part and some estimates of heritability have dropped as a result. In these four studies of separated identical twins a total of 122 were reported but 53 of them were in the reports of Burt alone. In spite of this, the involvement of genes in the determination of the I.Q., whatever the test measures, appears to be well established.

Twin studies have been applied to many human traits, some with more solid results than others. It is a technique that has been recognized for a long time. Francis Galton advocated the use of twins to distinguish between the effects of heredity and environment in the middle 1800s, when he collected data on various traits in twins for statistical studies. This method of analysis may serve to provide some valid answers about the effect of nongenetic factors on phenotypes, especially where polygenes are involved. Twin studies are particularly valuable in helping to estimate the relative roles of heredity and environment on the variance. To this extent, it is sometimes possible to provide a reasonable evaluation for the heritability of a trait in a population. One should be aware, however, that the technique is of no real significance in working out the genetic mode of action or the number of gene pairs involved.

A nice example of the use of twin studies to ascertain the involvement of environmental factors in phenotypic development is found in the study of **dermatoglyphics** or fingerprint patterns. It is a relatively simple matter to make ridge counts for all 10 fingers and then make comparisons with other persons, related or unrelated. It is possible to group large numbers of individuals into pairs to run a statistical analysis called a **correlation coefficient.** If the numbers of ridge counts bear no relationship to each other, as might be expected among pairs grouped randomly from the general population, the correlation coefficient would be 0 or close to 0. If there were complete correlation, as might be expected in identical twin pairs, the coefficient would be 1.00 or close to 1.00, indicating the ridge counts were the same within all pairs examined.

When such a study was summarized by Sarah Holt in 1968, it was seen that the predictions were essentially correct. The correlation between unrelated pairs (husbands and wives) was 0.05 (essentially 0 in this type of analysis). Correlation coefficients between parents and children, nonidentical twins, and siblings were close to 0.50 in all three categories. This might have been predicted, since the proportion of genes shared in all three types of pairs is one-half.

The correlation coefficient for a comparison of identical twins in the Holt study is 0.95. This figure is interesting because it indicates that identical twins do not always correspond exactly in their fingerprint ridge counts. If the trait were determined exclusively by genetic factors, the coefficient should be 1.00. Because it is not, one may assume that the embryonic movement and organization of cells can be affected by external factors during the first 4 months of life, even if only slightly. Notice, however, that 0.95 is as close to 1.00 as 0.05 is to 0, and that figure was found where 0 was expected earlier. Because experimental counting errors could account for both of these differences, one should be cautious about an interpretation involving environmental factors affecting ridge counts in identical twins.

HUMAN POLYGENES AND FUTURE STUDIES

Even a brief look at polygenic systems helps to explain why it is often so difficult to develop an understanding of hereditary traits in humans. If there were no effects of environment, the phenotypes of these quantitative traits would still be spread over a spectrum whose categories would be difficult to recognize. Where small human families have made genetic analysis difficult with simple traits, they make analysis of polygenic systems almost impossible. The presence of a large number of quantitative characteristics does not indicate a bright future for the analysis of some of the most interesting and "human" traits.

It is likely that some of the most often mentioned characteristics in the work of those who would like to "select" future offspring fall into this category. Although it is possible that such topics as behavior will never be analyzed genetically, many geneticists and psychologists are confident that continued research will provide answers. Musical ability, creativity, congeniality, manual dexterity, and language ability might be the sorts of characteristics found on everyone's list of preferred traits. They all are likely to have strong genetic components, but since there is a spectrum of phenotypes throughout all populations, they may be expected to be polygenic.

All traits of this sort are particularly subject to the effects of environment. Thus, the problem of resolution and understanding becomes overwhelming in the varied conditions of life that exist.

The absence of clear-cut genetic ratios and the influence of environmental conditions may make analysis impossible, but it does not mean that Mendelian principles are not a consideration. In spite of the complexity of analyses, the fundamental underlying mechanism is still the genetic basis seen repeatedly. It is important to keep this fact in mind and not be tempted to confer an aura of mystery to human inheritance simply because the problems cannot be reduced to their lowest common denominator.

ADDITIONAL READING

BODMER, W. F., and L. L. CAVALLI-SFORZA, "Intelligence and Race," *Scientific American*, (1970), 223, No. 4, 19–29.

______, *Genetics, Evolution, and Man*. San Francisco: W. H. Freeman & Company Publishers, 1976.

FALCONER, D. S., *Introduction to Quantitative Genetics* (2nd ed.). New York: Longman, 1981.

FELDMAN, M. W., and R. C. LEWONTIN, "The Heritability Hang-up," *Science* (1975), 190, 1163–68.

HOLT, S., *The Genetics of Dermal Ridges.* Springfield, Ill.: Chas. C Thomas, 1968.

JENSEN, A. R., "Race and the Genetics of Intelligence: A Reply to Lewontin," *Bulletin of the Atomic Scientists* (1970), 26, No. 5, 17–23. (Also reprinted in Baer, A. S. (ed.), *Heredity and Society* (2nd ed.). New York: Macmillan, 1977.)

KAMIN, L. J., *The Science and Politics of I.Q.* Potomac, Md.: L. Erlbaum Associates, 1974.

LEWONTIN, R. C., "Race and Intelligence," *Bulletin of the Atomic Scientists* (1970), 26, No. 3, 2–8. (Also reprinted in Baer, A. S. (ed.), *Heredity and Society* (2nd ed.). New York: Macmillan, 1977.)

SCARR-SALAPATEK, S., "Unknowns in the I.Q. Equation" (book review), *Science* (1971), 174, 1223–28.

STERN, C., *Principles of Human Genetics* (3rd ed.). San Francisco: W. H. Freeman & Company Publishers, 1973.

SUTTON, H. E., *An Introduction to Human Genetics* (3rd ed.). Philadelphia: Saunders, 1980.

WADE, N., "I.Q. and Heredity: Suspicion of Fraud Beclouds Classic Experiment," *Science* (1976), 194, 916–19.

REVIEW QUESTIONS

1. In studies involving twins if the two members of a pair both manifest a particular phenotype, they are said to be concordant for the trait. If they do not both show the same phenotype, they are said to be discordant. In a study involving a number of twin pairs it is possible to report the amount of concordance as a percentage, which represents the fraction of the total sample of twins who match each other for the trait.

    The following chart summarizes a compilation of hypothetical data for three traits studied in identical (monozygotic) and nonidentical (dizygotic) twins. Based on this information, which trait is the most likely to be most influenced by hereditary factors? Which is most likely to be influenced by environmental factors? What evaluation might be placed on the remaining trait in regard to the concordance differences between identical and nonidentical twins?

	Concordance for Twin Pairs (%)	
Trait	*Identical twins*	*Nonidentical twins*
I	85	34
II	95	91
III	24	16

2. Assume that the human skin color trait responsible for phenotypic differences between so-called black and white persons is based on five pairs of polygenes. Assume further that the situation represents an ideal example of typical polygenes: equally acting codominant alleles, cumulative in effect, independent and not linked.

(a) If two individuals with the genotype *AaBbCcDdEe* (each with one black parent, *AABBCCDDEE,* and one white parent, *aabbccddee*) are mated, what would be the chance for a black child (*AABBCCDDEE*) to be born? For a white child (*aabbccd-dee*)? For a child of the same intermediate skin color as the parents?

(b) Discounting the effects of environment, if six different gene pairs were involved, how many different classes of skin color should be distinguishable?

3. Assume that human height is controlled in part by four different pairs of alleles acting in cumulative fashion as polygenes. Assume also that each capital letter gene contributes 2 inches to the basic height of 4 feet 10 inches; each lowercase letter contributes no height increment. Thus, a person with one capital letter allele would be 5 feet tall; with two such alleles, 5 feet 2 inches; three, 5 feet 4 inches, and so on. Disregard environmental effects in this question.
   (a) What would you anticipate to be the shortest and tallest person in a population based on these polygenes? What would you anticipate to be the mean height?
   (b) Diagram a mating (genotypes) in which the parents are both 5 feet 6 inches tall and the only possible height for the children is 5 feet 6 inches.
   (c) Diagram a mating (genotypes) in which one parent is 5 feet 6 inches tall and no child can be any taller than 5 feet 6 inches.
   (d) Diagram a mating (genotypes) in which one parent is 5 feet 6 inches tall and all children will be at least 5 feet 6 inches or taller.
4. Fingerprint ridge counts are made by counting the ridges that cross a straight line drawn at key points on each fingerprint. All 10 finger counts are summed to give a single number. Suppose that a large collection of data from a population gives a range of ridge counts that runs from 0 to 260. The mean count is 130 and the variance is calculated at 70. Twin studies indicate that fingerprint ridge count heritability (based on a crude formula, p. 241) in this hypothetical population is 0.90 (on a scale of 0.0 to 1.0).
   (a) What is the variance due to genetic factors?
   (b) What is the variance due to environmental factors?
5. Analysis of the hypothetical population in Question 4 indicates that 1.6% fall in the range of zero to 37 ridges and 1.6% are found in the range of 223 to 260 ridges. Ridge counts of 110 to 150 are found among approximately 30% of the population. Based on these data, estimate the number of genes controlling the ridge count trait.
6. Suppose that a hypothetical population of organisms is found growing in a pond and is analyzed for the time required for mitosis to occur. The population is sampled randomly and includes organisms from deep and shallow areas, shady and sunny spots, areas of low and high nutrients and oxygen, high and low temperatures, and so on. The analysis reveals that mitosis times run from 30 to 150 minutes with the mean being 90 minutes. The variance is calculated to be 30.

   The same organisms taken into the laboratory and grown on the same carefully controlled nutrient medium, at the same temperature, and under the same light intensity are also analyzed for mitosis times. In the controlled environment the time range is 60 to 120 minutes with the mean at 90 minutes. The variance is calculated at 15.

   Calculate the heritability for the rate of mitosis trait using the simple formula given on page 241.
7. List several reasons why an "all or nothing" genetic trait or disease in humans, such as albinism, would be easier to analyze, predict, and treat than a quantitative one.
8. Why is the concept of heritability difficult to utilize in human studies?
9. Give some reasons for criticism of the use of I.Q.'s in comparing the genetic intellectual

endowment of randomly chosen individuals from a population or for the comparison of two different populations.

10. To study the effects of heredity and environment on variations in I.Q. measurements, a classical technique involves the comparison of I.Q.'s of pairs of individuals. When a data list is compiled for pairs of persons of a given relationship, a statistical test which produces a correlation coefficient (c.c.) is run. If the I.Q.'s of the pairs in the list all match exactly, the c.c. is 1.00 (100% are the same). If there is no correlation between pairs at all, the c.c. is equal to zero. If half the pairs are correlated and half are not, the c.c. equals 0.50, and so on. Thus, a list of identical twins' I.Q.'s with the same lifetime environment and experiences should produce a c.c. of close to 1.00 (presumably any deviation must be due to slight environmental differences). Randomly chosen pairs of unrelated individuals from different environments should yield a c.c. close to zero.

Results from several different studies have been combined to produce correlation coefficients for various degrees of kindred and environments. Among such data are the followig c.c.'s:

Identical twins reared together	0.88
Identical twins separated	0.77
Dizygous twins reared together	0.63
Siblings reared together	0.58
Unrelated pairs of persons reared together (adopted children)	0.23

Explain what sort of inferences you might be able to draw from this compilation of data. Justify any conclusions about the comparative effects of heredity and environment on variations in the I.Q.

Attempt to estimate a heritability value for I.Q. from these data and assign relative values to the environmental and genetic components. What sort of errors might be inherent in your estimate? What interpretative precautions could you provide about the estimate's application to other I.Q. studies?

CHAPTER SUMMARY

1. Some inherited traits do not exhibit clear-cut discontinuous Mendelian ratios. They are due to interacting pairs of genes that produce cumulative effects on the phenotype, resulting in continuous variation across a spectrum.
2. It is usually difficult to determine the number of gene pairs controlling a quantitative trait. Inadequate means of distinguishing various classes in the spectrum and environmental interaction with the phenotype often contribute to the difficulty.
3. Where fully heterozygous parents are involved, the binomial distribution describes the classes within a polygenic spectrum and their frequency.
4. Heritability is a complex concept that is used to describe the amount of variation from the population mean that is due to genetic variation, as opposed to environmentally caused variation. Heritability figures are theoretically applicable only to the population from which derived.
5. An example of a human polygenic trait is human skin color, for which there may be five or six controlling pairs of genes.
6. Most human polygenic traits present unusually difficult problems of analysis. Environmental factors are so significant and the genetic mechanism is so elusive that there is little reason to expect a quick solution to the problems. Human stature, fingerprint ridge counts, and intelligence are examples of polygenic traits. Although large quantities of numerical values for height, ridge counts, and I.Q. can be analyzed, it has not been possible to estimate the number of gene pairs involved. Human life span seems to be under genetic control with a polygenic system involved.
7. Various estimates of I.Q. heritability have been made, with the general conclusion that environment and genes are both significant. Figures for I.Q. heritability usually fall in the range of 0.40 to 0.80.
8. Human twins have provided good samples for research on polygenic traits where it is helpful to have individuals with the same genotype, as opposed to those having different genotypes in the same environment. Several important studies of identical twins separated at birth have provided information on polygenic traits, such as I.Q. and fingerprint ridge count.

CHAPTER

11

GENES IN POPULATIONS

All Communities
divide themselves into the few and the many.

Alexander Hamilton,
Debates of the Federal Convention (1787)

Although a little noticed facet of his work, Mendel actually reported on work that contained the seeds of a study of gene frequencies in populations. In one section of his 1866 paper he noted, "The proportions in which the descendants of the hybrids develop and split up in the first and second generations presumably hold good for all subsequent progeny."* It seems clear that Mendel recognized that gene frequencies could be maintained automatically at a given level within a distinct population. He carried some of his crosses through six generations to demonstrate the balanced situation.

Mendel apparently was able to predict mathematically a changing genotypic ratio while the relative number of genes remained constant. A complex formula was derived for the proportion of genotypes after n generations of self-fertilization in peas which showed that there was a regular, predictable tendency for the proportion of homozygotes (*AA* and *aa*) to increase, while heterozygotes (*Aa*) decreased. His formula, $(2^n - 1)\ AA{:}2\ Aa{:}(2^n - 1)\ aa$, applied only to a situation in which self-pollination was the mechanism of reproduction; Mendel did not generalize in regard to the frequency of genes within a pool of genes in the population. It is doubtful that he conceived of his populations or progenies in this fashion, but it is clear that he had a better grasp of the flow of genes from one generation to the next within populations than some scientists in the early 1900s.

It is interesting that ideas about **population genetics** had their birth in an attempt to apply Mendelian principles to studies of human heredity. Wilhelm Weinberg,

* Mendel, G., "Experiments in Plant-Hybridization," *Verh. naturf. Ver. in Brunn, Abhandlungen* 1865, iv. (Reprinted in English with comments in Peters, J.A. (ed.), *Classic Papers in Genetics*. Englewood Cliffs, N.J.: Prentice-Hall, Inc., 1961.

a German physician, was interested in Mendelian principles in relation to humans. In 1908 he published on the generalization of the equilibrium of genes in populations that Mendel had noted. Although he was a self-taught mathematician, Weinberg extended the equilibrium principle to human studies through the use of algebra and probability theory. His publications obviously qualify him as a founder of the study of population genetics. He extended his theoretical explanations in papers of 1909 and 1910, and even provided formulas for population studies of multiple alleles. As so often happens, his work was largely unnoticed until after his death in 1937.

In an often repeated pattern of science, in 1908, the same year as Weinberg's pioneer publication, Godfrey Hardy, an English mathematician, published a very brief paper explaining his derivation of a formula which generalized the equilibrium principle only implicit in Mendel's observations (Fig. 11.1). He was stimulated to publish in response to some critics of Mendelian particulate genetics who cited human **brachydactly** (short fingers) as proof that the principles were invalid. Their thesis was that, since the trait was based on a dominant gene, it should eventually result in a 3/4:1/4 ratio among humans, if Mendel were correct. Of course, the gene for brachydactly is rare and no such tendency to overwhelm the recessive can be detected.

Figure 11.1. Godfrey Hardy (left) and Wilhelm Weinberg (right). In 1908 the principle of gene equilibrium in populations ($p^2 + 2pq + q^2$) was expressed mathematically. The principle was proposed independently by the English mathematician Hardy and the German physician Weinberg. (Hardy photograph courtesy THE BETTMANN ARCHIVE/BBC Hulton Picture Library; Weinberg photograph courtesy *Genetics*.)

It often comes as a surprise to a beginning student that dominant alleles are not in the process of swamping recessives in populations. Somehow the "typical" monohybrid cross with its 3/4:1/4 ratio leads to the erroneous conclusion that with each succeeding generation the dominant trait should increase in frequency. However, it is possible to stand back from genetic crosses involving small numbers of persons and consider instead populations of individuals. It is obvious that PTC non-taste ability persists generation after generation, as do attached ear lobes, even though both are determined by recessive genes.

The mathematical basis for all population genetics can be found in the papers of Hardy and Weinberg. **Gene equilibrium** under conditions of **random mating** in a population has therefore come to be known as the **Hardy-Weinberg principle.** The original formulas have been expanded and embellished to cover many aspects of population studies in a science that is sometimes highly theoretical, but an understanding of the simple underlying mathematical basis is essential for a rational view of gene distribution in human populations.

NORMAL DISTRIBUTIONS IN POPULATIONS

The **binomial expansion** was introduced in Chapter 10 to describe the distribution of various phenotypes in quantitative inheritance. This same procedure can also be used to describe the so-called normal curve for populations in other genetic situations. As a matter of fact, the underlying principle of the technique sits right at the heart of population studies in heredity. With some careful observations one can see the correlation with Hardy and Weinberg's analysis. First, however, the binomial equation will be reviewed briefly. The purpose now is to consider frequencies in all or nothing genetic situations, where clear-cut distinctions in phenotypes are involved.

Binomial Expansion to Predict Family Sex Ratios

To demonstrate the use of the binomial expansion, it is most helpful to choose an example which is not controlled by a single pair of Mendelian alleles but still follows a similar "either-or" pattern. Such an example is the **sex ratio** in human families. As was pointed out in Chapter 3, sex determination is based on a theoretical 1:1 system for males and females because of the XY-chromosome mechanism. While this 1:1 ratio is not strictly true in actual birth results, to simplify one can assume that it is valid for a demonstration of the method of analysis. It is possible to utilize the 106:100 ratio in the same fashion. Even though sex determination may not be based on a single gene, it is clearly genetic because of the chromosomal control.

Often such statements as "This is their fourth child and they have two girls and a boy, so this is bound to be a boy," or, "They have four girls, so it appears that they can have only female children," are heard. These statements take no account of the statistical basis for distribution of sex ratios, and more importantly, they disregard the basic underlying fundamental that each conception and birth is an event which is **separate** and **independent** from all others. It is possible to assign statistical figures to such statements, based on the important fundamental consideration that sex ratios in families follow a **normal distribution curve.**

In the case of sex determination, as in many hereditary characteristics, there are two distinct phenotypes. The mechanism involved produces a result similar to a cross between a homozygote and a heterozygote. This situation easily allows application of the binomial expansion. Furthermore, it permits a simple statistical analysis, if assumed that there is a one-half chance for each of the two phenotypes. Thus, the abstract symbols x and y can be used to symbolize males and females, respectively, with one-half chance for each to occur at every birth.

It should be obvious that at each birth a child can be either a male (x) or a female (y). Therefore, all the possibilities for gender at each birth are represented by $x + y$. The possibilities for two births in terms of the sex of the children would be $(x + y)^2$; for three children it would be $(x + y)^3$, and so forth.

Employing the method of binomial expansion used in Chapter 10, the following can be seen:

$$(x + y)^3 = x^3 + 3x^2y + 3xy^2 + y^3$$

This can be interpreted as predicting the chance for a family of three boys (x^3), two boys and one girl ($3x^2y$), one boy and two girls ($3xy^2$), and three girls (y^3). Thus, by inserting the chance for boys ($x = 1/2$) and the chance for girls ($y = 1/2$), predictions can be made about the combinations of boys and girls in a family of three children or about the distribution of sex types among families with three children in a population.

Notice that even without the binomial expansion the chances for different combinations in a family of three children could have been worked out. Three boys can occur in such a family in only one way. The first child must be a boy, the second must be a boy, and the third must be a boy. Each has a one-half chance of being a boy, so mathematically the chance would be $1/2 \times 1/2 \times 1/2$. The chances for two boys and one girl are three times as good, however, since such a combination can occur three ways. The first child could be a boy, the second a boy, and the third a girl. Alternatively, the second could have been a girl, with the first and third boys, or the first might have been a girl, with the second and third boys.

The binomial expansion predicts that if a family has three children, there is 1/8 chance of having three boys, a 3/8 chance for two boys and one girl, a 3/8 chance for one boy and two girls, and a 1/8 chance for three girls. To say the same thing another way, if all families with three children are examined in a population, 1/8 of them would have three boys, 3/8 would have two boys and one girl, 3/8 would have one boy and two girls, and 1/8 would have three girls.

Remember that if a couple already has one boy and one girl and are about to have a third child, the chance of it being a boy is one-half (106/206) and the chance of it being a girl is one-half (100/206). Regardless of the number and sexes of preceding children, each birth still has the same chance as the others. This is the fundamental basis for the other calculations and cannot be overlooked. Considering family units in a population is something quite different than separate births involved in the origin of these family units.

In the family with two girls and one boy or in the family with four girls the next child has one-half chance to be of either sex. But is a family of three girls and one boy unusual, or is a family of five girls really rare? The binomial expansion says that there should be one-fourth of the families with four children that have three girls and one boy. Likewise, one in every 32 five-children families would be expected to have five girls.

Binomial Expansion for Other Genetic Traits

The binomial expansion can be used to predict the distribution of single-gene controlled traits in a fashion similar to that for sex distribution in families. However, it is usually necessary to specify carefully a particular segment or group within a population, since the use of the technique is limited to a situation where all crosses being considered will produce the two given phenotypes with a known frequency. This is similar to sex distribution, where it was assumed that all matings produced boys and girls with equal frequency.

It is possible to use the binomial expansion to analyze a pair of dominant and recessive alleles in a population, if the chances for each kind of phenotype to be produced are known. One might wish to examine the frequency of free and attached ear lobes in families by looking only at a specific segment of the population. For instance, one might choose to make predictions about the distribution of **ear lobe shape** in families where one parent with free ear lobes was heterozygous (*Aa*) and the other had attached ear lobes (*aa*). In such matings the chance for each child to have free ear lobes would be one-half and the chance for each to have attached ear lobes would be one-half.

To use the binomial with these phenotypes, the symbol x can be assigned to free ear lobes and the symbol y to attached ear lobes. The distribution of ear lobe shape can be described in families of four children.

$$(x + y)^4 = x^4 + 4x^3y + 6x^2y^2 + 4xy^3 + y^4$$

The chance for a family of four children to have two with free ear lobes and two with attached ear lobes, when the parents are as specified, equals 3/8 ($6x^2y^2$).

If one chooses to analyze the families of marriages between two heterozygous parents for ear lobe shape ($Aa \times Aa$), the situation would change because the chance for free ear lobes at each birth would be 3/4 and for attached ear lobes, 1/4. In such a segment of the population the chance in families of four children for two with free ear lobes and two with attached ear lobes would equal 27/128 ($6x^2y^2 = 6(3/4)^2(1/4)^2$).

If the chances for the two different forms of a trait are unequal, the distribution of such phenotypes is skewed more toward one end of the spectrum than the other. When the population is graphed it appears as an **asymmetrical curve,** rather than the normal distribution curves that are most familiar. As an example of such differences in normal distribution, the results of the expansion of $(x + y)^6$ for both $x = 1/2$ and $x = 3/4$ can be compared:

$$(x + y)^6 = x^6 + 6x^5y + 15x^4y^2 + 20x^3y^3 + 15x^2y^4 + 6xy^5 + y^6$$

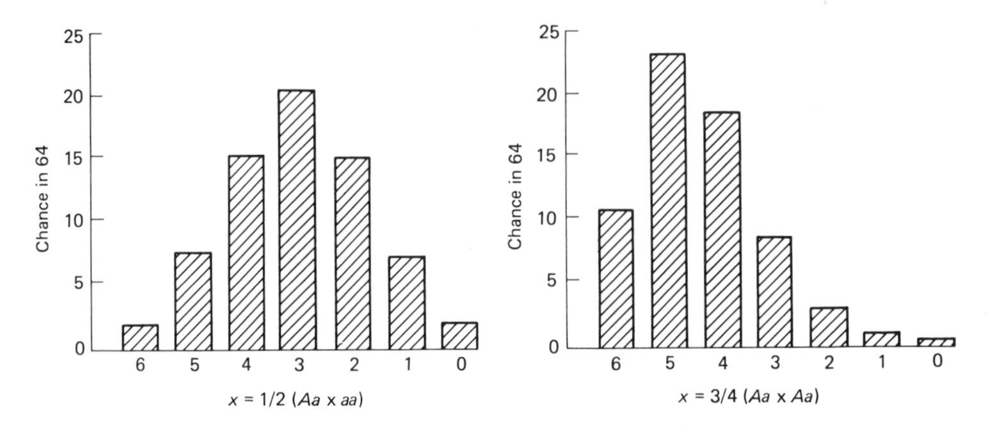

Number of offspring showing dominant phenotype in families of six children

A quick survey of the two situations reveals that in the normal curve based on a one-half chance for both traits, families with three children of each type are the most common combination. However, in the other distribution, where the chance for free ear lobes is better than for attached, the most common type of distribution in six-children families is that five have free ear lobes and one has attached ear lobes (probability or chance of 729/2048).

THE HARDY-WEINBERG EQUILIBRIUM

In spite of the proposal that rare dominant alleles increase in a population with each succeeding generation up to the level of 3/4 of the individuals, Hardy and Weinberg both recognized that this was not the case. They were able to demonstrate mathematically that, under given conditions, once a gene was established in a population at a given frequency, it is maintained in the population at that same frequency generation after generation.

Requirements for a Hardy-Weinberg Population

It is important at the outset to point out that the Hardy-Weinberg principle is based on a set of ideal conditions in the population. A deviation or interference with any of these conditions will cause a distortion in the balance of the alleles and thereby raise or lower the frequency of one of them in succeeding generations. Much can be explained about actual populations by the highly theoretical Hardy-Weinberg law, even where the ideal conditions are not met. The principle serves as the springboard for a considerable body of investigations concerning selection and evolution. In addition to providing the foundation for basic studies in this area it is the fundamental source of calculations utilized in modern plant and animal breeding programs.

It is doubtful that any situation truly meets the requirements for a population following the law of Hardy and Weinberg. The following requirements are necessary for the equilibrium to be established:

(1) The population must be of **large size** (theoretically of infinite size) and it must be a stable, distinct group made up of interbreeding individuals;
(2) There must be **no selective migration** into or out of the population of individuals with certain traits;
(3) There must be **random mating** within the population, which implies that no preferential mating of particular types to each other can occur;
(4) There must be **no differential survival** or **reproductive advantage** of one type over the other;
(5) There must be **no mutation** of the genes controlling the trait in question.

It should be obvious that a deviation from any of these requirements could change the frequency of a gene by adding to or subtracting from the **gene pool** within the population.

In human populations it is impossible to find a situation that meets all these requirements; however, the Hardy-Weinberg principle can be used as a working formula to estimate and predict **gene frequencies** with reasonable accuracy. Today it is difficult to find a truly stable, nonmigrating group. Often populations cannot be distinguished along geographical lines, since there are ethnic groups and social reproductive barriers within a larger population. There is invariably some flow of genes from one group to another within such situations. It may well be that there are **reproductive isolation** mechanisms at work in any human population that are not recognized. Modern societies generally allow fairly free rein in terms of mate selection and a broad overview might suggest that mating is random. However, under such circumstances it is probably inevitable that there is a certain amount of choice involving selection of similar interests, intellectual levels, and possibly even physical features.

Naturally, the impossible criterion for a balanced Hardy-Weinberg population is the requirement that no mutations can be occurring. No matter how low the frequency, mutations do occur. For most common characteristics, the mutation rate is so low as to be insignificant, but in very rare traits, the mutation factor can be an important consideration. As a matter of fact, in rare genetic disease situations, where the trait is detrimental and results in a much lower survival rate, the mutation rate may be almost as high as the frequency of the gene and account for its maintenance in the population.

Realistically, no human population truly meets the standards necessary to apply the Hardy-Weinberg law strictly. However, the calculations and formulae that it provides are valuable in making a thorough study of gene frequencies and flow in populations. Such studies are highly instructive about the operation of genes in populations and have considerable use in analyzing the concepts of gene frequencies, evolution, and race.

Calculation of Gene Frequency

The first calculations in this chapter involved the frequency and distribution of phenotypes. Throughout all genetics studies the ultimate interest resides in an analysis and prediction of phenotypes. However, it is obvious that genotype determines phenotype and the underlying basis of calculation must center on the genotype. In any population analysis one must go back to the calculation of **gene frequencies** to understand Hardy-Weinberg calculations. Notice that the principle states that once a *gene* is established in a population at a given frequency, it is maintained at that same frequency generation after generation. This does not necessarily mean that the phenotypic frequency is always the same generation after generation, but usually it is.

A simple hypothetical example can be used to explain this anomaly. Imagine that a population consists of persons who are all homozygous, one-half men with the two dominant alleles (*AA*) and one-half women with two recessive alleles (*aa*). The frequency of the dominant allele (*A*) would be one-half, as would that of the recessive allele (*a*). When mating occurred in this population, each child would have an *AA* father and an *aa* mother and would therefore have the genotype *Aa*. No matter how many children were produced, they would all have the same phenotype, whereas the preceding generation was half dominant phenotype and half recessive. Significantly, in this second generation the frequency of the two genes (*A* and *a*) is still one-half each.

One can determine the third generation because all matings among second-generation individuals would be between two heterozygotes and would produce the familiar 3/4:1/4 phenotype. The genotypes in the population would be one-quarter *AA*, one-half *Aa*, and one-quarter *aa*. Look carefully at this distribution:

1 *AA*

2 *Aa*   4 *A*'s + 4 *a*'s

1 *aa*

This shows simply that the frequency of each allele in the third generation is still one-half.

Although it is much more involved to follow this example to future generations in this basic fashion, it can be done. One must assume random mating among the various genotypes, as well as equal production of offspring from each type of mating. If this is done, it can be shown that the frequency of the two alleles is maintained at one-half in each subsequent generation. Furthermore, from this point on, the **phenotypic frequency** will be maintained with 3/4 showing the dominant trait, 1/4 the recessive.

It is not necessary, however, to utilize such a cumbersome analysis to demonstrate the basis for the Hardy-Weinberg law. The same sort of mathematical abstractions found helpful before can be used, much in the style originally suggested by the discoverers of the principle. The Hardy-Weinberg equilibrium can be expressed as a formula in which p represents the frequency of one allele (say, *A*) and q represents the frequency of the other allele (*a*).

First, it should be apparent that the two frequencies are dependent on each other. If one allele is not present, the other one must be. Thus, if the entire gene pool in the population is represented as unity (all or 100%), the total gene population may be expressed as the figure 1:

$$p + q = 1$$

From this two more simple formulas can be derived:

$$p = 1 - q \qquad \text{or} \qquad q = 1 - p$$

If half the alleles are A, the frequency of the allele A is equal to 0.5. One can then insert the frequency of 0.5 in the preceding formula:

$$q = 1 - 0.5$$

$$q = 0.5$$

Likewise, if it were known that the total gene pool contained 25% a, then its frequency would be expressed as 0.25 and the equation would be as follows:

$$p = 1 - 0.25$$

$$p = 0.75$$

Such formulas deal only in frequencies of genes, however, and it is necessary to turn to an expansion to represent genotypic frequencies in the population. Notice that each individual in the population has two loci for each gene and that each can carry either A or a. The chance at each locus is represented as $p + q$, so that multiplying the two independent chances yields $(p + q)^2$. As with a simple binomial expansion, this equals $p^2 + 2pq + q^2$ and is known as the **Hardy-Weinberg equilibrium** equation. The whole population is then represented by the fundamental formula:

$$1 = p^2 + 2pq + q^2$$

To interpret in more meaningful terms, the entire population (unity) is represented by AA homozygous dominants (p^2), plus Aa heterozygous individuals ($2pq$), plus aa homozygous recessives (q^2). Real figures can be inserted in the formula, if the gene frequencies are known, to give the genotypic frequencies. Thus, if the frequency of A and a is 0.5,

$$1 = (0.5)^2 + 2(0.5)(0.5) + (0.5)^2$$

$$1 = 0.25 + 0.50 + 0.25$$

One-fourth of the population (0.25) is AA, one-half (0.50) is Aa, and one-fourth (0.25) is aa.

If the frequency of A were 0.75 and a were 0.25,

$$1 = (0.75)^2 + 2(0.75)(0.25) + (0.25)^2$$

$$1 = 0.56 + 0.38 + 0.06.$$

One should expect to find 56% of the population to be AA and 38% Aa. Thus 94% of the population would exhibit the dominant phenotype and only 6% would be homozygous recessive.

As in all mathematical procedures, if the genotype frequencies can be determined from the gene frequencies, it should be possible to reverse the process and ascertain the gene frequencies from the genotype frequencies. The key to this manipulation is in being able to recognize the **homozygous recessive** individuals. Once recognized, it is then possible to apply the Hardy-Weinberg equilibrium to the determination of gene frequencies. With known gene frequencies, one can calculate the other genotype frequencies in the population.

At this point note again that it is virtually impossible for any population to meet all the requirements for an ideal Hardy-Weinberg situation. In conditions where heterozygotes cannot be recognized, it is necessary to assume that the population is in a Hardy-Weinberg equilibrium. The use of their formula is an approximation and should be used only for short-term predictions.

An example of the Hardy-Weinberg analysis might be the frequency of **tasters** and **nontasters** in a population. It has been reported that in the United States 70% of persons tested are tasters of P.T.C and 30% are non-tasters. Although at first glance it might appear that the *T* allele is more common than the *t* allele, calculations show otherwise. Notice that nontasters are represented by q^2, which would be equal to 0.30; therefore $q^2 = 0.30$. It is a simple step to convert this to $q = \sqrt{0.30}$ and find the square root. In this case $q = .55$. When this is known, one can determine the frequency of *T*.

$$p = 1 - 0.55 = 0.45.$$

The *t* allele is thus more abundant in the gene pool than *T* allele.

From the frequencies of *T* and *t* (0.45 and 0.55), the frequency of the three genotypes, *TT, Tt,* and *tt*, can be determined:

$$p^2 + 2pq + q^2$$

$$(0.45)^2 + 2(0.45 \times 0.55) + (0.55)^2$$

$$0.20 + 0.50 + 0.30$$

Homozygous dominant individuals are found only 20% of the time, while heterozygotes account for 50% of the population.

The technique, often most valuable in analyzing the frequency of very rare human genes, has been used to study tyrosinase negative albinism (see Fig. 8.1, p. 203). Some studies have indicated that the frequency of tyrosinase negative albinism in the U.S. population is one in 35,000 persons. The frequency of 1/35,000 represents *aa* individuals and equals q^2 in the equilibrium formula. Thus, $q^2 = 0.000028$ and $q = \sqrt{0.000028} = 0.0053$.

If one wishes to know the chance that a person is a **heterozygous carrier** of the gene, it is necessary simply to determine what $2pq$ equals. It is known that $p = 1 - 0.0053$ or 0.9947. This figure is so close to 1 that it can be assumed essentially to be 1. This implies practically no chance of encountering anything other than *AA* individuals in the population. There is, however, a slight chance of finding *Aa* carriers. The chance is equal to $2\,(0.9947 \times 0.0053)$ or 0.0105. This means that about one in every 100 persons is a carrier in a population where albinos occur with the frequency of 1/35,000.

What is the chance for two carriers to mate and produce an albino child? Two independent individuals each have a chance of 0.0105 of being *Aa*. The product of these separate probabilities gives the random chance of their mating

(0.0105 × 0.0105 = 0.00011). The probability of an albino child is one-fourth:

$$Aa \times Aa$$
$$\downarrow$$

3/4 Normal

1/4 Albino

The product 1/4 × 0.00011 gives the chance for a tyrosinase negative albino child in marriages of normally pigmented persons in the population. The chance is 0.0000275.

Notice that this is practically the same as the frequency of such albinos in the population, indicating that their primary source is this sort of mating with virtually none being produced from *Aa* × *aa* or *aa* × *aa* marriages. The calculations here have been rounded off somewhat, but the chance for a very rare recessive trait to be produced by any combination other than by two heterozygotes is so rare as to be of no statistical consequence. As one might suspect, there are indications that persons with like genetic traits, such as albinism, do tend to choose each other as mates more frequently than random choice would predict. Naturally, this may result in offspring manifesting the trait in question. In very rare traits such matings generally do not constitute the major source of homozygous recessive members of the population.

GENE FREQUENCY CALCULATIONS IN SEX-LINKED TRAITS

As seen, some of the commonly investigated human traits, such as **colorblindness**, occur at a higher rate among males than females. It is also true that some severe hereditary diseases in males, such as **hemophilia**, are practically nonexistent among females. The reason for this situation is of course that the genes controlling such traits are carried on the X chromosome. It should be obvious that such genes cannot follow the Hardy-Weinberg equilibrium among both males and females. In fact, such genes do follow the Hardy-Weinberg principle for females in a population, since they carry two X chromosomes. On the other hand, males have only one allele for each X-linked trait, and the calculations are much simpler. Such calculations are based as other population studies, on the Hardy-Weinberg fundamental: Genes established at a given frequency are maintained at that same frequency generation after generation, provided no selective or migratory forces are at work.

It is probably apparent that it is necessary to analyze males and females separately for population studies where **sex-linked traits** are concerned. An important first principle is that the gene frequency is the same throughout the population among males and females alike. The major difference is, of course, the frequency of phenotypes. Actually, it should be no problem to see that among males the gene frequency and the frequency of individuals showing the recessive trait is exactly the same. That is simply another way of saying that males exhibit the trait whenever they have the recessive allele, because there is no dominant gene to overshadow its presence.

The next logical step, and an important consideration, is to recognize that it is a simple procedure to determine the frequency of sex-linked genes. Because the frequency of males exhibiting the trait is the same as the gene frequency, one simply needs to know what proportion of males displays the characteristic. The gene frequency is the same figure throughout the population.

Because females carry two genes for each sex-linked trait, it should be obvious that the Hardy-Weinberg equilibrium applies to their segment of the population. Thus, the distribution of the two genes is shown mathematically by $(p + q)^2$. Once the frequency of the genes is known for the male segment (q), the proportion of homozygous and heterozygous females from $p^2 + 2pq + q^2$ can be ascertained.

Hemophilia in Human Populations

An example of a rare sex-linked recessive, is illustrated in the chance of encountering a hemophiliac female in a population where one in 10,000 males has the disease (a common estimate of its frequency). Notice that the frequency of *persons* in the population who are hemophiliac is about 1/20,000 or 0.00005, whereas the frequency of hemophiliac males is 1/10,000 or 0.0001. This means that the frequency of the recessive allele for hemophilia (q) is also 1/10,000 or 0.0001 and the frequency of the dominant allele for normally clotting blood (p) is 0.9999.

To determine the genotypic distribution among females, one simply inserts these figures in the binomial expansion:

$$p^2 + 2pq + q^2$$

$$(0.9999)^2 + 2(0.9999 \times 0.0001) + (0.0001)^2$$

This indicates that hemophiliac females should occur in the population with a frequency of 0.00000001 or one in 100,000,000 females. Notice that their occurrence is predicated on the reproduction of a hemophiliac father, since such females would have two *h* alleles. Past studies have indicated that such men produce only one-fourth or one-fifth of the children that "normal" men do. Thus, the anticipated frequency of hemophiliac females would be more like one in 500,000,000 females.

The probability of hemophiliac females is so low that the possibility of their discovery was seriously questioned. However, in 1951 a family with several females who were *hh* was discovered. Apparently, these women had been able to survive because the interplay of another gene helped to mediate the effects of their homozygous condition. Since this first discovery only a few other hemophiliac females have been identified.

The frequency of *Hh* females in the population would only be 2 per 10,000 women. They serve as the major reservoir for the *h* allele in the gene pool, and many hemophiliac males have apparently opted not to reproduce and pass on their *h* gene. In the past many did not survive to reproductive age. As late as the 1950s, the mortality rate for hemophiliac males was high and only about one-quarter even reached the age of 16 years. Now probably 95% attain this age and well beyond.

It has been found that hemophilia can be effectively treated with self-administered injections of **antihemophilic globulin (Factor VIII)**. Obviously, males who formerly would have died or chosen not to pass along genes to the next generation can now survive in a healthy state. They may live to attain normal old age through the use of the antihemophilic globulin. Whether or not they opt to reproduce is an interesting situation.

If the treatment of the disease allows a more normal life, will it result in the passage of a gene which was normally selected against in the past? Keep in mind that unless the mother was a carrier, a male hemophiliac would not produce hemophiliac children. There would be only carrier females, and only with a one-half chance. In the next generation, however, these females could produce hemophiliac sons.

Apparently some carrier females, who can be identified by a fairly sophisticated blood test, have chosen to produce only daughters. It is now possible to identify a male or female fetus in utero. There is a one-half chance that the son of a carrier female will have hemophilia. Some women have chosen the option of abortion for male fetuses, rather than risk having a hemophiliac son. Even with the possibility of overcoming the disease symptoms with injections, the financial drain for such treatments is enough burden to suggest this alternative.

The **mutation rate** for the change of the dominant allele to the recessive h has been found to be 2.7/100,000 or about 0.00003. This is about one-third of the frequency of the gene in the population, a very high relative proportion. In a rare gene of this type the mutation rate becomes a significant factor in the maintenance of the gene frequency. In the case of some extremely rare genes that are strongly detrimental to survival the gene frequency is about balanced by the mutation rate. With hemophilia, the mutation rate has about balanced the reproductive selection against hemophiliac males in the past.

If hemophiliac males begin to reproduce in significant numbers, or female carriers abort males and give birth only to females, the number of carrier females will increase in the population. With the mutation rate continuing at its normal rate, as selection against the gene is eliminated, the gene frequency would increase. This is another situation presenting a moral dilemma to be solved by future generations and is a direct result of expanded understanding and practical genetic research.

Colorblindness in Human Populations

Not all recessive sex-linked alleles are as rare as hemophilia. Although such alleles may be found homozygously among females, such women manifest the trait much less often than men. Remember that the frequency of males showing the recessive trait is the same as the gene frequency, and females only exhibit the trait if they are homozygous. It is apparent that the mathematical relationship is based on the square of the frequency of affected males (q) being equal to the frequency of affected females (q^2). In such a relationship the rarer the gene, the wider the difference between numbers of affected males and females.

An example of a sex-linked recessive allele that is not rare is red-green colorblindness. Although red-green colorblindness was treated as a single-gene controlled characteristic while solving problems in Chapter 7, the trait is controlled by at least two pairs of genes. The effect of this can be seen rather easily in a population analysis.

About 8% (0.08) of the Caucasian male population in the United States is red-green colorblind. This means that the frequency of the recessive allele is the same, or equal to 0.08 (q). Females with the homozygous recessive genotype should be found with a frequency of q^2 or 0.0064. Thus, about 0.64% of the female population should consist of colorblind individuals. In fact, surveys reveal that somewhat less than 0.5% of females are colorblind.

This anomaly can be explained by the fact that two gene pairs are involved. If it is assumed that the frequency of the green-blind allele is 0.06 and the frequency of the red-blind allele is 0.02, a new estimate of the frequency of colorblind females can be made. Notice that some would be colorblind due to defective green vision (0.06^2), some would be colorblind due to defective red vision (0.02^2), and some would be colorblind because of the possession of both recessive alleles homozygously ($0.06^2 \times 0.02^2$). The expected frequency of red-green colorblind women then would be 0.36% + 0.04% + 0.0001%, or about 0.4%, somewhat closer to the actual value.

The situation, however, is more complex: There are two mutant alleles affecting green vision, one more severe than the other, and two alleles similarly affecting red vision. Furthermore, there are other factors affecting color vision in homozygous recessive females. As a result, deviations from expected values are found and these are especially noticeable with the small frequencies involved.

GENETIC DRIFT

Variation in gene frequencies in a population due to chance fluctuations is called **genetic drift**. The term implies that there is a tendency for a value to "drift around" from generation to generation. Because only a small segment of the available gametes are ever utilized in the production of progeny, it is obvious that a **sampling process** is involved in the formation of each generation.

This sampling process is analogous to a procedure one might follow to ascertain the number of black and white beads in a box. If half the beads were white and half were black, a sample could be withdrawn assuming that it would be representative of the whole population. If there were 500,000 black and 500,000 white beads and only 10 were withdrawn, the chances of picking eight black and two white are not too bad. The binomial expansion would estimate that one might find deviations this great or greater about 10% of the time.

If the sample of beads somehow could reproduce by dividing and produce like-colored beads, the frequency of beads would have drifted in one generation from 50:50 to 80:20. Likewise, due to random sampling errors, the frequency of genes may show more variation than would be expected from a strict interpretation of the Hardy-Weinberg principle. The smaller the population size involved, the greater the effect of random fluctuations. There are situations in the history of a population when numbers drop down to a small number of individuals before building back up to a large size (comparable to the hypothetical 10-bead sample from a box of 1,000,000). Geneticists label this phenomenon a **bottleneck effect**, and note that genetic drift is likely to be most pronounced under such circumstances.

A corollary is found in the fact that the entire population is not relevant to the genetic drift phenomenon. Within any population, only a certain number of individuals are involved in reproduction. There may be a number of factors that contribute to the selection of the **reproductive subpopulation**. Most obvious is probably age. A rough approximation would probably categorize about one-third of the total population as providing the source of genes for the next generation.

FOUNDER EFFECT AND MIGRATION

If a small group of individuals or families fragments away from a larger population to form another population in another geographic area, it is apparent that a nonrepresentative sample may be involved. At some later date an examination of the gene frequencies in the new population would reflect the distribution of genes of the founders of the splinter group more closely than that of the parent population. This result has been labeled by geneticists the **founder effect**. Remember that the gene frequencies may represent those of the "founders" of this population, but the distribution is renewed in each generation. Thus, every generation is the founder of the next.

Frequently, what appears to be a large population actually consists of numerous distinct local populations that are more or less isolated from each other. If there were complete reproductive isolation from the larger population in such small groups, genetic drift would result in distinctly different gene frequencies among the various smaller units.

If families migrated away from their original homes into an area populated by another group, they would carry with them genes at a different frequency than in their new residence. Subsequent intermarriages between the two populations would result in a change in the gene frequencies. When intermigration occurs among subpopulations, the resultant flow of genes from one group to another has a tendency to counteract the effects of genetic drift in the subpopulations. In studying large populations it is necessary to recognize any isolated or semiisolated subpopulations and consider the overall effects of drift and migration.

MUTATION EFFECTS

Gene frequencies in small populations or subdivisions of populations show greater effects of mutations than in large populations. Even though **mutation frequency** is the same in any group, if one occurs, it has a more pronounced effect on the frequency of a rare allele in the small population than in a large population. For example, one might consider the royal families of Europe as shown in the Queen Victoria pedigree to represent a distinct reproductive subpopulation (Fig. 9.1, p. 230). The hemophilia mutation carried by Victoria resulted in a much higher frequency in the royal subpopulation than in the overall European population.

SELECTION MECHANISMS

Gene frequencies clearly are partially dependent on **selection processes** that have conferred various advantages or disadvantages on certain individuals. **Environmental conditions** enhance or hinder the ability of some individuals to survive and reproduce as effectively as others. Notice that this action of specific environmental factors is on the phenotype of individuals. The phenotype is controlled by the genes, but it is individuals carrying genes that are at an advantage or disadvantage, not the genes themselves. A relative increase or decrease in gene frequency may thus be associated with the particular **advantage** or **disadvantage** of certain **phenotypes** in a given environment.

It is sometimes difficult to understand why some genes confer a selective advantage in a given environment. Past environmental conditions are often difficult to evaluate, let alone the manner in which a genotype responded. It is fairly easy to see the advantage to **hemoglobin mutants**, such as those described in Chapter 4. Because the heterozygotes were resistant to malaria, they had a better chance to survive and reproduce than persons without the mutant gene.

In this regard an interesting question about **cystic fibrosis** concerns its rather high incidence among persons of European ancestry. Although it has been highly and effectively selected against in the past through the death of virtually all homozygotes prior to reproductive age, why does the gene persist with a frequency of about 2% (0.02)? This is too high a frequency to be accounted for by mutation rate. Many population geneticists believe that heterozygotes for cystic fibrosis must enjoy some sort of selective advantage over homozygous dominant persons. In fact some studies have indicated that a larger number of offspring are born to heterozygotes (*Cc*) than to homozygotes (*CC*), but these data are not conclusive. The situation requires considerable study and there is no doubt that careful analyses will be made.

In Chapter 8 it was noted that **Tay-Sachs** disease occurs at a much higher frequency among persons of northern European Jewish descent (Ashkenazi Jews). The frequency of the gene has been estimated among such persons at about 1.6% (0.016). This is considerably higher than can be accounted for by the normal mutation rate and it is doubtful that the mutation rate is higher in this particular population. It is difficult to explain the presence of such a severe allele at this high frequency. It is invariably lethal in the homozygous state. Can genetic drift or the founder effect explain its high frequency, or might there have been some selective advantage conferred on heterozygotes?

CONCEPT OF HUMAN RACES

For various reasons, groups of humans are found associated in populations throughout the world. The reasons underlying the formation of such groups of individuals are not always clear and speculation about the causes is often without much basis. Clearly, there is generally a familial aspect to such populations. It is conceivable that in many instances minor groups within a large population migrated to a new area, becoming geographically and reproductively **isolated** from the original population.

As discussed with the founder effect, the migrating group might have carried with them a particular combination of genes in higher or lower frequency than existed in the original population. The particular gene would thus be found in both populations, but at significantly different frequencies.

Following fragmentation of a population and separation to a different geographical area, requirements for survival in the new environment might exert new selective forces on combinations of genes which were not particularly advantageous in the old environment. In addition, genetic drift is more pronounced in smaller populations. It is not difficult to envision that, over long periods of time and many generations, a new population might become distinct from the original in gene frequencies and resultant physical traits.

Even though humans all belong to one biological **species**, capable of **interbreeding**, the difference from one population to another can be obvious. It is commonly accepted that humans all share common ancestors in antiquity. However, many anthropologists have studied differences among various populations to distinguish certain groups as **races**.

Anthropologists often differ about how many races of humans actually exist and have postulated from three to 20 or more different races. The number of races into which *Homo sapiens* is divided usually depends on how much detail is involved in describing the differences among populations. Notice that the populations are often not clearly isolated from each other geographically or reproductively, due to expanding numbers of individuals and migrations. Thus, clear-cut distinguishing differences may not exist and there are no "pure" races. Any classification is bound to be arbitrary and imperfect.

The concept of differences in humans along with casual observation of races has led to some of the most serious problems confronting society. It is true that there are differences from one population to another, but these differences are based on gene frequencies. Genes found in one population probably exist in other populations, but the frequencies can be extremely different from one group to another. More importantly, the low frequency of a gene that is considered either advantageous or unfavorable may be more than compensated for by other genes in high frequency.

Unfortunately, it has become the vogue within some social structures to associate particular characteristics with a so-called race and then apply broad judgments to everyone categorized as belonging to that race. Even if certain genes are found at a high frequency within a population, it should be obvious that **individual differences** still exist within the population.

Possibly a more insidious use of racial concepts, coupled with basic genetic information, has been the proposal of racial inferiority in such areas as native intelligence. Based on implied lack of genetic intellectual quality, some have predicted a failure of educational efforts in certain races.

Persons who espoused this line of thinking have even proposed that educational systems be organized along racial lines. The basis for their analyses and proposals has been the acceptance of a strong genetic component to intelligence. Other workers who have looked closely at the situation feel that the environmental effect on intellectual ability has been neglected in such studies. There is no doubt that various racial groups segregated within our society have been exposed to different philosophical and practical approaches to intellectual endeavors.

That different levels of genetic ability may exist seems undebatable where individuals are concerned. To suggest that educational endeavors be forfeited for various groups of persons based on their ancestry or on a handful of obvious phenotypic traits seems ludicrous. Although the educational establishment could be improved through some restructuring, a system that finds and emphasizes the strengths of all persons, without regard to racial background, seems like a more logical route.

EUGENICS

While Mendel was establishing his understanding of the genetic mechanism, an active and interested scientist was thinking of ways to emphasize the "good" inherited traits in humans and eliminate the "bad." Francis Galton, a cousin of Charles Darwin, proposed in 1865 that conscious efforts be made to encourage the reproduction of some persons and discourage the reproduction of others (Fig. 11.2). He named his new study **Eugenics** and categorized it as a field of science. It is interesting that

Figure 11.2. Francis Galton, a cousin of Charles Darwin, originated the study of eugenics in England in 1865.

Darwin recognized this proposed breeding program as evolution under the direction of humans. He complimented Galton on his proposal, but thought he was a bit too optimistic about its implementation.

Unfortunately, Galton would be classified as a racist of the worst order today. He believed in the supremacy of Caucasians and pointed out the differences in his writings. Of course, any selective breeding program for humans requires a judge for "good" and "bad" traits. Galton thought that he could recognize the difference, as did some before and after him. The efforts of Adolf Hitler in the 1930s and 1940s to eliminate or enhance perceived human characteristics have been cited as an organized program in eugenics. Scientists often differ with such opinions, since the Nazi effort was grounded in hatred and had little scientific basis.

Eugenics programs can be thought of in two ways. **Positive eugenics** involves the encouragement of reproduction by those possessing desirable characteristics. **Negative eugenics** is the discouragement or prevention of reproduction of those with defective genes. A majority of states have laws, although they vary from state to state, that allow sterilization of individuals for a host of nebulous genetic reasons, primarily mental deficiencies.

There have been many supporters of diverse eugenics plans, and many persons of recognized scientific and political stature still support such plans in one form or another. Such plans have become increasingly popular as the problems created by the expanding world population have become more visible. No doubt there will be attempts to legally formalize some aspects of these programs, or, if not legally, by more subtle means, such as the fine print in health insurance policies. With the knowledge to prevent the birth of various identifiable genetic maladies through **abortion**, governments may impose a requirement for such action without individual choice.

Some eugenics schemes involve the prevention of reproduction of persons who are homozygous for serious and rare recessive traits. They hold out the promise of eliminating or significantly lowering the incidence of such maladies in future generations. These plans often have considerable popular appeal because it could mean a large financial saving of institutionalization at public expense. The attempt to eliminate such genes from a population may not be worthy of consideration. Plans to alleviate or overcome the phenotypic effects are frequently more plausible. Programs in which an attempt is made to correct the phenotypic defect have been categorized as **euphenics**. Some feel that any attempt to improve, eliminate, or overcome genetic effects is a form of tampering and not worthy of human efforts. Their arguments are probably futile, since it is characteristic of human curiosity to investigate scientific matters underlying human nature. Once such information is obtained, it inevitably leads to practical plans to improve the plight of humans.

MATHEMATICAL BASIS OF SELECTION PROCESSES

It was Charles Darwin and Alfred Wallace's realization in 1858 that there are natural forces operating in the environment which tend to select for or against certain phenotypes. These phenotypes are controlled by genes that tend to make the individual who possesses them more or less fit to reproduce. As a result of this selection process, whether complete or not, a gene tends to increase or decrease in a population with subsequent generations. This process of selection by natural conditions is the basis of **evolution** or change of living things through time.

Artificial selection is no less effective than **natural selection** and through conscious effort crops or domestic animals can be improved for food production, among other things. No doubt Mendel recognized this selection process with his agricultural background, but he did not synthesize it into a broad concept of evolution as Darwin and Wallace. He probably knew of their work, but no comment from him relating his genetic knowledge to their broad theory can be found. It is possible that in his role as a monk such a subject was taboo and he simply did not think much about it.

Utilizing the mathematical principles of Hardy and Weinberg, it is possible to derive equations that will produce accurate numerical data regarding the selection process. Extremely complicated formulas and problems can be proposed, taking many factors into account. Sometimes detailed population studies, involving various degrees of selection for or against certain traits, require the use of computers. To show the means of implementing such extensions of the Hardy-Weinberg equilibrium, two equations will be examined. For purposes of this text, it is not necessary to go through the mathematical derivations; the formulas will be used to observe the processes involved and the type of analysis possible.

It is easiest to consider only processes involving selection against certain alleles, although mathematical consideration of positive selection might also be investigated. Furthermore, to simplify, only **complete selection** against a trait will be examined. "Complete selection against" means that homozygous recessive individuals are prevented from producing offspring (lethal selection); however, heterozygous individuals are not identified and they continue to reproduce. It is not too difficult to modify such calculations to demonstrate partial selection. For those interested in learning the formula for partial selection against homozygous recessive individuals and in seeing an example of this process, try Review Question 15 at the end of this chapter.

In general, it is necessary to solve selection equations for two different figures: the number of generations involved in the selection process and the frequency of genes. Mathematical abstractions must again be used in such studies. Notice that the convention of using n has been employed to symbolize the number of generations. The symbol q is once more used to represent the numerical frequency of the recessive gene; q_o is the frequency of the allele in the population at the outset of the study, and q_n the frequency of the allele after n generations of selection against it.

The following equation is probably the easiest to use when the two allele frequencies are known and the **number of generations** involved is desired:

$$n = 1/q_n - 1/q_o$$

If one knows how many generations of complete selection against a trait are involved, the easiest way to learn the **frequency of a gene** after selection has taken place is to use the formula

$$q_n = \frac{q_o}{1 + nq_o}$$

To apply the first of these, assume that a naive eugenicist, or possibly a politician hoping to save money, has proposed a plan to reduce or eliminate a hypothetical genetic defect caused by a single gene. If the cost of caring for such defectives is borne by the state at a cost of $100,000 per year per individual, and the frequency of persons showing the trait is one per 20,000, a sizable sum is involved in a state having only 4 million inhabitants. There would be 200 defective persons for a total cost to the taxpayer of 20 million dollars per year.

Suppose that, although the politician is an idealist, he or she recognizes that the disease cannot be completely eliminated by preventing the reproduction of homozygous recessives. Suppose that he or she proposes only to cut the number and the bill in half by preventing the production of offspring by homozygous recessive persons. Remember that laws already exist in many states permitting sterilization of some defective individuals, so the scenario may not be as bizarre as it first appears.

The frequency of defective individuals (0.00005) is the square of the allele frequency (q^2). Thus, the frequency of the recessive allele (q) is 0.007. To lower the frequency of defectives by half would be to change their frequency to 1/40,000 persons in the population (gene frequency = 0.005) and save $10,000,000 per year, once the new frequency was reached. How long would it take?

Using the preceding formula, one can insert the gene frequencies at the outset and at the completion of the selection process:

$$n = 1/q_n - 1/q_o$$

$$n = 1/0.005 - 1/0.007$$

$$n = 200 - 142$$

$$n = 58$$

The number of generations of complete selection or prevention of reproduction by homozygous recessives would be about 58. If it is assumed that about 30 years are involved in each human generation, somewhat more than 17 centuries are involved! Of course, this is a ridiculous consideration in terms of saving money. Plans to find a means of overcoming the effects of such genes would be more meaningful.

A more realistic proposal might be made in terms of a gene that is not as rare as the last example. Suppose that complete selection against a recessive trait were made in a population, one-quarter of which exhibited the phenotype. This 25% of the population would be unable to pass along the gene to their offspring, but the heterozygotes would be as fit to reproduce as the homozygous dominant individuals. One might ask how effective a number of generations of selection might be and whether or not the allele could ever be eliminated from the population.

In such a case, where the gene frequency and the number of generations involved is known, the second equation is easier to use. Because the frequency of homozygous recessive individuals is 0.25(q^2), the frequency of the gene is 0.5 ($\sqrt{0.25} = q$) and can be inserted in the following formula:

$$q_n = \frac{q_o}{1 + nq}$$

1 generation of selection $\longrightarrow q_n = \frac{0.5}{1 + (1 \times 0.5)} = \frac{0.5}{1.5} = 1/3$

2 generations of selection $\longrightarrow q_n = \frac{0.5}{1 + (2 \times 0.5)} = \frac{0.5}{2} = 1/4$

3 generations of selection $\longrightarrow q_n = \frac{0.5}{1 + (3 \times 0.5)} = \frac{0.5}{2.5} = 1/5$

It is apparent that gene frequency can never reach zero in the population, but tends toward an infinitely smaller value under the conditions of selection. It should also be obvious that, as the frequency gets smaller, the chance of eliminating it also diminishes (1/2 → 1/3 → 1/4 → 1/5 → 1/6 . . . as compared to 1/401 → 1/402 → 1/403 → 1/404 → 1/405, etc.).

Such exercises show the fallacy of selection against a trait by eliminating homozygous recessive individuals from the reproductive pool. However, it is important to note that there are many situations in which a careful selection process could work to produce a given result in a positive or negative program. For instance, a dominant gene could be eliminated from a population in one generation, if only homozygous recessive individuals were allowed to reproduce. Likewise, if heterozygous individuals can be identified, selection against a recessive trait can be completely effective. In this case it would be necessary to allow only homozygous dominants to transmit genes to the next generation.

If one considers carefully what is happening in the complete selection process against recessive alleles, as a recessive gene is selected against and the frequency becomes smaller, the proportion of the recessive alleles in heterozygotes versus homozygotes becomes larger. This can be seen mathematically by quickly considering four different allelic frequencies for a recessive gene—0.33, 0.25, 0.20 and 0.17 (or in fractions, 1/3, 1/4, 1/5, 1/6). The Hardy-Weinberg equilibrium for these values, using fractions instead of decimal equivalents, would be:

$(2/3)^2\ AA + 2(2/3)(1/3)$ or $4/9\ Aa + (1/3)^2$ or $1/9\ aa$

$(3/4)^2\ AA + 2(3/4)(1/4)$ or $6/16\ Aa + (1/4)^2$ or $1/16\ aa$

$(4/5)^2\ AA + 2(4/5)(1/5)$ or $8/25\ Aa + (1/5)^2$ or $1/25\ aa$

$(5/6)^2\ AA + 2(5/6)(1/6)$ or $10/36\ Aa + (1/6)^2$ or $1/36\ aa$

As the allele frequency decreases from 0.33 to 0.25 to 0.20 to 0.17, the proportion of heterozygotes compared to homozygotes goes from four, to six, to eight, to 10 times greater. Such an example demonstrates that the great reservoir of very rare recessive genes is to be found among heterozygous carriers and not among those showing the trait.

A knowledge of the Hardy-Weinberg equilibrium and its mathematical extension to the effects of selection is important to understand gene distribution in human populations. To answer questions concerning eugenics proposals and the basis for differences among human populations, the formulas derived from the original 1908 studies are essential tools. Sophisticated and complicated models requiring computers have been utilized in population genetics, but, as in the work of Mendel, there exists a basis of underlying fundamentals that can be grasped and used for a broad understanding of human genetics.

ADDITIONAL READING

ALLEN, G. E., "Science and Society in the Eugenic Thought of H. J. Muller," *Bioscience* (1970), 20, 346–53.

BODMER, W. F., and L. L. Cavalli-Sforza, *Genetics, Evolution, and Man*. San Francisco: W. H. Freeman & Company Publishers, 1976.

FALCONER, D. S., *Introduction to Quantitative Genetics* (2nd ed.). New York: Longman, 1981.

GLASS, H. B., "The Genetics of the Dunkers," *Scientific American* (1953), 189, No. 2, 76–81.

HARDY, G. H., "Mendelian Proportions in a Mixed Population," *Science* (1908), 28, 49–50. (Reprinted with comments in Peters, J. A. (ed.), *Classic Papers in Genetics*. Englewood, Cliffs, N. J.: Prentice-Hall, 1961.)

LOEHLIN, J. C., G. Lindzey, and J. N. Spuhler, *Race Differences in Intelligence*. San Francisco: W. H. Freeman & Company Publishers 1975.

MATSUNAGA, E., "Possible Genetic Consequences of Family Planning," *Journal of the American Medical Association* (1966), 198, 533–40. (Reprinted in Baer, A. S. (ed.), *Heredity and Society* (2nd ed.). New York: Macmillan, 1977.)

METTLER, L. and T. Gregg, *Population Genetics and Evolution*. Englewood Cliffs, N.J.: Prentice-Hall, 1969.

STERN, C., "The Hardy-Weinberg Law," *Science* (1943), 97, 137–38. (Reprinted in Levine, L., *Papers on Genetics*. St. Louis, Mo: C. V. Mosby, 1971.)

______, *Principles of Human Genetics* (3rd ed.). San Francisco: W. H. Freeman & Company Publishers, 1973.

WALLACE, B. (ed.), *Genetics, Evolution, Race, Radiation Biology: Essays in Social Biology, Vol. II*. Englewood Cliffs, N.J.: Prentice-Hall, 1972.

WEINBERG, W., "On the Demonstration of Heredity in Man," *Jahreshefte des Vereins für Vaterländische Naturkunde in Wurttemberg, Stuttgart* 1908, 64, 368–82. (Reprinted in English with comments in Boyer, S. H., IV (ed.), *Papers On Human Genetics*. Englewood Cliffs, N.J.: Prentice-Hall, 1963.)

REVIEW QUESTIONS

1. In a randomly chosen family with six children, what is the probability that all six will be girls? What is the probability that two children will be boys and four will be girls? If five girls have been born, what is the chance that the sixth will be a boy?
2. Suppose that a new human population is established on a desert island when 200 pioneers arrive. One hundred of these pioneers are males homozygous for the free ear lobe trait and 100 are females with attached ear lobes. There will be 100 marriages in the first generation and each marriage will produce four children, two boys and two girls.

What are the frequencies of the A and a alleles in the parental population at the outset? What are the frequencies of A and a among the 400 children? What are the phenotype (free and attached) frequencies in the two generations?

3. Assume that there are 200 marriages among the members of generation II in Question 2, and that each results in four children. Among the 800 children in this third generation, what would be expected proportions of genotypes? What would be the phenotype frequencies? What are the frequencies of the A and a alleles?

4. Carry on the population of Questions 2 and 3 for two more generations by following the same pattern (random marriages for all members of each generation with four children produced in each family). Each generation will contain equal numbers of males and females and the genes will follow ideal Mendelian distribution. There will be 400 marriages among generation III and 1600 children produced.

   Hint: Among 800 persons there are 400 males (100 AA, 200 Aa, 100 aa) and 400 females (100 AA, 200 Aa, 100 aa). In random mating the marriages would be as follows:

Number	*Males*		*Females*
25	AA	×	AA
50	AA	×	Aa
25	AA	×	aa
50	Aa	×	AA
100	Aa	×	Aa
50	Aa	×	aa
25	aa	×	AA
50	aa	×	Aa
25	aa	×	aa

   There will be 800 marriages in generation IV resulting in 3200 children, following the same pattern.

   What will be the frequencies of genotypes in each of the two generations (IV and V)? What will be the phenotype frequencies? What will be the frequencies of alleles A and a?

5. Assume that a tryrannical government arises on the island in generation III and decides to eliminate the attached ear lobe trait before any children are born in generation IV. This is to be done by preventing the reproduction of those showing the trait (aa) in generation III. This means that 600 persons will produce families in generation III instead of 800 (200 AA and 400 Aa). Use the same rules to calculate the results in generation IV (random marriages among AA and Aa persons, four children per family, equal sex distribution).

   Hint: Now the marriages will be as follows:

Number	*Male*		*Female*
34	AA	×	AA
66	AA	×	Aa
67	Aa	×	AA
133	Aa	×	Aa

What will be the genotype and phenotype proportions among the children of generation IV? What will be the frequencies of the *A* and *a* alleles? What do you think the frequencies of *A* and *a* would be after one more generation of this type of selection?

6. What factors could cause gene frequencies in a population to change or deviate from the Hardy-Weinberg prediction?
7. Suppose a population was found in which 16% lacked middigital hair on their fingers. Calculate the genotype frequencies and the frequencies of the dominant and recessive alleles for the population. What is the chance for an $Aa \times Aa$ mating in this population?
8. Cystic fibrosis occurs with a frequency of 1/2500 (0.0004) in the U.S. white "population." What does this indicate the frequency of the gene to be? What is the chance that a person is the carrier of the gene? What is the chance that two carriers will marry and that their child will have cystic fibrosis in this population?
9. How many generations of complete selection against reproduction of cystic fibrotic persons would be required to reduce the incidence of the disease by one-half (1/5000)?
10. Cystic fibrotic individuals have not been able to reproduce in the past, since the disease was generally fatal before reproductive age (lethal). Do you think the gene diminished as quickly as your calculation in Question 8 predicted? Explain.
11. What effect on the frequency of the gene for cystic fibrosis may an effective treatment for the disease have? Explain.
12. Suppose that a survey of an African population is made by examining blood samples to determine whether a person has sickle-cell anemia (Hb^SHb^S), sickle-cell trait (Hb^AHb^S), or normal blood (Hb^AHb^A). The survey is categorized according to age classes with the following numbers of individuals in two groups.

	Infants under year of age	*Adults over 35 years of age*
Hb^AHb^A	295	572
Hb^AHb^S	97	211
Hb^SHb^S	8	1

Using the Hardy-Weinberg equation, determine whether or not these two groups are in equilibrium. Explain any discrepancies that your analysis reveals.

13. Suppose that medical research indicates through experiments that persons who are carriers of the recessive Tay-Sachs allele (*Tt*) are resistant to tuberculosis. Using this hypothetical information, can you construct an explanation for the above average frequency of the Tay-Sachs allele among Ashkenazi Jews?
14. Expand the binomial equation to the power of 10: $(a + b)^{10}$. Use this as a model to describe the different combinations of equal-sized, well-mixed red and blue marbles that might be drawn from a large container, 10 at a draw. Assume that on the first draw there are 500 red marbles and 500 blue marbles in the container. Based on the binomial expansion, what is the mathematical probability of drawing eight of one color and two of another, *or* a greater deviation from the 5:5 norm (e.g., 8:2, 9:1, or 10 of one color)? How might this situation serve as a model for the founder effect?
15. It is possible to analyze the effect of partial selection, where the homozygous recessive condition is not completely lethal and a certain percentage may survive to reproduce. To

determine the recessive allele frequency after one generation of selection, the following formula is used:

$$q_1 = \frac{q_O - kq_O^2}{1 - kq_O^2}$$

where

q_1 is the new allele frequency (one generation of partial selection)

q_O is the initial allele frequency

k is a fraction (from 0 to 1) representing the proportion of homozygous recessives eliminated in each generation (coefficient of selection)

Assume that 25% of the population carry the homozygous genotype for a recessive gene (*aa*) responsible for a condition in the body fluids that is not invariably detrimental to survival and reproduction. In only about 25% of the homozygotes is it detrimental, and then it results in sterility (thus, the gene cannot be passed on to the next generation). Disregard any mutation effects and calculate the frequency of the allele in the progeny of this generation. What will be the distribution of *AA, Aa,* and *aa* genotypes among this next generation (after one generation of partial selection)? After two generations of partial selection what will be the frequency of the *a* allele, and what will be the distribution of genotypes *AA, Aa,* and *aa*?

CHAPTER SUMMARY

1. Godfrey Hardy and Wilhelm Weinberg in 1908 described independently the principle of gene equilibrium in populations. Their formula for genotype distribution, $p^2 + 2pq + q^2$, has become the basis of sophisticated analyses of populations.
2. For a population to meet the ideal standards of the Hardy-Weinberg equilibrium, it must be a very large, stable, randomly mating population. There must be no selective migration into or out of the population and no differential survival or reproductive advantage for the genes being studied. Mutational change in gene frequency will disrupt the equilibrium. Probably no natural population satisfies the ideal requirements.
3. The binomial expansion, or normal curve, can be utilized to describe and predict the classes and frequencies of single-gene-controlled traits in a family or population. Frequencies of genotypes and phenotypes are dependent on the frequencies of alleles.
4. Gene frequencies can be calculated on the basis of the equilibrium formula ($p^2 + 2pq + q^2$). Because q^2 represents homozygous recessives, the recessive allele frequency is equal to q and the frequency of the dominant allele is equal to $1 - q$.
5. In sex-linked traits the frequency of the recessive allele is equal to the frequency of males showing the recessive trait and the frequency of the dominant allele is equal to the frequency of males showing the dominant trait.
6. Variation in gene frequencies in a population due to chance fluctuations is called genetic drift. When small groups leave a larger population and form a new population, a similar sampling error occurs. The new and different gene frequency is said to be due to the founder effect.
7. Analysis of very rare and lethal genes often indicates that equilibrium is maintained in a population with the gene frequency about equal to the rate of mutations producing it. Some seriously detrimental genes are maintained at a level well above mutation rate. This phenomenon often depends on heterozygote advantage, as in the malaria resistance of sickle trait heterozygotes.
8. Human races are distinguished genetically by a number of distinctive gene frequencies in variously isolated populations.
9. Francis Galton invented a proposal for the improvement of human populations by selection of those to reproduce. His plan was called eugenics and it has been invoked in various ways. Some types of eugenics plans can still be found. Unfortunately, such plans require human judges to select "good" and "bad" traits, a highly questionable process for maintaining the biological diversity characteristic of living organisms.
10. An analysis of gene frequency in very rare recessive traits helps to explain the difficulty of lowering the incidence of such traits through eugenics proposals to prevent the reproduction of homozygous recessives.
11. Equations that show the effect of selection have been derived from the Hardy-Weinberg formula. It is possible to predict the number of generations required for a given change in frequency or to predict the amount of frequency change in a given number of generations of selection.
12. As the frequency of a recessive gene diminishes in a population due to selection, the proportion of heterozygotes relative to the homozygous recessives increases.

CHAPTER

12

EPILOGUE AND PROLOGUE TO THE FUTURE

If you do not think about the future, you cannot have one.
John Galsworthy, Swan Song *(1928)*

Practically everyone would agree with efforts to cure genetic diseases by surgical methods. When such programs are launched, however, information is provided that may as easily be applied to genetic traits that are not disabling. It would not take an extensive survey to show considerable differences of opinion about what measures should be taken to influence human hereditary qualities that are not disabling or clearly defective.

Almost incredible means of altering or influencing the normal human inheritance process have been proposed in science fiction. Such schemes are at least partially based on factual scientific knowledge or an extension of techniques already available. In most cases the methods that may ultimately be used to change hereditary patterns are only in their infancy, with truly amazing and sophisticated developments yet to come.

The future may produce a number of techniques for improvement of the human condition through manipulation of the normal, randomly operating genetic mechanism. Whether or not such activities constitute an improvement is debatable, but it is essential to consider the possibilities and be prepared to aid in making decisions about the utilization of new techniques.

Some of these future techniques are already in use, if only in a relatively limited fashion. Other methods are speculative, but some preliminary experiments have been done. Regardless of the acceptability of such experiments or techniques by the majority of current standards, rest assured that somewhere there will be advancement of such ideas. The biggest problems will be involved with sorting out the acceptability of various proposals and the formulation of ethical codes or legislation controlling their application.

To make the important decisions required, knowledge will be the most valuable attribute for all concerned. Knowledge now, as in the past, will depend on persons with vision.

CORRECTION OF GENETIC DEFECTS

The most obvious technique is the correction of defective genes through their replacement with normally functional genes. This sort of maneuver is probably still some distance into the future, but it must be the ultimate thought in any repertoire of future techniques.

The groundwork for this possibility has already been laid and many areas of active molecular biological research are opening up means of providing a workable technique. Practically all work with gene transplants, as well as other molecular gene studies, utilizes bacteria as experimental organisms. It is a long jump from unicellular, simple bacterial cells to multicellular complex human organisms. In spite of this, it is important to remember that many of the concepts of gene functioning at the molecular level were originally discovered and worked out with bacteria as models. Later studies have shown many of these same mechanisms to operate in much the same way among higher organisms. There are fundamental and intricate complications with the genetic mechanism of higher organisms that do not exist in the bacteria, but bacterial techniques at least provide a starting point.

An area of scientific research that has occupied more public attention and concern than almost any activity in recent memory is with **recombinant DNA**. The basis for this type of study is the laboratory growth of viruses and bacteria. It was first discovered that viruses could pick up foreign DNA segments, even human genes, and transfer them to bacteria. The foreign DNA was sometimes incorporated into the bacterial genetic material and often produced its normal products there.

However, much more elegant and precise techniques for transplanting genes became available with the discovery of **restriction endonucleases**. As seen in Chapter 5, such enzymes can be used to excise chromosome segments at specific positions, leaving "sticky ends" which will join with similar sticky ends on bacterial DNA.

Bacterial cells have single circular DNA strands that constitute their hereditary material. Most species also often have additional DNA in the form of a small circle with one to a few genes. Such circular DNA segments are called **plasmids**. These plasmids do not carry essential bacterial genes, but often are responsible for such traits as resistance to antibiotics. They replicate freely inside the bacterium and are transferred naturally among cells. It is a fairly simple matter for a skilled molecular biologist to cut open one with a restriction endonuclease and insert a foreign gene excised with the same enzyme. The matter of returning the plasmid to the bacterium where it will replicate has also become routine, and in some cases the foreign gene's product will be produced in abundance. When specific genes are grown and maintained in the laboratory in this manner, they are said to be **cloned** (grown from a single common ancestor).

To add to the sophistication of the technique, some laboratories have artificially synthesized genes (and even modified them to create "improvements"). One human gene that has been synthesized and then implanted in bacteria via plasmids is the 514 base pair gene for **interferon**. Interferon is a suspected anticancer, antipathogen, human product which may have great economic value.

As discussed, it has been possible to transfer human chromosomes to mouse cells growing in culture. Using the restriction endonuclease techniques, genes have been transferred to plants; a beginning has been made in transferring genes to mammalian cells in culture. Amazing things have been done with **tissue cultures**, but as yet it has not been possible to transfer genes or chromosomes into the cells of a human. Such experiments, tried without success, have been met with generally severe criticism because of their tentative nature.

One attempt was through the use of a viral carrier, hoping that the "infection" might transfer DNA segments to human cells where they could become incorporated into the human genome. Another was to try to confer the ability to produce normal hemoglobin in individuals by removing bone marrow cells to tissue culture conditions, where the normally functioning gene could be inserted and then returning the cells to the marrow. It was hoped that the cells would take up residence there and begin to produce red blood cells carrying normal hemoglobin. The experiment had not been successful on mice, but was nonetheless attempted on two humans in 1980. Considerable furor, both inside and outside the medical-biological community, was raised and steps have been taken to prevent such experiments on human subjects. Nevertheless, considerable hope remains that a means of replacing or adding genes to the cells of those with defects will be discovered.

Medical researchers in recent years have expended much effort attempting to ascertain the specific action of disease-causing genes. Because their goal is the treatment or cure of such diseases, it is essential that the mode of action be known. Once the specific site and mechanism of operation of a gene is found, it is possible to think about methods of controlling the defective function. However, this is not to say that an effective treatment or a therapeutic technique will be readily available. Some modes of action have been well known for years and, even though several agents have been tested, no effective treatment is yet available.

Of course, the ultimate goal of modern medical genetics is a means to **cure** gene-caused diseases. Unfortunately, probably nothing short of genetic "surgery," in which the defective gene could be excised and replaced by a normally functioning one, could provide a true cure. Needless to say such a technique is probably years away from a practical reality. Until such a technique is available, several general methods of **treatment** or **therapy** may be utilized, depending on the mode of action of the offending gene.

Review some examples of human genetic disease and note where it might be possible to prevent the development of symptoms by either eliminating an agent in the environment, supplying a missing gene product, or removing or inactivating a gene product. Such techniques constitute therapy and they are not cures. However, they may be significant medical techniques to alleviate suffering and disability, and they are based on a knowledge of the problem. In many cases such therapeutic techniques will act as stopgaps until more permanent approaches can be discovered and utilized.

An area of research that bears watching has developed in the study of **cancer**. Several separate experiments have shown that DNA removed from cancerous cells has the ability to cause cancer in experimental laboratory animals. The idea that some forms of cancer might be caused by genes is in itself important. Possibly more significant are the analytic tools that such knowledge provides. As the suspected genes are defined and their products identified, the group of diseases known as cancer will be better understood.

There can be little doubt that, with further research and investigation, many more human defects will be shown to have a genetic basis. There is clearly familial association with some well-known diseases, such as **diabetes**, but the genetic mechanism is not clear. Many other disease situations that are not directly attributable to genetic factors have a hereditary component. Even infectious diseases display varying responses in individuals, based on different gene products present in the body.

It is obvious that genetic knowledge will become more significant to the future of medicine. This will be true not just for practitioners but for laypersons who must take decisions in regard to the risk of having unhealthy offspring.

There will very likely be proposals in the future for means of eliminating defective genes from a population through organized efforts, possibly supported by legislation. Many states already have laws which permit sterilization of individuals to prevent the birth of defective children (usually referring to mental defects). It is even possible that more subtle types of pressure will be applied, as in the case of fine print in health insurance policies that might require abortions to prevent lifelong expensive institutionalization.

No doubt refined methods will eventually lead to a realistic approach to the attack on genetic defects. True cures for genetic diseases could be realized if a missing or defective gene could be replaced in a person's cells with something like a minor viral infection or vaccine.

The problems with such a technique arise when one considers what most people would look on as misuse of a valuable tool. What sort of controls could be anticipated for a technique that might find use among those who would like to change hair color, complexion, or some other minor trait? Suppose that genes for intelligence are identified. Would it be appropriate to raise everyone's intellectual expectations as far as possible by making such genetic material generally available?

What sort of protection would there be if various behavioral genes are identified and isolated? Although remote, the possibility may exist for control of populations by the spread of an infectious virus that could lead to behavioral modification and manipulation. Could a sort of genetic correction of the criminally insane be justified by a future society? Could an advanced society justify not developing or using such a technique?

GENETIC SCREENING

Many problems resulting from inherited diseases could be alleviated if members of the population were screened for the presence of defective genes. It is possible to survey potential parents to determine whether or not they are carriers of seriously detrimental genes. When such knowledge is available prior to the birth of a child with a genetic disease, options are available that may prevent much suffering. In such cases the persons involved can be counseled and assisted in making enlightened decisions. It is also possible to survey newborn infants for defects, often in time to prevent the development of detrimental symptoms.

It has become a virtual obligation of society to make such **screening programs** available, and in some cases even a law, for the newborn. It is possible to screen for defects prenatally, as well as among children or adults. Obvious advantages are the enlightenment of the public in general and individuals in particular about risks, in addition to the identification of those requiring therapy.

Most states have laws requiring the screening of newborn infants for **PKU.** A number of other conditions, such as **maple syrup urine disease**, **galactosemia**, and **sickle-cell anemia,** are also tested for in some states. While some consider such mandated testing to be an invasion of privacy, the benefits to family and state are clearly significant. The precedent for such mass screening was established in 1962 in Massachusetts with the PKU program. It appears that more programs will be added in the years ahead as technology is developed.

Testing newborn infants is somewhat easier to implement than other mass screening programs. Attempts have been made to require preschool screening for such traits as sickle hemoglobin, but such programs have generally been unsuccessful. Programs for adult screening would appear to be even more difficult to administer. Some mass screening programs have shown varying degrees of success when they were applied to specific segments of society. Thus, screening for sickle-cell trait in various black communities has been moderately successful. Screening of Jews of northern European descent for **Tay-Sachs disease** has met with some success, primarily among those contemplating marriage and pregnancy.

Does the future hold the possibility of mass screening for the identification of those with genetic diseases or carriers of defective recessive genes? An effective public health service might well recommend such a program. Mass screening for treatable genetic defects would be very important to individuals who might be able to avoid severe consequences by the timely intervention of therapy. Not only could painful consequences be forestalled, the money saved in comparison to the cost of the program would probably make it worthwhile. Considerable suffering and unhappiness might be remedied if the knowledge provided to the public were used to prevent the birth of children with untreatable genetic illnesses. Furthermore, mass screening could be important economically for the same reason, where persons with serious genetic defects require institutionalization at public expense.

The future reaction of the public cannot be accurately assessed, but apathy, fear, or rejection of an invasion of privacy might doom such programs from the start. There are definite problems in gaining acceptance and applying the programs. Good public relations and information programs coupled with involvement of many persons in such screening endeavors would help. It is difficult to see from a scientific or medical viewpoint how screening programs would be detrimental.

GENETIC COUNSELING

Genetic counseling for potential parents already exists and can be expected to expand in the future. The technique is an important adjunct in progressive medical institutions involved in the study of hereditary problems and there are presently more than 400 genetic counseling clinics in the United States. Active and sophisticated counseling programs are in daily operation in medical centers throughout the world. There exists a need for more widespread and available units in local institutions and the demand will surely increase. However, there is no reason for couples, who have need of counseling in regard to the risk of genetic defects in their families, to feel that such services are unavailable to them.

As more genetic problems are understood and recognized, the ability to communicate the sophisticated knowledge in an understanding way to laypersons will become more significant. It will be most important as identification of problems becomes more routine. Most significant will be the need for counseling of prospective or potential parents as to the risks, consequences, and options involved. In a relatively simple fashion persons with average educational backgrounds will have to be led to an understanding of sometimes complicated medical and genetic knowledge. It is probably true that one counselor is not sufficient to deal with all aspects of a situation. Counseling teams are the best approach in the final analysis and are now used by most clinics.

Goals and Features

The major goal of genetic counseling is to provide accurate, understandable information to potential parents of children with genetic defects. The type of information must include explanations of the medical consequences of the situation, the genetic basis of the defect, and the statistical chances for a defective child. It is also important that the options that are available and feasible be explored thoroughly with the counselees.

In short all pertinent information should be presented to the parents so that a rational decision can be reached. There are definite techniques of presentation that will provide such data to the concerned persons and at the same time permit a calm, sensible approach to develop.

In the recent past genetic counseling was utilized only after an unfortunate genetic event or defective birth had occurred. With modern techniques of analysis, however, it has become easier to counsel potential parents prior to such an unpleasant experience, and provide information on the option of abortion.

Diagnosis of possible gene or chromosome defects in the parents of other children should be a subsidiary goal during counseling with the provision for referral services and advice about treatment. A good counseling program will set other goals that center on the problems incurred in pregnancy that are not necessarily genetic.

Multiple abortions or failure to conceive are areas that require the attention of counselors. There should be an alertness to problems that may arise in regard to exposure to drugs or other agents during pregnancy that may result in birth defects, so that referrals can be made.

There is little doubt that a need exists for much more genetic counseling than is presently available. Many couples who need and could benefit from counseling are not even aware of the existence of such programs. It is likely that the most successful counseling programs will be associated with large medical centers where the programs can be run on a group basis and provide a spectrum of talent and knowledge. But this does not preclude the widespread involvement of smaller local clinics.

Techniques and Factors Affecting Counseling

Throughout the counseling process, the counselor attempts to maximize the options available to each couple. Along with solid, understandable information, each couple should be able to come to a decision that is their own, not the counselor's.

Although patients will generally look to the counselor for guidance, the final decision must rest with the potential parents. The counselor must attempt to provide all the data possible without actually swaying the decision toward a personal bias. In this respect a rather impersonal relationship may be necessary; at the same time, however, personal knowledge and understanding are likely to be required. A wise counselor might well recognize a person's true feelings better than the person or spouse.

Some factors of prime importance are the educational and emotional backgrounds of both patient and counselor, as well as the motivation of all concerned in regard to the counseling process. The attitudes of counselors and physicians usually must be carefully assessed to provide an effective atmosphere.

In conducting the counseling process the patients' prior knowledge and background must be determined accurately as a reasonable starting point. It is then possible to explain the consequences of having a trait, as well as the underlying mode of operation of the genetic mechanism. The patients' understanding of the situation must be checked periodically throughout the process.

The emotional impact of the knowledge must be assessed constantly, to prepare for crises that may require assistance from another counselor. Seldom has it proven worthwhile to attempt genetic counseling without full involvement of both potential parents.

Results

The information that counselors receive about their own genetic situation is the most important result of counseling. This is information that they can understand and share with other knowledgeable persons. Based on this knowledge and rapport with others, they can make rational decisions in regard to producing future progeny. Thus, they may avoid unnecessary shock or surprise, as well as unfortunate defective births in their family.

On a broader scale, effective and readily available counseling programs can help to decrease the incidence of genetic diseases and birth defects. A savings in economic terms can be realized for individual families and the tax-paying public.

An intelligent approach to widespread and readily available genetic counseling will be important in the future. Coupled with more effective screening programs and sophisticated techniques for treatment, as well as widening options for prospective parents, there is no substitute for the personal attention and information dissemination that counseling can provide.

BIRTH CONTROL

To say that there has been an immense amount of thought and effort devoted to the control of fertilization and pregnancy in a majority of human societies would probably be an understatement. Throughout recorded history individuals and institutions alike have debated, studied, and worried about the natural human efficiency of procreation. The reasons for such intense interest vary from the prevention of childbirth out of wedlock to the prevention of mass overpopulation in the face of dwindling natural resources.

In most instances interest in **birth control** has centered on numbers of offspring. Families in practically every modern society attempt to control the number of offspring through very personal efforts at **contraception**. Reverend Thomas Malthus, an English clergyman, wrote *An Essay on Population* in 1798 that lucidly pointed to the geometric growth of human populations in the face of arithmetic growth of food supplies. Today the predictions are more dire, immediate, and accurate.

Regardless of the reason for an interest in birth control, it is obvious that a thorough understanding of the fundamentals of human reproduction are essential if one is to propose a means of interfering with the process. Knowledge of human reproductive physiology allows medical researchers to analyze the process in regard to logical points for interference with its natural culmination. A diverse array of devices, physiologically active compounds, and surgical procedures have been studied and are available for use.

In spite of the knowledge and skill of modern science, it is strange that no universally effective and acceptable technique is available, although there should be no problem for anyone who is determined to prevent conception in a reasonably affluent modern society.

Techniques for Selecting Genes in Future Offspring

While most attention to birth control has centered on the limitation of quantity of offspring, there is at least an implied goal of maintaining and improving the quality of future generations. Such an aspiration is probably more apparent at the level of institutional and governmental planning. Within families, however, it is likely to be true that those planning to have children, regardless of the number, wish to have the healthiest, most intelligent children possible.

In the past the only efforts to attain such ends were devoted to providing the best environment, education, health care, and nutrition possible. Little could be done to insure the production of offspring with the best possible genotype. This is essentially still true, but some methods are beginning to develop. It is true that most people would consider the choices currently available to be basically negative options, but sometimes the alternatives are often so much worse that the chance to make the choice is welcomed.

The Control of Gene Transmission

From time to time a kit is advertised which purportedly allows one to choose the sex of a future child. Sometimes "recipes" are described that have the same effect—the conception of either a male or female offspring. Usually these gimmicks are not long on the market nor are they taken very seriously, since no one is prepared to guarantee their efficacy. More serious, and apparently more effective, efforts to allow the selection of a child's gender are offered by some reproduction clinics. **Artificial insemination** with variously treated semen has been utilized, but results have not been completely as predicted (usually reported as about 70% effective). It is true that X- and Y-bearing sperm cells can be influenced externally, most effectively in an electrophoretic field, where migration rates or direction can be influenced differentially. The routine clinical approach has supposedly dealt with the differences in size and mass of X and Y chromosomes and the resulting differential sedimentation qualities of sperm bearing them. A patent on sperm separation was awarded in the late 1970s; one can anticipate increasing interest in the use of such techniques.

In 1983 it was announced by Japanese researchers that X-chromosome-carrying sperm cells could be isolated and utilized for artificial insemination in laboratory animals with 100% assurance. The technique, based on slightly different electrical charges on X- and Y-bearing sperm, consisted of repeated introduction of the sperm cells into media of varying electrical charges. While completely effective in isolating X-bearing sperm, the technique was only 85% successful in isolating male-producing Y-bearing sperm.

The Japanese scientists predicted that, by refining their procedures, the rate of success for producing media containing only Y-bearing sperm would soon reach 95%. Furthermore, it was stated that the technique should be available for use in humans in the near future.

Even if perfected, there will probably be limited use of techniques for controlling the sex of offspring. Such techniques would still be a long way from the selection of individual genes for children. However, where severely disabling **X-linked genes** are carried by a person, the selection of sex could be a significant aid in the prevention of the disease. If a male with a trait such as **hemophilia** chose to produce only male children, the transmission of the gene could be prevented. If a female carrier of the same gene opted to have only female children, none would have the disease. Clearly, there is at least a start toward gene selection in this concept of sex selection through the artificial segregation of X- and Y-bearing sperm cells. One can envision the evolution of very sophisticated techniques of discriminating sperm cells carrying particular genes and then selecting these for exclusion or use in fertilization.

In the case of a sex-linked gene it would be possible to prevent the transmission of a defective allele from father to child, such as the recessive sex-linked allele for a trait like hemophilia. This procedure would permit only sons to be born, but it would prevent transmission of the allele to daughters.

Likewise, such a technique could prevent the birth of a hemophiliac son to a heterozygous mother by selecting only X-bearing sperm cells. One-half the daughters would be carriers of the recessive allele, but none would be hemophiliacs. Under these circumstances, the technique could be extremely important to the family involved.

Procedures of this kind will be effective almost exclusively through artificial insemination. While this process of mechanically implanting viable sperm cells into a female's reproductive tract is of importance to the future, it is a method that has been utilized routinely for a number of years. Most often it has been applied in cases where a couple has been unable to produce a child, usually because of male sterility. Physicians have most often utilized semen from donors unknown to the couple; it has been estimated that 250,000 U.S. citizens have been conceived by the technique.

Research has also shown that an ovum can be fertilized outside the body and then implanted into a woman's uterus, where it will develop normally through fetal stages to birth. The first verified case of an **in vitro fertilization** and implantation of the embryo which led to the birth of a child was reported in England: Louise Brown was born to Mrs. Lesley Brown in Lancashire, England, in 1978. Researchers Patrick Steptoe and Robert Edwards worked for 12 years, making over 100 attempts before finally succeeding. The goal of their research is a medical one, in that it allows women with oviduct abnormalities or infertile ova to carry a fetus and give birth in otherwise normal fashion.

With such techniques several possibilities for careful manipulation may evolve. It is obvious that genes carried by either one or both partners in a family may be selected against by utilizing gametes from other persons. The procedure might even provide a means of selecting particular gametes from either parent. If the ability to identify undesirable genes in sperm or eggs through chemical techniques were developed, it could be that identification and elimination of gametes carrying these genes, followed by in vitro fertilization with accepted gametes, would become a method of choice.

To date no one has devised a plausible method for selectively choosing genes to be incorporated into the gametes or a zygote. No doubt there are those who might like the idea of pressing buttons on an instrument to select eye color, hair color, height, sex, and so on for their future child, but such a fantasy will probably never become reality.

The real measure of the success of modern genetics will be the ability to communicate to future citizens an understanding of the workings of the human genetic mechanism and the hereditary patterns that develop by chance. An effective and useful program of counseling for prospective parents will be more important than all the machinations resulting from genetics research.

PRENATAL ANALYSIS

At several points in preceding chapters reference has been made to the use of prenatal diagnosis. While the transmission of particular genes or chromosomes through gametes remains essentially uncontrollable in the foreseeable future, it is possible to determine that certain chromosomes or genes have been transmitted long before the birth of the child. The ability of medical science to analyze chromosomes and genes of a human fetus without inducing any harmful effects to mother or child is an achievement of great significance to genetics counseling programs. Without doubt, the technique will carry even greater significance in the future.

The most frequently used technique of analysis, **amniocentesis,** is based on the removal of a small amount of fluid contained in the amnion, or sac in which the fetus is suspended (Fig. 12.1). The procedure, once considered very risky, has prov-

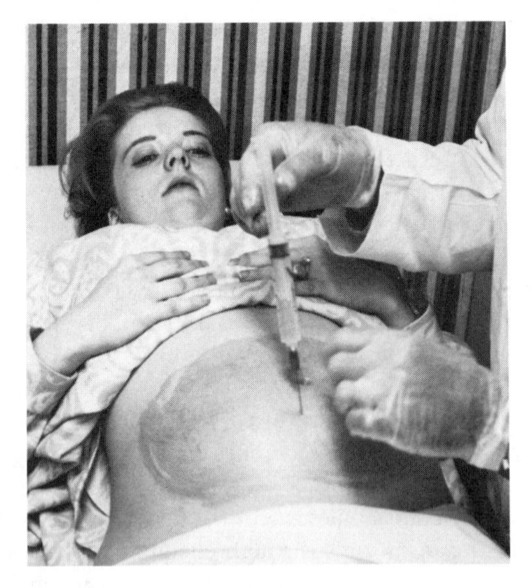

Figure 12.1. Woman undergoing amniocentesis. The needle of the syringe will penetrate the amnion, a sac of liquid holding the fetus in the pregnant woman's womb. A sample of the fluid will be removed aseptically so that fetal cells can be grown in tissue culture and analyzed for genetic and chromosomal defects. (Courtesy Leonard McCombe, *LIFE Magazine*©1971, Time Inc.)

en to be safe when performed by competent personnel. After determining where a needle can be placed safely clear of the fetus, the sterile needle is inserted into the mother's abdomen through the amnion and fluid extracted (Fig. 12.2).

Ultrasound maps of the fetus in the mother's body are usually used to determine exactly where contact with the fetus can be prevented (Fig. 12.3). The "picture" of the fetus seen on a television screen is a projection of dark and light areas resulting from the variable penetration and reflection of ultrahigh frequency wavelengths of sound beamed through the different textured tissues of the mother and fetus. Ultrasound is well on the way to becoming a diagnostic technique in its own right. It was reported in late 1983 that the sex of a fetus could be determined accurately as early as the fourth month of pregnancy.

Amniocentesis is based on a feature of epithelial tissues, the covering layers of cells in the human body. As a normal protective mechanism, these tissues are constantly sloughing off the surface and are shed into the environment. In the fetus many

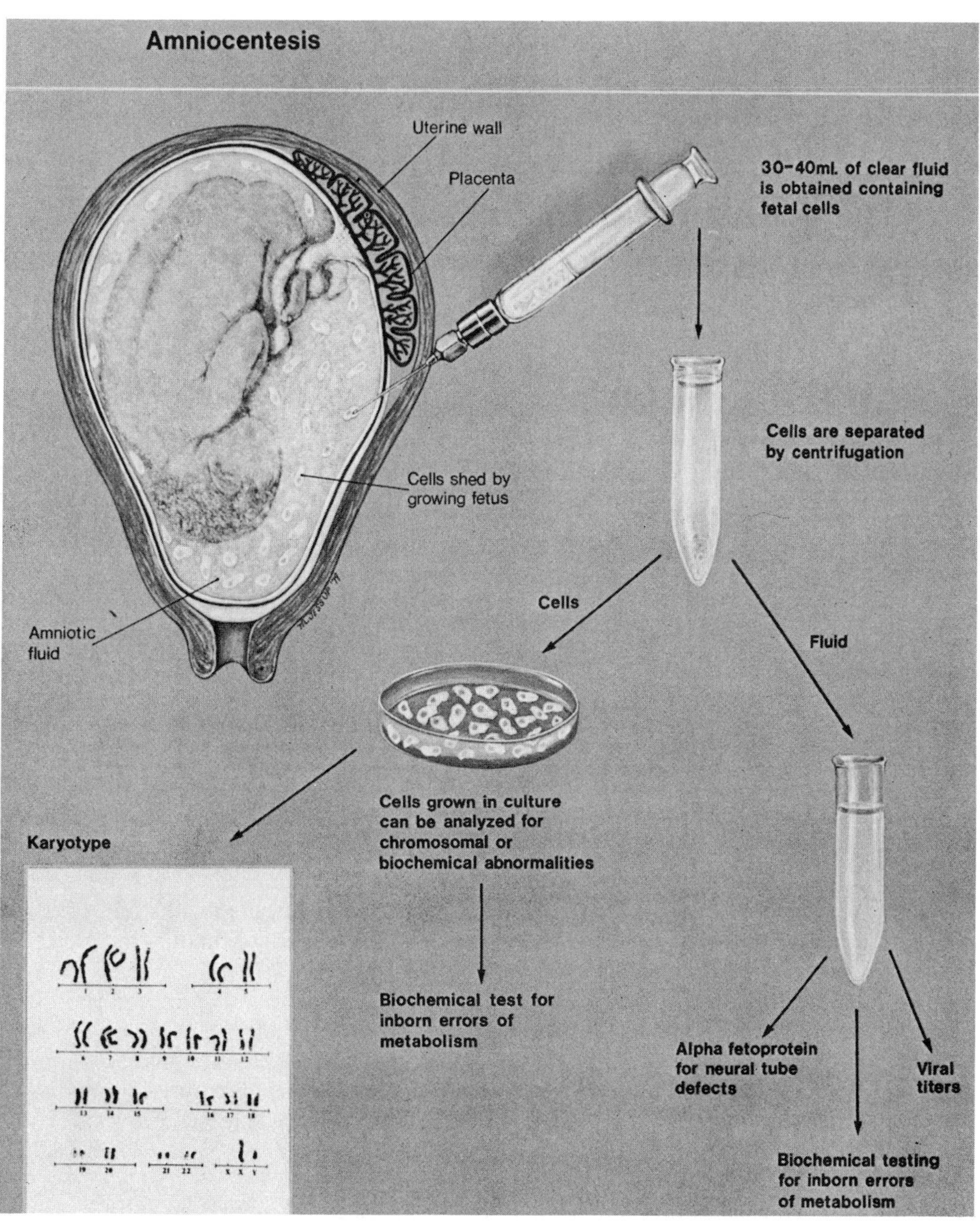

Figure 12.2. Diagram illustrating the protocol for removal and analysis of amniotic fluid and fetal cells during amniocentesis. Biochemical tests on cells grown in culture depend on the synthesis of gene products, but recent advances in technology have made possible an additional type of analysis. The chromosomes from the cultured cells can now be analyzed for the specific presence of some genes. DNA base sequences for such genes can sometimes be determined. (Courtesy Michael K. McCormack and *American Family Physician*)

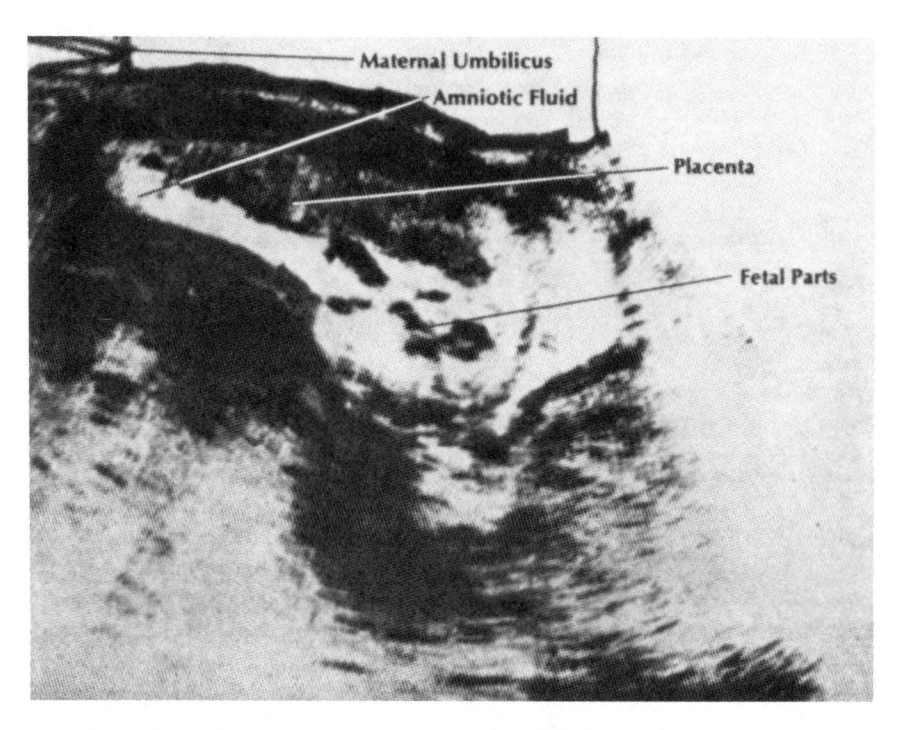

Figure 12.3. Ultrasonograph showing fetal position. Ultrasonography prior to amniocentesis shows obstetrician precisely where to insert needle to avoid placenta and fetus; it also localizes deepest pool of fluid. This longitudinal scan was made in seventeenth week of pregnancy by Division of Diagnostic Ultrasound, Department of Radiology, Downstate Medical Center, Brooklyn, NY. (Courtesy Henry L. Nadler and *Hospital Practice*)

of these cells remain alive in the amniotic fluid and are removed as the fluid is withdrawn in amniocentesis. There is no apparent harm associated with fluid removal and it is probably replaced in a short time. Of course, such techniques can never be totally free of risks and no competent practitioner would utilize the method without sound justification. Experience indicates a 1 to 2% risk to the fetus, and this must be balanced against the risk of not obtaining the information.

As in so many other modern techniques, the ability to conduct analyses through amniocentesis is dependent on the ability to grow cells in tissue culture. The amniotic fluid is centrifuged to remove fetal cells and they are then grown in test tubes for further analysis. Because the cells are dividing by mitosis, it is relatively routine to analyze for number and form of chromosomes.

Abnormal numbers of chromosomes can be easily spotted, as in the 47 of a **Down syndrome fetus**. It is a relatively simple matter to determine **genetic sex** by the presence of XX or XY complements. While it is somewhat more difficult, **chromosomal aberrations** may be identified, such as translocations, inversions, or deletions. Utilization of **Giemsa staining** techniques now permits a careful analysis of well-known banding patterns.

Neural tube defects are also identifiable with amniocentesis. These defects result from the failure of the tubelike structure which forms the spinal column and brain to close properly. One such defect is **spina bifida** (literally, split spine). The condition is detectable because a protein (alpha fetoprotein) that is not normally present in the amniotic fluid accumulates there at a high level. This is generally thought to be due to leakage from the open spinal column. Spina bifida exhibits a pattern which indicates that it is **polygenic**. It occurs at the rate of about one in 200 births and has a recurrence rate of about 5% for parents having a child with the defect.

Children with the defect often survive for 12 or more years, but they have a range of disabilities including mental retardation, paralysis, and lack of body function control. Surgery to close the neural tube is often performed after birth, but its prognosis is generally not good. Identification of the condition can be made immediately from an amniotic sample without culturing cells, allowing for a timely **abortion** if the parents elect. Because of the very serious nature of abortion, it is essential that amniocentesis be coupled with a thorough counseling program.

Amniocentesis provides a means of analysis for some genetic defects beyond chromosomal abnormalities or the physical presence of chemicals in the fluid. Because many genes are functional within the cells being cultured in test tubes, it is sometimes possible to analyze for gene products, such as **hexosaminidase A**, the enzyme that is lacking in a Tay-Sachs infant. At least 40 or 50 gene defects can be identified routinely, and more are being added to the list constantly. Generally, only genes that result in changes in the type or amount of some product have been analyzable. As more analyses are devised and become routine, one can anticipate research to reveal elegant methods for determining the presence or absence of numerous other gene products. A discovery in 1982 using restriction endonucleases to identify the sickle hemoglobin (HbS) gene in amniocentesis probably is the beginning of a series of tests that will greatly expand the list of defects amenable to the procedure.

As seen in Chapter 5, the restriction endonucleases can be used to cut the genome of fetal cells in the amniotic fluid into identifiable segments. Then, by using known mRNA or DNA probes, the presence or absence of different genes may be determined. In an interesting discovery it was found that a particular endonuclease (Mst II) recognized and cleaved DNA at the exact point where the triplet CTC specified glutamic acid for hemoglobin A. When a mutation changes the sequence to CAC (for valine in hemoglobin S), the endonuclease no longer recognizes or cleaves the DNA (see Review Question 14, p. 124).

Because of this, it is possible to recognize with only a small number of cells the presence of the following: (1) normal—two genes for HbA (all small DNA fragments); (2) sickle cell trait—a gene for HbA and one for HbS (some small and some large DNA fragments—uncut); and (3) sickle cell anemia—two genes for HbS (all large DNA fragments). Because only a small number of cells are required, tissue culture is not necessary and the analysis time is cut from about 5 weeks to 2 weeks.

Prior to the discovery of this new means of identifying sickle hemoglobin genes with endonucleases, another means of prenatal diagnosis was necessary. The inherent risks are so much greater than with amniocentesis that very careful consideration of the possibilities is needed. The merits must be ample to justify its use. In this technique a sample of fetal blood is removed from the placenta by using a **fetoscope**, a sort of manipulative microscope inserted through a small surgical opening in the

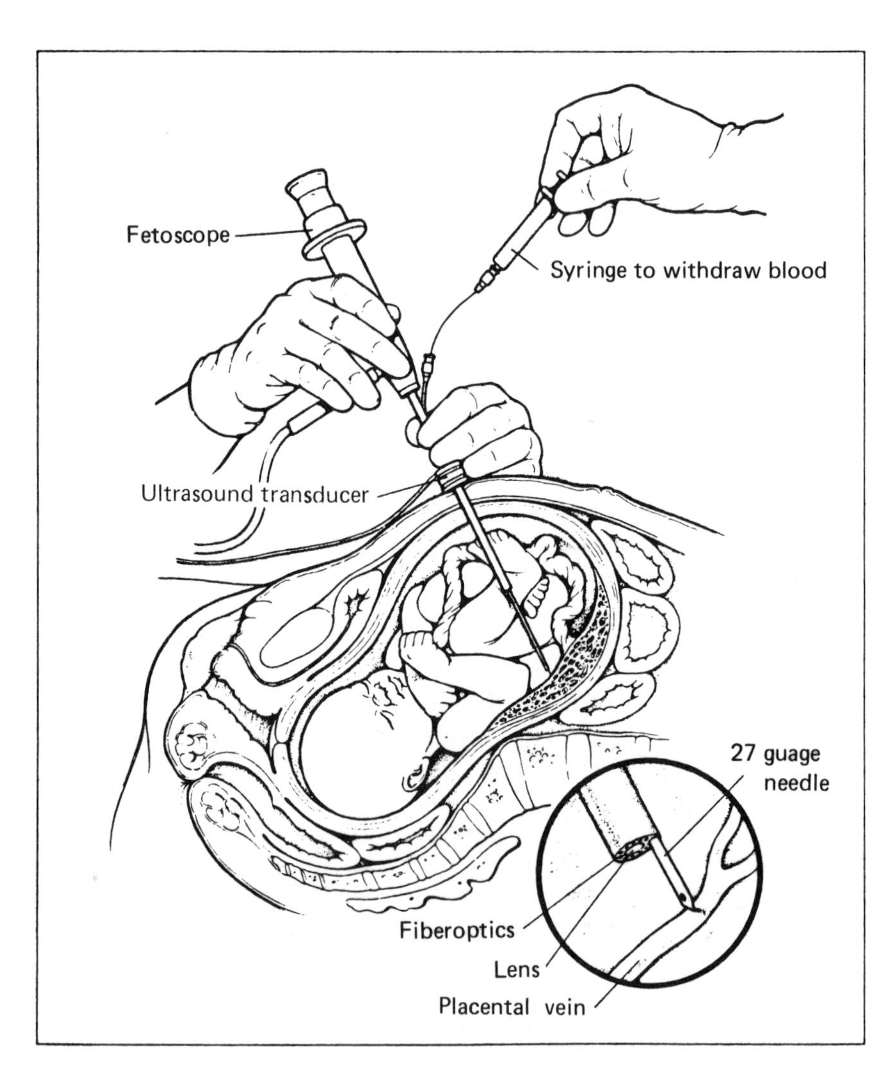

Figure 12.4. Fetoscopy to obtain blood sample from placental vessels, as performed by J. C. Hobbins at Yale University, is shown here in an artist's diagram. Detail at lower right shows needle about to enter vessel. Ultrasonography is used first to locate placenta and fetus. (Courtesy Henry L. Nadler and *Hospital Practice*)

abdomen of a pregnant woman, to place carefully a syringe and aspirate a small blood sample (Fig. 12.4).

This technique can provide information about some traits that amniocentesis has not been able to provide. Hemoglobin defects, such as those seen in sickle cell anemia and the thalassemias, have been measured directly in the hemoglobin-containing red blood cells. In these situations blood cells were necessary, since the tissue culture cells do not produce hemoglobin. Thanks to the endonuclease technique, the allele for sickle hemoglobin and other special cell type products may now become identifiable routinely by amniocentesis.

In truth, **prenatal diagnosis** is probably of practical value only when coupled with abortion, even though some problems arise. There are not enough fetal cells and amniotic fluid to make amniocentesis useful until about the fifteenth week of pregnancy. One to four weeks are required for cell growth and chromosome analysis of the culture and it is seldom possible to opt for an abortion under these circumstances until about the eighteenth week of pregnancy. Biochemical analysis of gene products require longer than chromosomal analysis, and elective abortion may not be possible until about the twenty-first week.

Good medical practice discourages abortions after about the twenty-eighth week, and it is best from many points of view to perform abortions as early as possible in a pregnancy. It is somewhat doubtful that the process of amniocentesis using gene products can provide much quicker results than at present, but, with restriction endonucleases and **gene probes**, there may be real improvements in the future.

Because abortion is so distasteful an option, it is often not considered in some cases. In such situations it may be difficult to justify prenatal diagnosis. If the option to abort the fetus is unacceptable after counseling has provided the knowledge and insured an understanding of the consequences, then the potential parents may not wish to avail themselves of prenatal diagnosis.

Indeed there have been, and will continue to be, many debates over the rights of both parents and fetus in the decision to have an abortion. As the practice becomes more widely acceptable, society will move out of the trying period of development toward logically organized legal rules and ethical codes regarding the uses of abortion.

In a precedent-making decision, the U.S. Supreme Court ruled on January 22, 1973, that is was unconstitutional for state laws to prohibit abortions during the first 3 months of pregnancy. The Court ruled that the states may regulate abortion procedures in ways reasonably related to maternal health after this period. A significant part of the decision ruled that subsequent to the **viability** of the fetus (last 10 weeks of pregnancy) the state could regulate and even prohibit abortion. Viability was defined as "capability of meaningful life outside of the mother's womb."* Whether "meaningful life" precludes such severely disabling and lethal traits as Tay-Sachs disease or Patau syndrome has not been tested.

Ultimately, of course, if the sort of genetic surgery that would allow the insertion of genes into the individual's cells becomes a reality, then amniocentesis could be extremely valuable in presenting information in conjunction with positive options. Although some observers of medical genetics envision the day when amniocentesis will become routine in this society for each pregnancy, with a mechanized, computer-style printout for a number of possible key gene products, such a scenario will probably remain a part of science fiction for many years.

Examples of treatment of defective fetuses in utero have drawn much attention recently. While the techniques are adaptable to only a few situations, at least the potential exists. Some symptoms and defects can be prevented from developing if the genetic condition is recognized. Such techniques will surely be perfected and extended to more situations with time and practice. In some cases then prenatal diagnosis may not necessarily be tied to abortion as the only option.

*Roe *v.* Wade, Vol. 410, U.S. Reports. *Supreme Court Reporter* (1973) 93, 705-39. St. Paul, MN: West Publ. Co., 1974.

CLONING

One of the commonly recurrent themes in science fiction involves the production of populations of humans in large numbers without sexual reproduction, a technique known as **cloning.** Of course, the individuals are exactly alike and selected for certain characteristics important to their function as virtual automatons. Fortunately, no one has yet devised a means of ''culturing'' humans in such fashion. It is conceivable that some person or government might decide to produce workers or soldiers of superior ability if such a technique were available.

The word *clone* comes from the Greek word for twig or branch. There is nothing new or unusual about the biological concept, since many plants show the ability. The fragile branches of willows have a natural ability to form new trees by rooting in damp soil and they are identical genetically to the ''parent'' plant. All the Lombardy poplars in the world belong to one clone, as this tree shows no sexual reproductive ability. Extensive cloning of this type carries real hazards. For instance, since all members of a clone are exactly alike genetically, they all carry the same defects and are all susceptible to the same diseases. The loss of a pool of diversity among humans through widespread cloning would be a detriment to future survival.

The process of experimental cloning of humans seemed one step closer to reality in the 1950s when botanists devised a technique of growing single separated cells from carrot roots into embryos and ultimately mature plants. A whole ''field'' of plants, genetically exactly like the original plant, could be started in a small dish of laboratory medium. Today commercial forestry companies are able to produce thousands of tree seedlings, all having the same superior growth characteristics in a very small space. Eventually, the many individual clone members can be transplanted to the final growing range.

While tissue culture techniques are now well known for human tissues, present indications are that a cloning technique similar to that for plants is not likely. Basically, the reason is not known, but human cells appear to have the ability to divide only a certain number of times. The discoverer of this phenomenon was Leonard Hayflick and the maximum number of cell doublings is called **the Hayflick limit.**

There appears to be the potential for only about 50 cell doublings in human soft tissue embryonic cells. Embryonic blood vessel cells will go through 70 doublings in culture before becoming senescent. After this the cells simply refuse to go on reproducing and eventually die. Current research on aging and senility is concentrating on this phenomenon for obvious reasons. If there is a built-in program which allows cells to divide only a set number of times, it appears that improved care and medication can do very little to extend life span.

Human tissue culture lines, such as **HeLa cells,** have grown in laboratories throughout the world for many years; such cells come from original cancerous tissue. **Tumor cells** have the ability to go on dividing in an uncontrolled fashion. It is as if the normal genetic control of cell division had been lost in malignancies. Although there is presently no known mechanism for restoring such control to the cells, it would be difficult to imagine a more useful discovery than a method to control cancer in its many forms by restoring the genetic control of cell division.

If the knowledge and ability to turn off and turn on the programmed control of cell division in human tissue is developed, it will be possible to maintain and grow "normal" tissue under standard laboratory conditions for indefinite periods of time. However, the ability to cause such tissue to form embryos and ultimately mature individuals is likely to present an even greater challenge.

Research efforts are underway now, as they have been for several years, to produce a human fetus outside the mother's body. To date, only very early stages have been attained before the development ceases. Human fetal development is much more complex and demanding than the early formation of carrot plants or pine trees. The development of artificial human uteruses to permit carefully controlled development of humans without the prenatal protection of a human parent may well be in the future, but, if so, it is probably so distant to be of little concern at present.

Because tissue culture of normal human tissue is limited, and normal fetal development does not succeed outside the female body, does not necessarily prevent cloning. If one forgets the potential for producing large numbers of offspring and concentrates on the aspect of cloning in which an identical genotype is perpetuated, another use for the process can be seen. Some persons feel that the perpetuation of certain individuals (or at least their genotypes) might be desirable. Supreme vanity might even be a reason to insure that one's own identical gene combination would be passed on to the next generation.

Cloning of this type may be within the realm of current technology. Good technicians can remove a nucleus from one cell and implant it in another, allowing the cell to go on living with a new set of genetic "instructions." It is also possible to implant a fertilized egg in a receptive uterus, resulting in a typical pregnancy. As discussed, such a technique is being perfected for use with in vitro fertilizations (p. 293).

Combining these two manipulations, a diploid nucleus from a mature **somatic cell** might be inserted into an egg and the diploid egg (zygote?) implanted in the uterus of a **surrogate mother.** Fertilization would be bypassed completely and the resultant fetus would have the same genotype as the "parent" nucleus.

In a classic experiment J. B. Gurdon in the 1960s using frogs' eggs was able to transplant nuclei from intestinal cells to the eggs after removing the egg nuclei. The transplantation process stimulated embryonic development and mature frogs were produced having the characteristics of the frog used to obtain intestinal cells. Similar experiments have been done with salamanders, but the development of amphibians takes place normally in water, without the necessity of a uterine environment and placental association with the mother.

In 1979 L. B. Shettles, a Vermont physician, claimed to have performed a similar experiment with human reproductive cells, but the scientific world has remained somewhat skeptical. Shettles reported that nuclei from human eggs were withdrawn and replaced by nuclei from human spermatogonia, the unreduced precursor cells of sperm. The resultant cells then proceeded to divide, reaching the blastula or hollow ball stage of embryo development in about 3 days. The embryo cells were not analyzed to determine that the nuclei were in fact genetically the same as the male's. The embryos were not implanted in an uterus and thus did not survive long.

If such manipulations can be performed, fertilization can be bypassed completely and a fetus might be produced using diploid nuclei from either male or female donors. The offspring would have the same genotype as the "parent" nucleus and represent a true clone.

The birth of a cloned offspring has already been claimed. David M. Rorvik wrote a book in 1978, *In His Image: The Cloning of a Man,* in which he described the production of a viable fetus through secret processes. The procedure and an allegedly successful outcome were supported by a wealthy man who wished to leave as an heir a duplicate of himself. The identities of the cloned man and the doctor conducting the operation were maintained as a strict secret.

Rorvik claims to have been called on to arrange the affair because of his past science news reporting and his writings on new developments in medical science. The book reads much like a fictional account and the scientific community places no credence in it. No convincing proof has yet been provided to support Rorvik's contentions.

The Rorvik book led to a legal action on the part of a geneticist whose cloning work on rabbits had been cited without permission. As a result, in 1981 a U. S. District Court in Philadelphia ruled that the book was "a fraud and a hoax."* In spite of the skepticisim, there was considerable revulsion over the idea.

What is to be said about the ability of a human cell to divide only a finite number of times? If the mechanism for programming this phenomenon resides in the nucleus, can such an offspring look forward to a shortened life span because the original "zygote" nucleus had already divided mitotically a given number of times? In contrast, does the cytoplasm of the egg confer a new program on the nucleus for a lifetime of divisions? Will the use of nuclei from spermatogonia, as suggested by Shettles, eliminate this possibility? Only time and experimentation can provide answers to such important questions.

How will cloning be accepted socially and morally in the future? Surely no early authors of religious codes could have anticipated this sort of manipulation. Undoubtedly, some societies and governments will wish to make laws controlling such processes. Sticky areas of conflict between the rights of individuals and modern codes of ethics will definitely be encountered.

GENE AND GAMETE BANKS

What seemed to be one of the ultimate ideas about genetic manipulation only a few years ago has come close to being a reality. Hermann J. Muller, a well-known American geneticist and Nobel Prize winner for his work with radiation effects, made a proposal in 1961 for the freezing of gametes, eggs and sperm alike, for the production of future generations. His idea was that all persons would be sterilized at birth and their gametes removed and stored. Obviously, in 1961, as at present, many technical details were needed to make his speculations feasible.

The goal of the plan was to judge after persons were dead whether or not their qualities were deserving of being passed on to future generations. In spite of Muller's sound scientific credentials and his superb knowledge of genetics, his foray into the

*Broad, W. J., "Court affirms: Boy clone is a hoax," *Science* (1981), 213, 118-19.

world of **eugenics** leaves some important questions unanswered. It is probably true that by selecting parents for desired inherited qualities one could "stack the deck" favorably. However, major problems exist in such proposals by the very nature of the genetic mechanism. Randomness and chance in the process practically guarantee that desirable clusters of genes will be broken up, while new and less compatible combinations will be created during fertilization. Because of the large number of factors involved and due to recombination of the various characteristics, such plans as Muller's have many pitfalls. Selecting successful, accomplished parents may not provide the talents required in the future. In fact, such selection processes might even tend toward less of the variability that is the essence of human nature. With all the proposal's difficulties and speculative techniques, its greatest weakness may have been Muller's plan to have a committee decide which gametes should be utilized to produce offspring!

Artificial Insemination

While the possibilities for a program like Muller's seem very remote, at least part of the technology necessary to the plan is now fairly routine. Frozen bull semen for artificial insemination is used routinely in cattle breeding and the technique is feasible for the storage and use of human sperm.

There are presently a number of **sperm banks** operating that provide an effective service for couples who wish to have children but have been unable to do so naturally. These banks keep a record of the physical characteristics of sperm donors and attempt to match as closely as possible the male member of the couple. From all reports, the sperm banks have compiled a record of success and satisfaction for their customers. The technique utilized may involve immediate use of the semen or freezing for artificial insemination use at an appropriate time during the female's estrous cycle.

It has been estimated that between 6000 and 10,000 children are born each year in the United States as a result of **artificial insemination by donor** (known by the mnemonic **AID**). The techniques involved (sometimes including freezing and storage of sperm) do not appear to have resulted in any more abnormalities than from conceptions following sexual intercourse. As a matter of fact, fewer birth defects can be anticipated with careful screening of donors.

AID presents the possibility for a number of unconventional uses. One of the more intriguing plans was formalized in 1980 when it was announced that an exclusive sperm bank had been formed in San Diego. It was called "The Repository for Germinal Choice," and its depositors were limited to Nobel laureates. The originator of the institution was Robert K. Graham, a wealthy businessperson who apparently planned to make the sperm available only to women with high I.Q.'s. Obviously, the underlying philosophy of the program was a sort of positive eugenics, with the goal of producing high intellect offspring. By 1982, when information about the progress of the plan began to appear, observers questioned its performance. The first reported pregnancy involved a woman who was reported to have served a prison sentence and who had lost custody of previous children because of child abuse. The second reported mother was not married, in spite of the requirement by the program. Both were above the average child-bearing age (the first, 39, the second, 40), and clearly ran a greater than normal risk of nondisjunction maladies. The women may have possessed superior qualities of intellect, however, and only time will demonstrate the effectiveness of the selection process.

The use of frozen sperm, gamete storage respositories, and even elitism in the admittance of donors does not seem to be unusual in the shadow of other ideas for manipulation of reproduction and selection of future gene combinations. Imagine the controversy such proposals would have engendered in the days of Darwin and Galton. Today such ideas are viewed as merely a small part of the spectrum of awe-inspiring possibilities that science has presented to humanity. Unfortunately, where human choice with its biases is the deciding factor, or where committees are responsible for decisions, long-term errors may be made in the interest of short-term gains. There is often great value to natural mechanisms of chance that operate in genetics.

In Vitro Fertilization

It has been shown that fertilization can be accomplished outside of the body by mixing donor sperm and eggs in a glass dish (in vitro fertilization resulting in what are often called "test tube babies"). By implanting such fertilized eggs in a receptive uterus, it has been possible to produce infants from routine pregnancies, as in the case of Louise Brown.

The first successful in vitro fertilization in the United States resulted in the birth of Elizabeth Jordan Carr on December 28, 1981. The 28-year-old mother had previously had both oviducts removed and her in vitro fertilization, pregnancy, and delivery were supervised by a clinic at Eastern Virginia Medical School in Norfolk, Virginia. The clinic, founded by Howard and Georgeanna Jones, provides similar services for couples who are unable to conceive for various reasons, and a number of pregnancies have been accomplished there.

The in vitro techniques have not been universally successful and are not as routinely accomplished as artificial insemination. However, by the time Elizabeth Carr was born about 20 **test tube babies** had been born, with probably 100 women pregnant from the process at the time. There are many tricky aspects to such a procedure, such as timing the fertilization to match the appropriate condition for implantation in the uterus. Freezing of an early embryo produced by in vitro fertilization, followed by cold storage, could allow the proper time sequence to be established. Frozen embryos have been utilized effectively in cattle production for years. S. Mukherjee and fellow researchers in Calcutta, India, claim to have used a frozen embryo for successful implantation and a normal human birth in 1978, but, controversy surrounds the situation.

In 1983 it was announced by Australian workers that a pregnancy had resulted from the use of a frozen human embryo. In this case multiple ovulation had been stimulated and four eggs were collected and fertilized in vitro. One was frozen as a back-up and the other three were placed in the patient's uterus. When the one embryo that had successfully implanted was later aborted as a fetus, the frozen embryo was then utilized as a replacement.

All cases of in vitro fertilization have been necessarily surrounded by secrecy and evasion. Thus, while the technique has apparently been utilized in a limited number of instances, notably in Great Britain, Australia, India, and the United States, it is a process which may someday be offered routinely to permit pregnancy using the mother's or a donor's ova. In the last several years more in vitro fertilization clinics have opened in the United States.

An approach that has been used in a few cases of female sterility or inability to produce viable babies has been the use of surrogate mothers. Artificial insemination has been utilized to produce children who are genetically related to the male member of the couple, but neither related genetically nor carried through pregnancy by the female. The surrogate mother legally waives all rights to the infant by contract. Of course, the biological mother is aware of her involvement and it is not easy to prevent all parties involved from knowing the identity of the others. Sperm banks do provide an easy mechanism for anonymity. Only time will tell how effective legal contracts are in separating the biological parents from later claims to parental rights. There surely will be judicial tests and procedures to establish precedents.

GENETIC ENGINEERING

Genetic engineering, without doubt, is a field of future research. Technically, the idea of "repairing" or "constructing" individuals through mechanisms of gene repair, replacement, or addition to a genome is not a reality at present. The fact that it is conceivable, coupled with numerous successes with experimental organisms other than humans, implies that someday such achievements are likely to be routine.

Such manipulations are certain to be of considerable criticism in a broad view. Some will consider the techniques to be tampering with a natural process, but there will be few meaningful arguments against the correction of severely disabling diseases through "engineering" methods.

Beginning in the 1970s genetic engineering began to move rapidly as various laboratories developed techniques for inserting specific genes into bacteria. Through the use of restriction enzymes or by employing viruses as intermediate carriers it became possible to insert genes that were not even remotely a normal part of the bacterial genome. It was clear that the possibility existed for inserting human genes or disease-causing genes into bacteria that are common inhabitants of human bodies. It is important to remember that the most often used, and most well-understood, experimental organism in molecular biology is *Escherichia coli,* an organism found overwhelmingly as the most prolific and common type of foreign organism in the human digestive tract. So reliable is the presence of *E. coli* in human feces that it is used as an indicator of polluted water. The organisms are ubiquitous in the environment and can easily assume residence within the human body from the surroundings.

Obviously, the potential for producing a serious pathogen—one that could spread and multiply rapidly if resistance to control measures were developed—poses a serious threat in the implantation of genes from humans in such organisms. A group of scientists in 1974 recognized this threat and a conference was held in Asilomar,

California, in February, 1975. The major result of this conference was a self-imposed moratorium on **gene-splicing** or implantation until satisfactory guidelines for the conduct of the research were drawn up in 1976. Most laboratories performing experiments with recombinant DNA were funded by grants from the National Institutes of Health (N.I.H.), and all such labs were required to follow the guidelines for containment of experimental organisms to maintain their grants. Generally, all other workers followed the stringent guidelines as well.

Of course, the matter was not completely settled. Some local governments attempted to add the force of law to the N.I.H. guidelines. Thus, in 1976 the City Council of Cambridge, Massachusetts, after prolonged and heated public debates involving lay citizens and university researchers, passed a city ordinance that converted the semivoluntary N.I.H. guidelines into law. Cambridge became the first and one of few governments to do this. Harvard and M.I.T. were thus forced to follow the guidelines, although in all probability such force was unnecessary.

When N.I.H. relaxed the rules somewhat in 1980, the Cambridge ordinance automatically followed the more relaxed rules. It has become generally accepted that the N.I.H. guidelines, while demonstrating a wise amount of caution in a new and unexplored area, were too stringent. At the outset, there was great concern over the production of a resistant and dangerous "Frankenstein's Monster" microbe that might become a worldwide scourge. In some quarters similar fears still exist, but most knowledgeable workers feel this is an extremely remote possibility.

Most recombinant DNA work now is being done in ordinary laboratories under relatively normal conditions. The strain of *E. coli* that is used most often is considered to be an enfeebled organism, incapable of surviving and multiplying outside the protection of a laboratory environment.

As one might suspect, with new and routine techniques developing, commercial enterprises have shown a strong interest in recombinant DNA work. With this commercial activity there has been renewed activity in the area of laws regulating such ventures. Therefore, in 1981 the Boston City Council began a new flurry of law-making activities designed to control and oversee what still could be an area with potential danger for the city's citizens.

The importance of recombinant DNA studies to commercial activities can be recognized when one remembers that a human gene successfully implanted in a bacterium may result in the bacterium's production of a human gene product (Fig. 12.5). This is an astounding concept with important results. It becomes theoretically possible to produce on a commercial scale large quantities of heretofore very rare substances, normally produced only in the human body or possibly synthetically. Penicillin is a well-known example of large-scale production of a compound by living microbes. To this extent, some major details of commercial production of biological compounds already exists.

It has been found that compounds like insulin, for the treatment of diabetes, or interferon, the rare human substance that may prove to be a potent anticancer wonder drug, can be produced by recombinant bacteria carrying human genes. Production of compounds is almost limitless and ranges from alcohol to hormones. Many companies in this emerging gene engineering industry were formed in the early 1980s and other established companies have expanded into the field. By the end of 1981

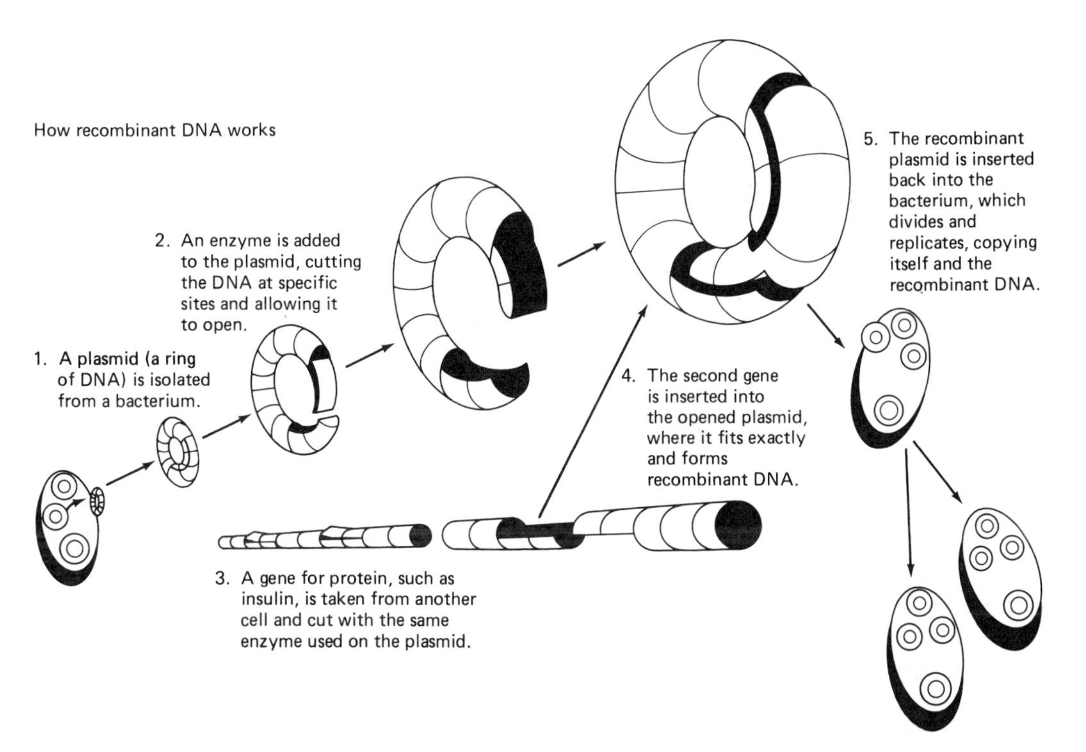

Figure 12.5. Recombinant DNA Technique. Shown graphically is a modern technique utilized for implanting foreign DNA into bacteria through the use of plasmids. Human genes can be inserted into bacteria in this fashion so that large amounts of human gene products can be produced by bacterial cells grown under controlled sterile laboratory conditions. (Courtesy Bob Conrad)

there were probably 200 companies engaged in such efforts. It is significant that in 1980 a landmark patent was awarded jointly to Stanford University and the University of California covering the techniques of gene splicing. Based on this patented procedure, in 1981 the holders of the patent began to offer licenses to commercial genetic engineering firms for use of the gene-splicing processes. An initial fee of $10,000 was charged, with an annual $10,000 fee, plus royalties of 0.5 to 1% of net sales of direct products. Noncommercial, basic research use of the techniques does not require a license.

More difficult legal questions than those of technique patents arise when specially engineered organisms are considered. The U.S. Patent Office has generally ruled that living things are not patentable, but bacterial cells with implanted foreign genes present a uniquely different situation. It was therefore ruled by the U.S. Supreme Court on June 16, 1980, that "a live human made microorganism is patentable subject matter."* The case was decided in favor of Ananda Chakrabarty in regard to a genetically engineered bacterium capable of breaking down crude oil multiple components. Such an organism has potential economic value in cleaning up oil spills.

*Diamond v. Chakrabarty, Vol. 447, U.S. Reports. *Supreme Court Reporter* (1980) 100, 2204-14. St. Paul, MN: West Publ. Co., 1982.

It will probably remain for individual corporations to guard and protect the organisms that they engineer through both secrecy in laboratory work and legal measures. Biological scientists are likely to find the adjustment to this new situation difficult at first, but it is not so different from many other scientific fields that are involved in protecting and utilizing scientific inventions.

In 1978 human **insulin** was produced by genetically engineered bacteria. This was a major accomplishment, since insulin is required by approximately 5 million diabetics throughout the world to replace the missing natural gene product in their bodies. The bacterially produced insulin is made under the direction of an artificially synthesized gene implanted in bacteria and it is identical to natural human insulin. It has been commercially named **humulin.** Its great advantage over insulin, obtained from the pancreas tissue of pigs and cows, is the lack of an allergic reaction to the product. In 1982 the U.S. Food and Drug Administration approved its use in this country.

PUBLIC AWARENESS: THE GREATEST NEED FOR THE FUTURE

Clearly, the future is impossible to predict or envision with any certainty. Great advances in the understanding of the genetic mechanism have taken place since 1865, involving the efforts of thousands of people. As in all scientific fields, one answer provides the fuel for several more questions and knowledge about minute details seems to accelerate faster than the public's awareness of them. All this basic knowledge ultimately leads to practical applications that are limited only by the creative ingenuity of modern workers.

While one hardly knows where the next significant finding or application will appear, the changes that take place will become familiar in one way or another. Some things may pose great and challenging decisions, while others may make life easier and more rewarding.

There are many proposals and plans for modifying or manipulating the human genetic condition. Although many segments of society accept such plans as accomplished fact, it is important for citizens in all societies to be alert to the many existing ideas. A basic level of understanding is required of everyone. Unscrupulous and unwise decisions may be made in a society that is not alert to the possibilities and shortcomings inherent in the potential for affecting the genetic future of humans.

Technologies that might be placed in the nebulous arena of the future are already present, simply lying in wait to affect one or a nearby person in some manner. One should be ready to face many trying issues involved with human heredity, legal and personal. Legislative and commercial efforts will require in-depth knowledge to influence the judgments and decisions required. No modern science text can provide the sort of information called for in such a situation. All literate people must become aware of the very basis of their beings—the genetic mechanism. A close interest in new developments must be maintained to emerge into a new and enlightened world.

ADDITIONAL READING

BAER, A. S., *Heredity and Society* (2nd ed.). New York: Macmillan, 1977.

BERGSMA, D. (ed.), "Intrauterine Diagnosis," *Birth Defects: Original Article Series VII* (1971), No. 5, National Foundation—March of Dimes.

______ (ed.), "Contemporary Genetic Counseling," *Birth Defects: Original Article Series IX* (1973), No. 4, National Foundation—March of Dimes.

FRIEDMANN, T., "Prenatal Diagnosis of Genetic Disease," *Scientific American* (1971), 225, No. 5, 34–42.

GOODFIELD, J., *Playing God: Genetic Engineering and the Manipulation of Life.* New York: Random House, 1977.

GROBSTEIN, C., "The Recombinant-DNA Debate," *Scientific American* (1977), 237, No. 7, 22–33.

HALACY, D. S., JR., *Genetic Revolution: Shaping Life for Tomorrow.* New York: Harper & Row, Pub., 1974.

HENDIN, D., and J. MARKS, *The Genetic Connection: How to Protect Your Family Against Hereditary Disease.* New York: Morrow, 1978.

HSIA, V. E., K. HIRSCHHORN, R. L. SILVERBURG, and L. GODMILOW, *Counseling in Genetics.* New York: Alan R. Liss, Inc., 1979.

LEVY, H. L., "Genetic Screening," *Advances in Human Genetics* (1973), 4, 1-104.

MERTENS, T. R., *Human Genetics: Readings on the Implications of Genetic Engineering.* New York: John Wiley, 1975.

NADLER, H. L., "Prenatal Diagnosis of Inborn Defects: A Status Report," *Hospital Practice* (1975), 10, No. 6. (Reprinted by The National Foundation—March of Dimes, Birth Defects: Reprint Series.)

OMENN, G. S., "Prenatal Diagnosis of Genetic Disorders," *Science* (1978), 200, 952–58.

REILLY, P., "Genetic Screening Legislation," *Advances in Human Genetics* (1975), 5, 319–76.

SHAW, M. W., "Perspectives on Today's Genetics and Tomorrow's Progeny," *The Journal of Heredity* (1977), 68, 274–79.

REVIEW QUESTIONS

1. Make a list of human genetic diseases or defects that are correctable through treatment or health maintenance programs. In each case briefly describe the treatment.
2. Who should be responsible for the cost of expensive treatment and maintenance programs for genetic diseases? If an agency of government supports such programs, should laws be made to control the birth of persons having genetic diseases?
3. Make a list of areas where laws should be made concerning genetic diseases or their transmission. Comment in each case on the constitutional rights of individuals, religious prerogatives, needs of society, and financial burdens involved.
4. Should mass screening for genetic defects or carriers of such genes be conducted by governments? What defects or diseases should or could be screened for? What use should be made of information gained in such screening programs?

5. Who should be selected for prenatal screening to identify fetuses with genetic defects? What defects or diseases should be screened for by amniocentesis?
6. Should genetic counseling be required for all at-risk parents in a modern society? Do you think a voluntary counseling program would be effective in modern society? If one defective child has been born, should genetic counseling be required for the parents? Is it appropriate to legislate concerning this type of educational program?
7. What value or advantage can you see in frozen egg and sperm banks? Should governments support such agencies, or should they be in the province of private enterprise?
8. Is it appropriate for research to be conducted in techniques of in vitro fertilization or cloning? Should government funds be used for such studies?
9. Are there any possible reasons for cloning of humans in regard to the needs of society?
10. List some areas where recombinant DNA techniques using human genes in bacteria might be worthwhile to medicine or society. Are there any dangers inherent in such techniques?

CHAPTER SUMMARY

1. Because of unforeseen developments that open up new areas of activity, it is always difficult to make predictions about the future. In human genetics it is probable that what is happening on a small or beginning scale now will become routine and much expanded in the years ahead.
2. It is probable that artificial manipulation of the human genome will become a reality. The correction or prevention of genetic defects will be of major importance, with some question about cosmetic or whimsical interference. Restriction endonucleases, nuclear transplants, viral gene transfers, and other new techniques are paving the way toward eventual genetic "surgery."
3. Although available and in widespread use in some areas, the future is likely to bring an expansion in genetic screening, not just in the number of individuals involved, but also for the number of traits examined. As more genetic defects are discovered and more persons carrying detrimental genes identified, a greater demand for genetic counseling will develop.
4. Good genetic counseling is available on a widespread basis, but such programs will become more accessible as they are better understood and an awareness develops.
5. Means of birth control in the future may be utilized as much to insure quality of offspring as to control quantity. Selection of gametes to be utilized in fertilization, artificial insemination, and in vitro fertilization is already a reality to some extent and may provide realistic options for insuring genetically "healthy" offspring.
6. Prenatal analysis by amniocentesis followed by fetal manipulation or abortion is already functional for many defects and the list is constantly expanding. Both chromosomal defects and defective gene products are identifiable in amniocentesis. Although the technique may never become routine for all pregnancies, it will surely be utilized widely to provide assistance for those at risk for genetically defective offspring.
7. Cloning of humans is often linked with the future and current techniques indicate that it is within the realm of possibility. However, many moral questions must be answered by society in regard to this sort of procedure.
8. Gamete banks or repositories already exist and may take on added significance in the future. How such facilities are to be used requires careful consideration by society.
9. Clearly, genetic engineering shows indications of widespread importance to future society. Not only may gene manipulation in humans become important, the production of human gene products, such as insulin and interferon, by other organisms carrying transplanted human genes will take on great significance.
10. Genetic literacy and awareness of the meaning of various proposals will be required by laypersons. In short, an understanding of the principles of human genetics will be essential for the public.

GLOSSARY

Abortion Early termination of pregnancy, either spontaneous or induced.

Acentric Chromosome or chromatid lacking a centromere.

Acrosome Enzyme packet at anterior tip of sperm.

AID Artificial insemination by donor.

Albinism Lack of melanin (brown) pigment in all body parts.

Allele One of the various forms of a gene created by mutation.

Amniocentesis Puncture by hypodermic needle of the amnion and removal of amniotic fluid to analyze fetus for defects.

Androgen Male sex hormone, responsible for masculine secondary sex characteristics.

Anemia Inadequate number of red blood cells.

Antibody Protein molecule that recognizes and inactivates foreign substances within living tissues.

Anticodon Triplet of nucleotides on transfer RNA molecule that is the complement of codon on messenger RNA.

Antigen Foreign substance that stimulates production of specific antibody in living tissues.

Artificial Insemination Placement of semen within female reproductive tract without copulation to effect conception.

Aster Array of microtubules radiating in all directions around centriole in mitotic or meiotic cells.

Autosome Chromosome other than X or Y (sex) chromosomes.

Bacteriophage Virus parasite capable of infecting bacteria.

Barr Body Condensed and largely inactivated (heterochromatic) X chromosome.

Bivalent Paired homologous chromosomes (each one consisting of two chromatids) seen in synapsis.

Blastula Hollow ball stage of the embryo at implantation in the uterus.

Cancer Disease characterized by uncontrolled cell division without differentiation of resulting cells.

Carcinogen Substance or agent capable of causing cancer.

Carrier Person not exhibiting a recessive trait but capable of passing it on to offspring (therefore heterozygous).

Centimorgan Unit of measuring map distance equal to 1% recombination.

Centriole Compact bundle of nine short triplet microtubules found at each spindle pole during mitosis and meiosis.

Centromere Primary constriction of chromosome and attachment point of spindle microtubules in mitosis and meiosis.

Cervix Constricted opening of uterus connecting to vagina.

Chiasma Microscopically identifiable point of crossing-over in a bivalent.

Chromatid Recently formed chromosome attached to its duplicate by the centromeres.

Chromatin Physical material of the chromosomes consisting of histone proteins and DNA.

Chromosome DNA-histone protein strand of hereditary material contained within the nucleus; each has one centromere and a specific sequence of genes.

Chromsome Band Stained or unstained constant region on chromosome length revealed by degree of affinity for various dyes.

Chromosome map Diagram of linear gene sequence within individual chromosomes, based on genetic or cytologic evidence.

Clone Individual produced without sexual reproduction, having the same genes as its sole parent.

Codominance Two alleles having equal and identifiable effect on the phenotype.

Codon Triplet of nucleotides in messenger RNA that binds to anticodon (complementary triplet) in transfer RNA molecule which is carrying a specific amino acid.

Coitus Sexual intercourse; copulation.

Conception Fertilization; the merging of sperm and egg to form a zygote.

Congenital Describing a condition which is present at birth.

Consanguineous Parents who are genetically related; having a recent common ancestor.

Contraception Prevention of conception by agent or device.

Copulation Sexual intercourse; coitus.

Correlation Coefficient Statistical analysis measuring the degree to which two factors vary in regard to each other (given as a real number from 0 for no correlation to 1 for complete correlation).

Crossing-Over The mutual exchange of chromosome segments between two homologous chromosomes at synapsis of meiosis.

Cytogenetics Study combining genetic phenomena with chromosomal structure and activity.

Cytology The study of cells and their organelles.

Cytoplasm Living material of a cell exclusive of the nucleus.

Deficiency Absence of a gene or genes from genome due to physical loss of chromosomal segment (deletion).

Deletion Chromosomal aberration consisting of a missing chromosomal segment (deficiency).

Deoxyribonucleic Acid (DNA) Self-replicating chemical constituent of genes within chromosomes.

Dermatoglyphics Study of fingerprints and palmprints.

Dicentric Chromosome or chromatid with two centromeres.

Dihybrid Organism that is heterozygous (hybrid) for two independent gene pairs.

Diploid Chromosome complement consisting of two of each chromosome (generally one of each from both parents).

Dizygotic twins Twins arising from two separate zygotes (fraternal, nonidentical).

Dominant Allele that exerts a phenotypic effect which masks the presence of the recessive allele.

Duplication Supernumerary presence of a chromosomal segment containing one or more genes within a chromosome.

Egg Female gamete or sex cell; ovum.

Ejaculation Forceful ejection of semen from penis.

Embryo Developing result of conception from zygote through the eighth week of pregnancy, when human form appears.

Enzyme Protein that is made by living cells and which assists chemical reactions without being altered itself.

Epistasis Overriding effect of the alleles of one gene pair on the action of a different, independent gene pair.

Euchromatin Chromatin containing functional genes, outstretched or relaxed at interphase; condensed and darkly stained at metaphase.

Eugenics Improvement of human population through selection of parents.

Eukaryotic Having hereditary material contained within a nucleus.

Euphenics Improvement of human population by modification of phenotypes through medicinal or therapeutic means.

Evolution Change of life forms over very long time periods.

Exon Integral segment of eukaryotic gene (DNA) that is necessary for manufacture of the gene product; transcribed to messenger RNA which dictates protein structure after excision of introns.

Expressivity Degree of manifestation of a gene in the phenotype; sometimes given as a proportion of full expression.

Fallopian tube One of a pair of tubes connecting the uterus with the ovaries; oviduct.

Fertile Producing viable gametes (sperm or eggs) capable of yielding a zygote on fertilization.

Fertilization Conception; merging of sperm and egg to form a zygote.

Fetus Developing result of fertilization from beginning of third month of pregnancy until delivery.

Follicle Swollen chamber in ovary containing single egg prior to ovulation.

Frameshift Mutation that causes sequence of gene DNA triplets to be disrupted by insertion or removal of nucleotides.

Fraternal twins Dizygotic or two-egg twins.

Gamete Sex cell; egg or sperm.

Gene Unit of heredity which is a segment of DNA that codes for a protein or other molecule (i.e., transfer RNA, ribosomal RNA, polypeptide).

Gene Splicing Technique of genetic engineering in which gene is removed from one genome and inserted into another.

Genetic Drift Random change in gene frequency from one generation to another.

Genetic Engineering Manipulation of genes involving gene synthesis, transfer from one genome to another, and production of gene products by microorganisms.

Genetics The study of heredity and variation.

Genetic Screening The analysis of individuals in populations to determine the presence of detrimental or disease-causing genes.

Genome All the genes of an organism; a full set of genes within a haploid complement of chromosomes (technically in humans 22 autosomes plus X and Y).

Genotype Describes the genes present in an individual (may be one pair or many).

Germ Cell One of the precursors of sperm or eggs.

Gestation Time from conception until birth; term of pregnancy.

Gonads The reproductive organs that produce gametes; ovaries and testes.

Haploid Single set of chromosomes; the number found in sperm and eggs.

Haplotype The combination of histocompatibility genes linked together in two groups in an individual.

HeLa Cell Human tissue culture cells representing a cervical cancer line obtained from Henrietta Lacks in 1953.

Hemizygous Having only a single allele of a gene naturally, as in males having single X and single Y chromosomes.

Hemoglobin Pigment in red blood cells that transports oxygen in body; composed of protein (two alpha globins and two beta globins) and heme (iron-containing porphyrin ring).

Hemolysis Breakdown of red blood cells and release of hemoglobin contents.

Heredity The science of inheritance and variation; genetics.

Heritability Mathematical estimate of the amount of variance in a trait which is due to genes.

Heterochromatin Chromatin that does not carry active or functional genes; late replicating; somewhat condensed in interphase and less compact, lighter staining than euchromatin at metaphase.

Heterozygous Condition in which two different alleles of a gene are present in the same organism (i.e., one dominant and one recessive).

Histocompatibility Genes Genes determining the antigens produced in the tissues of an individual that cause rejection of such tissues when placed in another body.

Histone Type of positively charged protein, rich in basic amino acids, associated with DNA in eukaryotic chromosomes (nucleosomes).

Holandric Found in males (on the Y chromosome).

Homologous Pair of chromosomes that are structurally similar; carrying a sequence of genes that affects the same traits.

Homozygous Referring to a pair of genes in an individual where the two alleles are identical (i.e., either both dominant or both recessive).

Homunculus A miniature human form, such as that postulated by early microscopists in the sperm head.

Hybrid Individual resulting from a cross of two different genetic forms; having two different alleles of a particular gene or genes.

Hybridoma Line of cultured cells (cancerous) representing two fused lines of cells from different sources or organisms.

Identical Twins Twins resulting from a single egg and sperm (monozygotic, maternal twins).

Idiogram Diagrammatic representation of the full complement of chromosomes of an organism.

Immunoglobin Antibody molecule; composed of four protein subunits under gene control (heavy chain = variable and constant domains, light chain = variable and constant domains).

Implantation Attachment of an embryo to the uterus wall about 8 days after conception.

Intelligence Quotient (I.Q.) Level of intelligence measured by test that determines "mental age" which is then divided by chronological age and multiplied by 100.

Intervening Sequence Sequence of nucleotides in gene that is not necessary to produce gene product; transcribed to messenger RNA and then spliced out prior to translation.

Intron Intervening sequence (located between exons).

Inversion Chromosomal aberration consisting of two breaks with segment between them turned 180° and reinserted.

In Vitro Fertilization Fertilization of egg by sperm outside the body (literally, "in glass").

Isochromosome Chromosome consisting of two identical arms (replicated strands) attached by centromere, with other arm missing.

Karyotype Pictorial representation of full chromosomal complement arranged according to centromere location and decreasing length.

Leukocyte White blood cell.

Linkage Condition existing where two or more gene loci are physically located in the same chromosome.

Locus The exact physical location of a gene within a chromosome.

Meiosis Reduction division; cell division involved in gamete production in testes and ovaries.

Melanin Dark brown pigment responsible for various pigmented parts of body (hair, eyes, skin).

Menopause Cessation of monthly cycle of ovulation in females.

Menstruation Breakdown of uterine wall in female following ovulation, if fertilization and implantation do not occur.

Metabolism Sum total of chemical reactions of life in cell or organism, especially in relation to release and utilization of energy.

Mitogen Chemical stimulant of mitosis.

Mitosis Nuclear or cell division resulting in two daughter cells having genetic properties identical to original cell.

Monoclonal Antibody Antibody produced in a culture of cells derived by mitosis from a single cell.

Monohybrid An offspring resulting from a cross of two parents differing in the alleles for one gene (heterozygous).

Monosomy Condition in which one particular chromosome is found only once instead of twice in a karyotype.

Monozygotic Twins Twins arising from a single fertilized egg; identical, maternal twins.

Mosaic Organism having sectors, regions, or areas with different genetic constitution, due to such factors as mitotic nondisjunction, inconsistent chromosome and gene inactivation, or somatic cell mutation.

Mutagen Agent or chemical that causes an increase in mutation rate.

Mutant Mutated form of gene or organism.

Mutation Heritable change in DNA that results in altered gene product.

Nondisjunction Nonseparation of certain chromosomes during anaphase in cell division; applied to paired homologues in meiosis I or to attached chromatids in meiosis II or in mitosis; yields products with either two of one type or none of that type.

Nucleosome Chromosomal structural unit consisting of a globular core of eight histone proteins with a DNA strand of approximately 140 nucleotides wrapped twice around the core.

Nucleotide Unit of molecular structure of DNA or RNA containing a phosphate, a sugar, and a nitrogen base.

Nullisomy Condition in which a particular chromosome is missing from a karyotype.

Octamer Containing eight units, as the eight histone molecules in nucleosomes.

Oncogenes Genes that cause cancer when activated.

Oocyte Cell undergoing meiosis in the ovary; gives rise to female gamete.

Oogonium Cell undergoing mitosis in fetal ovary; gives rise to oocyte.

Ovary One of the paired organs in female in which meiosis occurs and ova are produced.

Oviduct Tube in female carrying ova from ovary to uterus after ovulation; Fallopian tube.

Ovulation Release of ova from ovary.

Ovum Egg cell; female gamete.

Pedigree Chart Diagram of family used to trace inheritance pattern of a particular gene.

Phenotype Physical appearance of an organism under the control of genes interacting with environment.

Plasmid Small circular segment of DNA, peripheral to normal chromosome which replicates and functions within bacterial cells.

Polygenes Several different genes in a genome affecting the same trait in a cumulative and quantitative fashion.

Polymorphism Situation in which several different mutant forms of the same gene (alleles) exist and may be found with reasonable frequency in a population.

Polyploid Condition in which multiples of the haploid set of chromosomes (above diploid) exist in an organism (e.g., $4n$, $5n$, $6n$, etc.).

Proband Individual in a family pedigree who drew attention to the genetic trait under study; propositus.

Prokaryote Organism having no true nucleus and no membranous organelles in cells.

Propositus *See* Proband.

Protein Genetically determined chemical produced in living cells; composed of a number of amino acids linked together by peptide bonds.

Puberty Age of sexual maturity; onset of ovulation in females and of sperm formation in males.

Recessive Form of allele that is masked by the presence of dominant allele in heterozygous individual; expressed in phenotype only in the absence of the dominant allele.

Recombinant DNA DNA strand containing gene or genes that is excised from one genome or chromosome by restriction endonuclease and reinserted into a foreign genome (sometimes involves artificially synthesized DNA).

Recombination Loss of linked parental combination of certain alleles during gamete formation due to crossing-over in meiosis.

Replication (DNA) Precise enzymatically controlled production of two exact copies of double-helical DNA, based on nitrogen base pair complementarity.

Restriction Endonuclease General category of naturally occurring enzymes of bacterial cells that cut DNA double helix at precise points within strands (not at ends).

RhOGAM Injection used to prevent Rh sensitization of Rh-negative mother after delivery of Rh-positive fetus.

Ribonucleic Acid (RNA) Nucleic acid transcribed from DNA message; contains single continuous chain of ribose nucleotides (may form double chain with complementary sequences in same strand).

Ribosome RNA and protein containing particulate organelle in cell that is necessary to translate messenger RNA during protein synthesis.

Segregation Mendelian principle which states that paired genes of diploid organism separate and are passed on singly to offspring; genetic result of meiosis.

Semen Liquid containing sperm; released from male during ejaculation.

Serum Liquid portion of blood remaining after removal of clotting material and red blood cells.

Sex Linkage Genetic phenomenon observed when a gene is physically located on the X chromosome.

Sibling (sib) Brother or sister; having the same parents.

Somatic Cell Body cell; not one of the germ cells or gametes.

Species A distinctive taxonomic group of interbreeding individuals (i.e., *Homo sapiens*, the human species).

Sperm Male gamete produced by meiosis and differentiation in testis; spermatozoan.

Spermatocyte Cell undergoing meiosis in testis; gives rise to male gamete.

Spermatogonium Cell underoing mitosis in testis; gives rise to spermatocyte.

Spindle Chromosome-aligning and moving structure which appears during mitosis and meiosis; composed of proteinaceous microtubules.

Sterile Unable to produce reproductive cells capable of conception.

Synapsis Precise two-by-two pairing of homologous chromosomes seen in prophase I of meiosis; results in bivalent formation and permits crossing-over.

Testis Sperm-producing male reproductive organ.

Tetraploid Condition in which four full sets of chromosomes exist in one organism.

Tetrasomy Condition in which a particular chromosome is found four times instead of twice in a karyotype.

Therapy Treatment for disease or disability; serves to alleviate or modify condition, but does not necessarily cure it.

Transcription Synthesis of RNA under the direction of DNA.

Transduction Change of genetic characteristics through gene transfer by a virus.

Trihybrid Organism that is heterozygous (hybrid) for three genes.

Trimester One-third of the period of human pregnancy (first, second, or third 13-week period).

Triploid Condition in which three full haploid chromosome sets are found in one organism.

Trisomy Condition in which a particular chromosome is found three times instead of twice in a karyotype.

Tumor Cancerous growth.

Ultrasound Sound waves higher than audible range, capable of penetrating dense tissues.

Ultraviolet Light Light waves beyond visible range at violet end of spectrum; capable of inducing mutations through DNA breakage.

Variance Statistical measure of deviation from the mean in a population; calculated by summing the square of deviations and dividing by sample size.

White Blood Cell Cellular component of blood that has a nucleus and may be capable of cell division in tissue culture.

X Chromosome Sex chromosome found in both sexes, normally two in female, one in male.

Y Chromosome Sex chromosome normally found only in male.

Zygote Fertilized egg.

ANSWERS TO SELECTED QUESTIONS

CHAPTER 1 (Page 14)

4. A coin has two sides, thus a coin toss has two alternative results: either a head or a tail, determined solely by chance. Diploid organisms carry two genes for each trait and there are two alternative forms (alleles) in a heterozygote (dominant and recessive). During sexual reproduction one or the other, determined by chance, may be passed on to the offspring.

A coin when tossed can give only one of the two alternatives, not both. Two alleles in the same organism must separate and a reproductive cell can receive only one, not both.

5. Two coins tossed together are independent of each other, each has a one-half chance of landing either head or tail. Two separate gene pairs may be independent of each other (in different chromosomes); each has a one-half chance of being passed into a given reproductive cell.

The law of simultaneous independent events (such as two coins tossed simultaneously) states that the combined chance is equal to the product of the two separate chances. Thus, the chance of obtaining a head on two coins (homozygous dominant) tossed simultaneously is 1/4 ($1/2 \times 1/2$). Likewise, the chance for two tails (homozygous recessive) would be 1/4 ($1/2 \times 1/2$). The chance for a head and tail is based on the total of two combinations: a head on the first coin with a tail on the second ($1/2 \times 1/2$) and a tail on the first coin with a head on the second ($1/2 \times 1/2$) thus $1/4 + 1/4 = 1/2$.

Four different combinations or three kinds of results can be obtained with two coins:

Coin 1	*Coin 2*	
H	H	Homozygote
H	T	Heterozygote
T	H	
T	T	Homozygote

9. The trait is determined by a recessive gene, carried predominantly in the heterozygous condition.

10. A child receives 1/2 of his or her genes from the mother, thus shares these genes in common with the maternal parent (likewise sharing 1/2 in common with the father).

    Because the child has four grandparents, he or she shares 1/4 of the genes in common with each grandparent (1/2 the mother's are from her mother and she passes on 1/2 of these to her child, thus $1/2 \times 1/2 = 1/4$).

    The chance that siblings will receive a gene in common from a parent is 1/4. On a broader scale, this implies that siblings share 1/2 their genes in common—the same genetic relationship as parent-child in mathematical terms. However, the common genes are derived from both parents in siblings (1/4 from each parent).

CHAPTER 2 (Page 42)

4. Some of the proteins required for mitochondria are determined by the mitochondrial genome, but mitochondria require many more proteins that are under the control of nuclear genes. One can only speculate as to the significance of this compartmentation. One hypothesis that is espoused by some biologists views the mitochondria as primitive free-living prokaryotic organisms that invaded eukaryotic cells in early evolutionary stages. Both organisms benefited from the resultant symbiotic relationship and over long periods of time many genes unnecessary to the new conditions were probably lost or modified.

7. Maternal = ▨

   Paternal = □

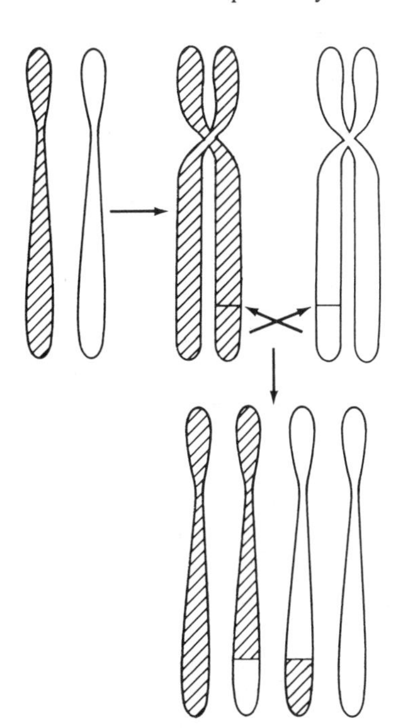

Four types of meiotic daughter cells

There would be 16 different types of meiotic daughter cells if there were two pairs of chromosomes with crossing-over at one point in each synapsis (4×4 types).

9. Some subjective interpretation is required in counting various stages in a micrograph,

especially in distinguishing interphase and prophase figures. It is best to make three or four counts and then take the averages.

(a) Total count of cells:

	Actual number	*Percentage of total*
Interphase	47	78.3
Prophase	6	10.0
Metaphase	1	1.7
Anaphase	2	3.3
Telophase	4	6.7
TOTAL	60	100.0

Such a count might not be representative of the actual frequencies, since the sample is probably too small to be a statistically significant representation of the whole organism.

(b). Time periods:

Interphase	78.3% of 720 min	=	9 hrs	24 min
Prophase	10.0% of 720 min	=	1 hr	12 min
Metaphase	1.7% of 720 min	=		12 min
Anaphase	3.3% of 720 min	=		24 min
Telophase	6.7% of 720 min	=		48 min

CHAPTER 3 (Page 62)

2. To increase the chances of encountering fertilizable female gametes, it is generally the practice to use fertility drugs to stimulate maturation of ova and ovulation. The surface of ovaries can be observed with a laparoscope that has been inserted into the body cavity through a small surgical opening. Mature ova can be observed by the swollen appearance of the follicle on the ovary surface. These ova may be removed with a syringe and maintained in a living condition outside the body.

3. The various options all involve the prevention of contact between sperm and ova in the female reproductive tract. Through the use of physical barriers (e.g., condoms or diaphragms) or chemicals (e.g., spermicides), viable sperm may be prevented from entering the oviducts. Other temporary procedures may involve the control of the presence of ova in the oviducts at the time of copulation (and up to 48 hours after). Examples of this procedure are found in the rhythm method or the use of chemicals that suppress ovulation (the Pill). Permanent means of prevention of sperm or ova release are seen in vasectomies and tubal (oviduct) ligation. None of the methods has been considered particularly detrimental to human health, with the exception of the Pill, whose use has been variously correlated with physical problems in some individuals, possibly even cancer.

Another means of birth control, which is not strictly contraception, is the prevention of implantation of an embryo after fertilization has occurred. A common method of accomplishing this is by the placement of a plastic intrauterine device (IUD) in the uterus.

It is a permanent method as long as the device remains in place. There is indication that such devices may cause physical problems in some women.

4. The process of meiosis in human males and females is the same in its basic essentials. There are minor differences involved in producing the two forms of gametes, but they do not alter the genetic outcome in the two sexes.

   Males:
   (a) Equal, symmetrical divisions produce four similar-sized gametes with very little cytoplasm.
   (b) Both division cycles are completed prior to sperm formation.
   (c) Meiosis begins at puberty and continues throughout life.

   Females:
   (a) Highly asymmetrical divisions produce one viable gamete with a large amount of cytoplasm for each meiosis.
   (b) Second division cycle arrested at metaphase II and not completed unless fertilized by a sperm.
   (c) Meiosis begins prior to birth, but is arrested at prophase I. Process resumes and continues in one oocyte per month, starting at puberty and continuing until menopause.

7. **(a)** The result would be fraternal or dizygous twins (nonidentical).
   (b) The result would be maternal or monozygous twins (identical). These twins would be genetically identical (100% of their genes shared in common), whereas the twins in part (a) would be no more closely related than siblings (50% of their genes shared in common).

CHAPTER 4 (Page 100)

5. Two amino acids—methionine and tryptophan—have only one codon. The codon for methionine serves as an initiation or start signal at the beginning of a message. In this start position it codes for formylmethionine. This unusual amino acid is subsequently modified or removed from the newly formed protein.

6. **(a)** Complementary DNA chain: G A T A T G C T T T C T A
   (b) mRNA produced by the segment: G A U A U G C U U U C U A
   (c) Aspartic acid—Methionine—Leucine—Serine

7. (G A A) forms instead of (G A U) in the first triplet position.

   Glutamic acid would be substituted in the first position for aspartic acid. The change might alter an enzyme's activity or specificity, but it might be a minor change and have little or no significant effect on the enzyme.

   In the second example the new sequence would be

   G A U A U C U U U C U A

   and would result in the amino acid sequence:

   Aspartic acid—Isoleucine—Phenylalanine—Leucine

   The sequence of amino acids would be changed completely from the point of the deletion to the end. Depending on the point of the deletion, it is likely that the gene product would be so different from normal that it would be ineffective in the expected role.

8. The sixth triplet would normally code for leucine in the protein, but in its mutated form it would become a termination signal. A sequence of five amino acids would be produced

and then terminated (Leucine—Isoleucine—Aspartic Acid—Leucine—Phenylalanine). This was called a "nonsense" mutation because the gene product was not effective in performing its intended role (the mutation resulted in nonsense).

In the second example the fourth triplet is changed from CUA to CUG. Both code for the same amino acid, leucine. Therefore, the mutation goes unrecognized; it is "silent."

9. AAA pairs with UUU, the codon for phenylalanine. UAC pairs with AUG, the codon for methionine. ACC pairs with UGG, the codon for tryptophan.

10. **(a)** The 1200 light chains could each be attached to any one of the 1100 different heavy chains, thus there would be 1200 × 1000 different combinations or 1,200,000 different antibodies.

(b) The functional and complete variable region is determined by the addition of the *J* region to the *V* region, before combining with the *C* region. Thus, if there are 5 *J* genes that combine with 200 *V* genes, there would be 1000 different combinations possible. If there were 1000 *V* gene combinations for the light chain, and 1000 *V* gene combinations for the heavy chain, there would be 1,000,000 different types of antibodies produced from only 412 genes (200 variable region genes (*V*), five joining genes (*J*), and one constant region gene (*C*) for both light and heavy chains).

12.

Child 1 Genotype		*Child 2 Genotype*		*Child 3 Genotype*	
3	12 (known)	3	12 (known)	9	8 (known)
11	?	2	5	11	12

Therefore the other parent's genotype is

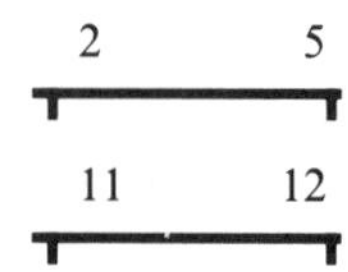

CHAPTER 5 (Page 123)

4. Gene order and distances in centimorgans:

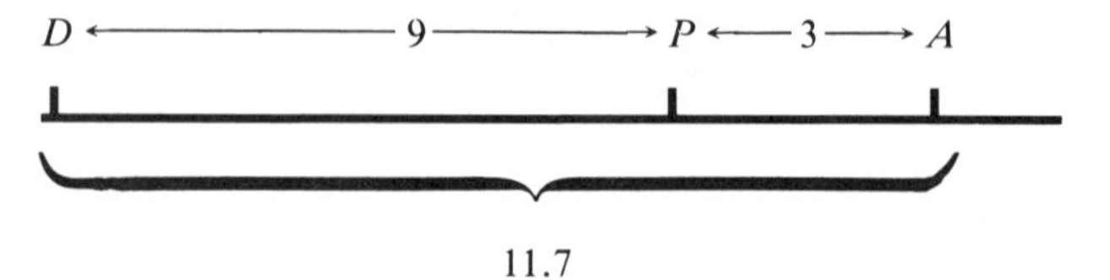

Because of distances involved, this is the only order possible, except the sequence can be reversed ($A \leftarrow 3 \rightarrow P \leftarrow 9 \rightarrow D$). The $D - A$ distance is lower than anticipated (12 expected) because of double cross-overs (one in each region simultaneously). These would be anticipated with a frequency of 3% of 9% (0.27%). *D* and *A* would not move in relation to each other, as they do with a single cross-over in either $D - P$ or $P - A$ regions. Thus, $12 - 0.27 = 11.73$ centimorgans.

6. 6PGD – Rh = 54%

Rh – UMPK = 44%

UMPK – PGM1 = 48%

PGM1 – AMY = 68%

Because of multiple cross-overs, i.e., one in 6PGD – Rh and one in PGM1 – AMY simultaneously, the total amount of recombination over relatively long distances can exceed 100%, even though 100% of the synaptic figures might not actually have a cross-over in one or both of the regions.

8. Daughter's genotype is *Ic ic*. Grandson has one-half chance of receiving *ic* gene from his mother and therefore of having the ichthyosis condition. It does not make any difference whether the husband has the *Ic* or *ic* gene, since he does not pass on his X chromosome to a son.

13. Genetic chromosome maps for the X chromosome must be based on recombination figures from females, since they have two X chromosomes that undergo synapsis and crossing-over. Males have only one X chromosome, thus synapsis, crossing-over, and recombination cannot occur for X-linked genes in males.

14. The mutation would prevent cleavage by the restriction endonuclease. Its presence could be detected by running the DNA in a centrifuge after enzyme treatment. If a person were homozygous for the mutation, all DNA would be in large units (heavy). If neither chromosome of a pair carried the mutation, all DNA would be in small units (light). If heterozygous, half the DNA would be heavy and half would be light.

    This mutation is the one responsible for sickle hemoglobin (glutamic acid changed to valine in β globin). By removing a small amount of fetal cellular material from the amnion and treating with the endonuclease (MST II), fetuses with sickle cell trait or sickle cell anemia can be diagnosed prior to birth.

CHAPTER 6 (Page 156)

1.

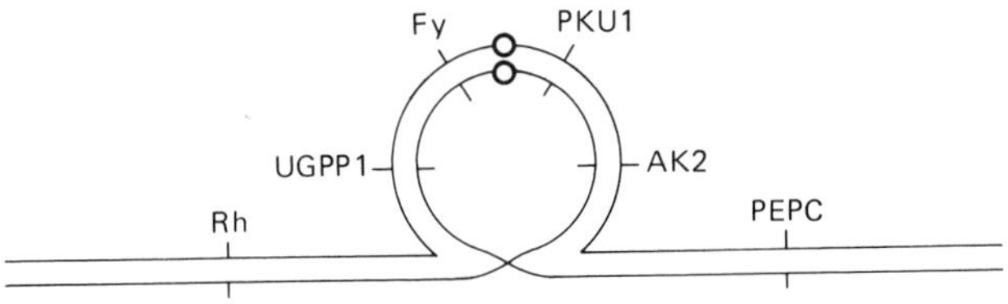

Fy PKU1
UGPP1 AK2
Rh PEPC

2.

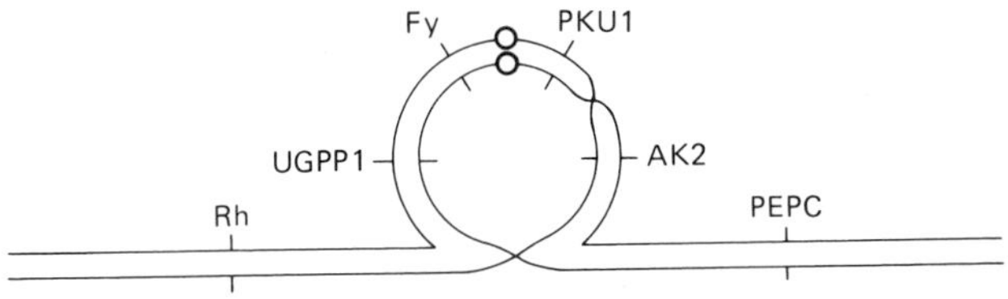

Fy PKU1
UGPP1 AK2
Rh PEPC

Rh UGPP1 Fy PKU1 AK2 Rh

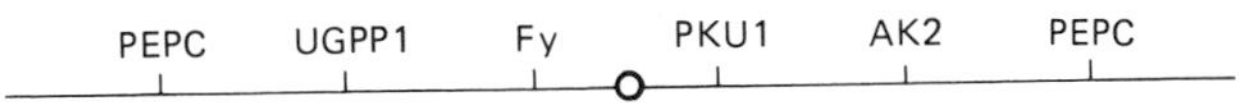

PEPC UGPP1 Fy PKU1 AK2 PEPC

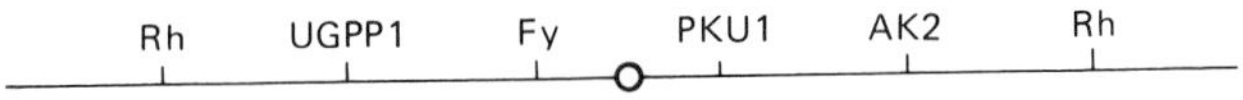

5.

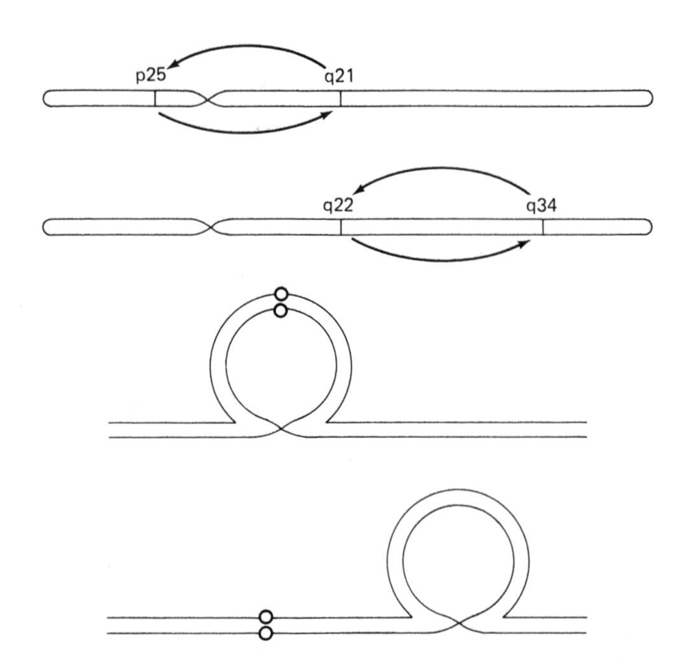

Duplications and deficiencies would result from crossing-over in the loop in chromosome 3 as shown in Question 2. Duplications and deficiences would result from crossing-over in the loop in chromosome 9. Dicentric and acentric formation would be the primary result. The dicentric chromosome would carry duplications and deficiencies as shown in the text example on page 144. The breaking of the dicentric, if centromeres moved in opposite directions, might further contribute to the deficiency situation.

6. The lack of significant genetic material in the short arms of chromosomes 13, 14, 15, 21, and 22 apparently allows the loss of these regions in Robertsonian translocations. The duplication of ribosomal RNA genes in these five chromosomes also permits their loss without detriment to cellular functions. Any two of the five appear to be good candidates for loss of short arms and centric fusions.

13. **(a)** XXXY Male Klinefelter syndrome
(b) XXYY Male Klinefelter syndrome
(c) XXXXY Male Klinefelter syndrome
(d) XXXX Female No syndrome (tetra-X female)

14. **(a)** XXY 1 Barr body **(d)** XXX 2 Barr bodies
(b) XO No Barr body **(e)** XXXY 2 Barr bodies
(c) XYY No Barr body **(f)** XXXX 3 Barr bodies

CHAPTER 7 (Page 193)

3.

	AP	*Ap*	*aP*	*ap*
Ap	*AAPp*	*AApp*	*AaPp*	*Aapp*
ap	*AaPp*	*Aapp*	*aaPp*	*aapp*

4. A monohybrid cross results in three different genotypes (*AA*, *Aa*, *aa*). Two simultaneous monohybrid crosses in the same organism (dihybrid cross) result in three combinations for each of the three genotypes, thus 3 × 3 = 9 genotypes (*AABB*, *AABb*, *AAbb*,*AaBB*, *AaBb*, *Aabb*, *aaBB*, *aaBb*, *aabb*). A trihybrid cross would produce 3 × 3 × 3 = 27 genotypes.
7. If neither parent shows a trait, but a child does, it is logical to assume that the trait (normal fingers in this case) is recessive. Thus, brachydactyly is due to a dominant gene.

$$B = \text{Brachydactyly} \quad b = \text{Normal fingers}$$

$$Bb \times Bb$$

$$\downarrow$$

$$bb$$

The chance for the second child to be brachydactylous is 3/4. The chance for the second child to have normal fingers is 1/4.

10. Man's genotype is *Aa cc pp*; woman's genotype is *aa Cc Pp*.

Child:	*Free*		*Crooked*		*Taster*	
	1/2	×	1/2	×	1/2	= 1/8 chance for all three simultaneously.

12. There is no chance for this type of child (zero probability), since it would be impossible to have a nontaster child, regardless of whatever other traits might be possessed.
14. **(a)** Type A (1/2), type B (1/2)
    (b) Type AB (1/4), type A (1/4), type B (1/4), type O (1/4)
    (c) Type B (1/2), type A (1/4), type AB (1/4)
15. **(a)** Father could be type B or type AB
    (b) Man 1 could be excluded (no I^a gene)
    (c) Man 3 could be excluded (only has *M* gene)
18. *As Ss* × *As Ss*; 1/4 (4/16) albino, 3/16 normal, 9/16 piebald spotting
20. Woman is *dd*; man is *Dd*; first child is *dd*. Chance for next child to be Rh positive is 1/2.
    Unless RhOGAM injection is utilized, there is a risk of HDN with subsequent Rh-positive children.
23. **(a)** Girls are either Xg positive (1/2) or Xg negative (1/2), boys either Xg positive (1/2) or Xg negative (1/2). Thus, there is a 1/4 chance for each.
    (b) Girls are Xg positive only, boys Xg negative only. Thus, there is a 1/2 chance for each.
    (c) Girls are Xg positive only, boys either Xg positive (1/2) or Xg negative (1/2). Thus there is a 1/2:1/4:1/4 chance situation.
26. *LL* × *Ll*; daughter will have long index fingers; son has 1/2 chance of having short index fingers.

CHAPTER 8 (Page 216)

1. To prevent galactosemia symptoms, the infant should be placed on a milk-free diet. Milk substitutes should be used which are free of galactose. The infant's genotype is *gg*. The parents are both *Gg*. The chance for another *gg* child is 1/4. Adequate amounts of the enzyme are produced by *GG* or *Gg* persons. There is 3/4 chance for the birth of a child without the *gg* genotype.
3. **(a)** The blood of phenylketonurics contains high levels of phenylalanine. Thus, the Guthrie test medium lacks phenylalanine, without which the bacteria will not grow.

(b) Galactosemic blood is detected from the blood samples by a direct assay for the missing enzyme, not the substrate.

6. (a) The pattern demonstrated is one of a dominant gene with full penetrance.
   (b) The pattern demonstrated for polydactyly is one of a dominant gene with partial penetrance (20/50 or 40%) and variable expressivity.

8. (a) Mother is *Dd*; Father is *d*__; Son is *d*__.
   (b) Chance is 1/2 for male children to have the disease.

9. (a) Males with Duchenne muscular dystrophy do not reproduce, thus it is not possible for a female with the homozygous recessive genotype to be produced.
   (b) If a segment of the X chromosome carrying the locus for Duchenne muscular dystrophy (*Xp* 12-21) were translocated to one of the autosomes, it is possible that the hemizygous presence of the gene could exist in a female:

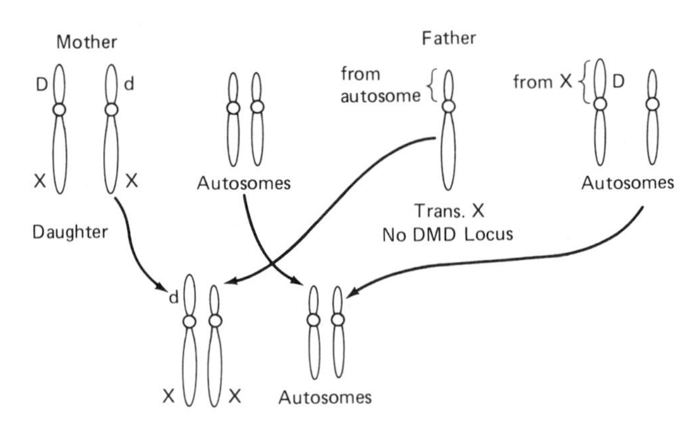

11. (a) Approximately one-half the cultured cells should show a deficiency of HGPRT. (She is heterozygous, one-half the cells have one X chromosome inactivated, one-half have the other inactivated.)
    (b) The chance is one-quarter. All girls would be disease-free, one-half the boys would have the disease.
    (c) By aborting all male fetuses, the birth of a child with Lesch-Nyhan could be averted. This technique would have a risk of one-half of aborting normal male fetuses. The lack of HGPRT can be identified in cultured fetal cells. Thus, the disease is amenable to amniocentesis, followed by abortion of fetuses lacking HGPRT.

CHAPTER 9 (Page 231)

2. (a) Autosomal recessive.
   (b) 1/9 ($2/3 \times 2/3 \times 1/4$)
   (c) Provide dietary substitute for milk which lacks galactose.
   (d) Measure amount of enzyme present in blood cells. Heterozygotes have about half the amount carried by homozygous dominant persons.

4. (a) Sex-linked recessive.
   (b) II – 2 = *Nn*
   II – 3 = *Nn*
   II – 6 = *Nn*
   III – 3 = *Nn*
   III – 10 = *NN*

(c) Risk of nystagmus is 1/8 (would affect only males, no chance for females). There is one-half chance that potential mother carries the recessive gene.

(d) Risk for sons with nystagmus is 1/4. Risk for daughters with nystagmus is 1/4. Thus, the risk for children with nystagmus is 1/2.

The chance for normal children is 1/2 (1/4 for daughters, 1/4 for sons).

6. II − 3 = 1
III − 1 = 0
III − 5 = 1/2

7. II − 2 = 1 (carries recessive gene)
II − 5 = 1 (carries recessive gene)
III − 4 = 1 (carries recessive gene)
III − 6 = 1/2 chance to carry
III − 9 = 1/2 chance to carry
The chance that the child of IV-2 and IV-7 will be albino is 1/64 (1/4 × 1/4 × 1/4).

10.

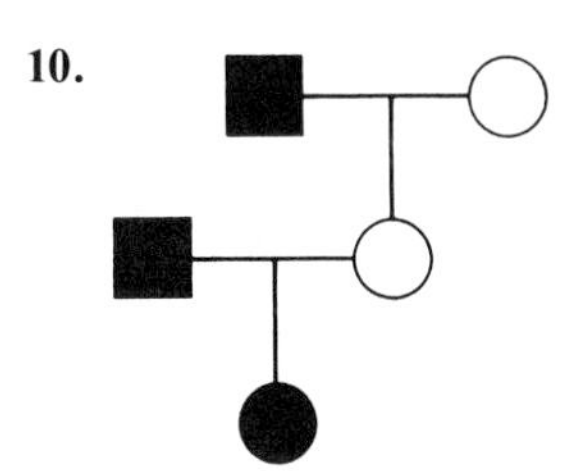

CHAPTER 10 (Page 253)

1. The trait most influenced by heredity is trait I, as shown by the pronounced difference in concordance in the two types of twins.

   The trait most influenced by environment is trait II, as shown by the similarity of concordance values in the two twin types.

   Trait III appears to have a tendency toward hereditary influence, but since there is a very low concordance for identical twins, only a tendency or proclivity is indicated.

2. Chance of 1/2048 for either a black or white child; chance of 63/512 for intermediate child.

   (b) There would be 13 different classes of skin color.

3. (a) Shortest = 5 feet 10 inches
   Tallest = 6 feet 2 inches
   Mean = 5 feet 6 inches

   (b) *AABBccdd* × *aabbCCDD* or *AABBccdd* × *AABBccdd* (or other similar combinations)

   (c) *AaBbCcDd* × *aabbccdd*

   (d) *AABBCCDD* × any other genotype

4. Genetic variance = 63
   Environmental variance = 7

5. 1/64 is approximately 1.6%. 1/64 indicates that three pairs of genes are operating. There would be seven categories of ridge counts with the median category (110 to 150) accounting for 5/16 (approximately 30%) of the population.

6. Variance due to genetics = 15
   Total variance = 30
   Heritability = 15/30 = 0.50

This figure can only be applied directly to the sampled population in its particular environment.

CHAPTER 11 (Page 279)

2.

	Phenotype Frequency	*Frequency of* *A*	*a*
Outset $AA \times aa$	1/2 : 1/2	1/2	1/2
400 children Aa	All free	1/2	1/2

3. Generation II $Aa \times Aa$ Frequency of genes

Generation III 1 AA :2 Aa :1 aa $A = 1/2$ $a = 1/2$

Number of children 200 400 200

Phenotype frequency = 3/4 free:1/4 attached

4.

Matings		*Children in generation IV*
25 $AA \times AA$	=	100 AA
50 $AA \times Aa$	=	100 AA, 100 Aa
25 $AA \times aa$	=	100 Aa
50 $Aa \times AA$	=	100 AA, 100 Aa
100 $Aa \times Aa$	=	100 AA, 200 Aa, 100 aa
50 $Aa \times aa$	=	100 Aa, 100 aa
25 $aa \times AA$	=	100 Aa
50 $aa \times Aa$	=	100 Aa, 100 aa
25 $aa \times aa$	=	100 aa
		Totals 400 AA, 800 Aa, 400 aa
		Gene frequency = 1/2 : 1/2

Totals for generation V = 800 AA, 1600 Aa, 800 aa

Gene frequency = 1/2:1/2

5. 34 $AA \times AA$ = 136 AA

66 $AA \times Aa$ = 132 AA, 132 Aa

67 $Aa \times AA$ = 134 AA, 134 Aa

133 $Aa \times Aa$ = 133 AA, 266 Aa, 133 aa

Generation *IV* totals = 535 AA, 532 Aa, 133 aa

Gene Frequency = 2/3 (1602/2400) A:1/3 (798/2400) a

Generation *V* totals = 1206 AA, 804 Aa, 134 aa

Gene Frequency = 3/4 (3216/4288) A:1/4 (1072/4288) a

In Generation *VI* the gene frequencies would be $A = 4/5$, $a = 1/5$.

7. $AA = 0.36$

$Aa = 0.48$

$aa = 0.16$

The chance for the $Aa \times Aa$ combination is 0.23 (0.48 × 0.48).

8. Gene frequency is 0.02 ($\sqrt{.0004}$).

Carrier frequency is 0.039 or 2 (0.98 × 0.02).

The chance for two carriers to marry and produce a child with cystic fibrosis is 0.0004 (0.04 × 0.04 × 1/4).

12.

Infants	*Adults*
0.737 *AA*	0.929 *AA*
0.243 *Aa*	0.069 *Aa*
0.020 *aa*	0.001 *aa*

Infant population is at equilibrium.

Adult population is not at equilibrium. There are more heterozygotes than anticipated (211 vs. 54) from the number of homozygous recessives. This may reflect the positive survival value of the sickle gene in the heterozygous state, because of its protection against malaria. Using the homozygous sickle gene condition as a measure of equilibrium may not be accurate, since it is often a fatal condition. However, the number of heterozygotes is still somewhat higher than expected (211 vs. 190) based on the gene frequency in infants.

15.

	AA	*Aa*	*aa*	*Frequency a*
Outset	0.25	0.50	0.25	0.5
1st Generation	0.28	0.50	0.22	0.47
2nd Generation	0.31	0.49	0.19	0.44

"Your problem is in the gene that makes antibodies, but since the Biophase Corp. now has a patent on that gene, I can't do anything for you."

Courtesy of Sidney Harris.

INDEX

C

D

E

F

G

H

N

O

P

Q

T

U

V